AGRONOMY

FACTS AND APPROACHES

THE AUTHOR

Dr. A.K. Ghosh has been working on senescence and source-sink relationship of different types of plants since 1985. Besides he has been working on pollution and ethnomedicine since 1993. With more than 19 years of teaching experiences in H.S. section, he is at present serving as Headmaster in Saraswati Vidyamandir. Dr. Ghosh has already published 40 original scientific papers in International Journals of India and abroad. He is the author of 7 books, good sportsman and a social worker.

He acted as a referee in Iraq and Pakistan Journal (PJSIR). Dr. Ghosh has been adjudged the best individual for his notable contribution in the field of Science Popularisation and Performance in disseminating science from the libraries and laboratories to the common people of West Bengal and abroad. He was awarded Gopal Chandra Bhattacharya Smriti Puraskar for the year 2004 by Department of Science and Technology, Government of West Bengal.

Dr. P. Das has been working as Agriculturist with more than 30 years of teaching experience in a H.S. school, and have many books to his credit.

AGRONOMY

FACTS AND APPROACHES

by

Dr. A.K. GHOSH

(M.Sc., M.Ed., Ph.D., Diploma in Naturopathy)

Swamiji Eco-Club, Midnapore–721101, W.B.

&

Dr. P. DAS

(M.Sc., B.Ed., Ph.D)

2025

Daya Publishing House®

A Division of

Astral International Pvt. Ltd.

New Delhi - 110 002

First Published, 2011
Reprinted, 2025

ISBN: 978-81-7035-698-1 (PB)

Published by : **Daya Publishing House®**
A Division of
Astral International Pvt. Ltd.
– ISO 9001:2015 Certified Company –
4736/23, Ansari Road, Darya Ganj
New Delhi-110 002
Ph. 011-43549197, 23278134
E-mail: info@astralint.com
Website: www.astralint.com

Laser Typesetting : **Twinkle Graphics**
Delhi - 110 088

Printed at : **Replika Press Pvt. Ltd.**

PRINTED IN INDIA

Dedicate to...

Late Abinash Chandra Ghosh
(Great Grandfather,
Former Headmaster during
1906-1914 of
Rajagram S.B. Raha Institution,
P.O. Rahagram,
Bankura which is previously known
as Rajagram A.S. School)
and to
Late Grandfather Bhutnath Ghosh
starter of Calcutta Tram Company
and to My Parents.

Acknowledgements

Authors would like to express their deepest sense of gratitude to Prof. Shyamapada Pal (V.U.), Prof. Sabyasachi Chatterjee, Joydev De (Asstt Director), Dr. Debal Deb, Dr. Subhasis C. Patra, Debdulal Bhattacharya, Narayan Santra (H.M) for their constant encouragement. Thank is also due to all the staff of D.I. office for their encouragement.

Dr. A.K. Ghosh wants to express his thanks to Sarathi Maity (A.I/S), Soma Ghosh and Biplab Ghosh for the patience shown by them during the preparation of the manuscripts. Dr. P. Das is thankful to his wife and children for their encouragement.

Dr. A.K. Ghosh & Dr. P. Das

Preface

Agronomy is one of the most important discipline of Agriculture. "Agronomy : Facts and Approaches" has been written to provide basic informations about different aspect of Agronomy. The book has been written in simple and lucid form under Indian context. Here attempt has been made to provide readily available informations. We do not pretent to claim credit for original contribution. Rather it should be viewed as a concise collection of various topic from different books, bulletins, journals etc. preliminary concept about growth and development of crops, manures and fertilizers, soil, agroforestry, etc. has been dealt in this book.

We shall consider our efforts successful and worthy if this book serves the purpose for which it is written. All suggestions and criticism for further improvement of this book will be thankfully acknowledged. Lastly we feel great pleasure in expressing our deep sense of gratitude to Mr. Anil Mittal for publication of this book so efficiently and promptly.

Dr. A.K. Ghosh & Dr. P. Das

Contents

List of Tables

List of Figures

1

INTRODUCTION

1.1 BASIC PRINCIPLES OF AGRONOMY

Agronomy is one of the most important discipline of agriculture. The term Agronomy has been derived from two Greek word-'agros' (*i.e.* field) and nomos (*i.e.* to manage). Literally it means the art of managing fields and technically it means the science and economics of Crop Production by management of farm lands. Norman (1980) has defined "Agronomy as the Science of manipulating the crop environment complex with dual aims of improving agricultural productivity and gaining a degree of understanding of the process involved." In recent times, Agronomy has assumed newer dimention and can be defined as a branch of agricultural science that deals with methods which provide favourable environment to the crop for higher productivity. In other words, it is the art and underlying science in the production and improvement of field crops with efficient use of soil fertility, water, labour and other factors related to crop production. Prior to recognition of agronomy as a district branch of agricultural science, Crop experimenters were usually botanist, chemist, general agriculturist etc. Later on new science of agronomy was built up by co-ordination of knowledge from the natural and biological sciences dealing with experimental works on crop production. The American Society of Agronomy was organised in 1908. Agronomy is a dynamic discipline. It deals with principles and practices of field management for the production of field crops. It is an integrated and applied aspects of different disciplines of pure sciences. The new technologies (*i.e.*, adoption of high yielding varieties, application of fertilizers, herbicides, insecticides, fungicides, scientific use of irrigation water, crop management techniques etc.) evolved so far are the major aspect of agronomy to boost up production and its usable products per unit land, time and inputs as follows :

(*i*) Production of field crops by adopting the technologies of LAB to LAND and *vis-a-vis*.

(*ii*) Factor affecting crop production and their inter-relationship.

(*iii*) Adoption of techniques of increasing use efficiency of inputs of crop production.

(*iv*) Evolution of technologies for better management practices of soil, water, plant nutrients, weeds and crop plant. Agronomy has three clean branches as follows:

(*a*) *Crop Sciences* : It deals with mainly the production of field crop.

(*b*) *Soil Sciences* : Soils and their origin and impact in crop production etc.

(*c*) *Environmental Sciences* : It deals only with the applied aspects.

Crop production is basically conversion of environmental inputs like solar energy, carbon dioxide, water and soil nutrients into economic products in the form of human or animal food or industrial raw materials. Ours is an agricultural country and seventy five per cent of total population depends mainly on agriculture. Population pressure is increasing but area under cultivation is static. Growing population is a great problem to the Agronomist for providing sufficient food to them. Agronomist is concerned with the production of food and fibre to meet the needs of the growing population by offering high yielding varieties of crops, Scientific method of crop production and proper plant protection measures. They are the only person who can help the farmers by providing technical knowledge of crop production. Scientific crop production follows breeding of improved varieties of crops along with adoption of improved agricultural practices including plant protection measures. Insect pest and diseases are a great problem to our agriculture now-a-days. Timely plant protection is an extra advantage of getting higher yields of cultivated crops. The traditional methods of cultivation should be avoided and new method of cultivation should be adopted to meet the requirement of the growing populations of our country.

There are some fundamental principles of agronomy as follows :

(*i*) Planning and executing measures for maximum utilization of production inputs like land, labour and capital.

(*ii*) Field management *i.e.*, preparation of land, levelling and adoption of other suitable land improvement practices and checking of soil erosion.

(*iii*) Selection of quality seed and crop varieties suitable for a particular area in a season.

(*iv*) Selection and application of manures and fertilizers in balanced way to increase the crop production.

(*v*) Water and its proper management for crop production.

(*vi*) Adoption of proper intercultural operations to get maximum yield.

(*vii*) Adoption of different types of cropping systems (*i.e.*, multiple cropping, relay cropping, mixed cropping, intercropping etc.) to ensure harvest even under adverse environmental conditions.

(*viii*) Adoption of need based plant protection measures to save crop from the attacks of crop enemies like weeds, insect, pest, diseases, rodents etc. causing the crop failure.

(*ix*) Adoption of suitable method for harvesting of crops in time to reduce field damage and to release the land for succeeding crops.

(*x*) Adoption of suitable post harvest technologies to make the produce suitable for human use and to reduce the losses of produce.

1.2 TWENTY POINTS WORKING SCHEDULE IN CULTIVATION

The following recommendations should be followed to get higher yield and extra profit in the cultivation of any crop :

(*i*) Crop rotation should be prepared according to the nature of climate and soil.

(*ii*) The crop suitable for a particular land and its high yielding variety (H.Y.V) should be selected and cultivated.

(*iii*) It is necessary to include at least one leguminous crop in the crop rotation.

(*iv*) Soil should be tested before starting the cultivation.

(*v*) Soil amelioration should be undertaken before the cultivation if the soil becomes acidic or alkaline.

(*vi*) Organic manures should be applied 3-4 weeks before sowing or planting of crops.

(*vii*) The balanced fertilizer should be applied according to the result of soil test.

(*viii*) The total amount of phosphatic and potassic fertilizers and half of nitrogenous fertilizer should be applied with the last preparatory tillage. The remaining nitrogenous fertilizers should be applied 2-3 times according to the growth of the crops as top dressing.

(*ix*) The applied fertilizers should be mixed with the soil just after application by tillage followed by planking.

(*x*) The crop should be planted by proper method, and in schedule time.

(*xi*) The proper spacing of the crop should be maintained.

(*xii*) The irrigation should be given according to the need of the crop.

(*xiii*) The proper drainage should be done.

(*xiv*) The measures should be taken to keep the land free from weeds. For this, weeding should be done in time.

(*xv*) Top dressing should be done in right time with accurate amount of fertilizers.

(*xvi*) Foliar application should be done to reduce the cost of fertilizer and also the losses of fertilizer due to leaching.

(*xvii*) Supervision of land should be the regular practice to determine the need of the crop and also to undertake the top dressing of fertilizer and crop protection measures.

(*xviii*) Protective measures should be undertaken as soon as the symptoms of the attack of insect, pests and diseases are noticed.

(*xix*) The crops should be harvested in due time to reduce the loss of the produce.

(*xx*) The advice of agricultural specialist should be sought if any problems appears in the process of cultivation of crops.

1.3 RELATION OF AGRICULTURAL SCIENCE WITH OTHER BASIC SCIENCES

The relation of agricultural science with other basic sciences is very much important. Because agriculture is said to be impossible without the help of other basic sciences. The relation of agricultural science with other basic sciences is as follows :

1. Agriculture and Biology : Agriculture is a biological sciences. The progress of agriculture is not possible without the proper knowledge of Botany and Zoology, the two branches of biological sciences. The botanist provides scientific knowledge about plant morphology, ecology, physiology, genetics etc. and this is helpful for progress of agriculture. We should have knowledge about plant pathology for detecting and controlling of plant diseases. The Zoologist provides knowledge about activities of different organism related to agriculture. Likewise in plant pathology, we should have knowledge of Entomology for identification and controlling of insect pests of crops.

2. Agriculture and Animal Husbandry : Animal husbandry is not considered as basic science. But the animal husbandry is an integral part of agricultural sciences. Different agricultural activities depend mainly on animal power now-a-days also though mechanisation of agriculture is increasing progressively. The healthy and productive animals are needed for the performance of different activities and this will be possible if we have sound knowledge about management and breeding of animals.

3. Agriculture and Geology : The geologist provides knowledge about the formation of soil from rocks. Soil is the medium for the production of crops. The maintenance of soil and crop cultivation according to the soil type is only possible if we have proper knowledge about the characteristics of soil. The knowledge of geology is considered helpful in this regard.

4. Agriculture and Chemistry : The agricultural chemist provides detail knowledge about soil formation and the constituents of soil, soil nutrients and their availability, soil management, soil organism and their role in agriculture etc. Besides these, the chemist became successful for producing the following inputs that are very much essential for production of crop as per our demand.

(*i*) **Fertilizers :** Fertilizers have a key role in the production of crops. Fertilizers are available mainly in three forms as follows :

(*a*) ***Nitrogenous Fertilizers*** **:** Sources of Nitrogen (N):

Example : Ammonium sulphate, Urea, Ammonium nitrate, Sodium nitrate, Ammonium Chloride, Calcium nitrate, Calcium ammonium nitrate (CAN) Ammonium sulphate nitrate etc.

(*b*) ***Phosphatic Fertilizers*** : Sources of phosphorus (P_2O_5).

Example : Super phosphate (Single, double and triple), Basic slag, Rock phosphate etc.

(*i*) ***Potassic fertilizers*** : Sources of Potassium (K_2O) :

Example : Muriate of potash or potassium chloride, potassium sulphate, potassium phosphate etc.

Besides these, complex fertilizers are also available in the market as follows :

Diammonium Phosphate (DAP), Ammonium nitrate phosphate, Urea Ammonium phosphate (Gromor 28:28:0) Gromor 14:35:14, IFFCO/Gromor/Paras/ Subarna-10:26:26. Sulphate 15:15:15, Sulphate 20:20:0 etc.

(*ii*) **Fungicides :** Chemicals used for controlling of fungi:

Example : Copper oxychloride 50 WP, Mancozeb 75 WP, Zineb 75 WP, Carbendazim 50 WP, Ziram 27 S.L., Captan 50 WP, Propiconazole 25 EC.

Quintazone 75 WP, Carboxin 75 WP, Hexaconazole 5 EC, Trisiclazole 75 WP, Tridimorf 75 EC, Copper hydroxide, Edifenfos 50 EC, Triamofos-48 EC, Metalyxsil 35 WP, Isoprothiolone 40 EC, Benomil 50 WP, Sulphur-80 WP, Bordeaux mixture etc.

Besides these some combine fungicides are also available in the market as follows :

Maneozeb 64% + Metalaxyl 8%, Carbendazim 12% + Mancozeb 63% etc.

(*iii*) **Seed Treatment Chemicals :** Chemical used for treating seed before sowing :

Example : Methoxyethyl Mercury chloride, Ethylmercury chloride, Thiram 75 WP, Carbendazim 25 and 50 WP, Mancozeb 75 WP, Captan 50 WP, Tricoderma Viridii, etc.

(*iv*) **Insecticide :** Chemicals used for controlling insect pests.

Example : **Dust**–Carbaryl 5 and 10%, Ethyl Parathion 4%, Methyl parathion 2%, Fenthoate 2%, Quinalphos 1.5%, Enlosulphan-4%, Chloropyriphos 1.5%, Lindane 1.3%, Fenvelaret-0.4% etc.

Granules : Phorate 10 g, Quinalphone 5 g, Carbofuran 3G', Lindane 6 and 10 g, Sevidol 4 g, Cartap hydrochloride 4 g, Aldicarb 10 g, etc.

Liquid : Acephate 75 SP, Alfamethrin-10 EC, Carbosulfan 25 EC, Cypermethrin-10 EC, Chloropyrifos-20 EC, Dichlorovos-76 EC, Diazinon 20 EC, Dimethoat 30 EC, Imidachloropid 17.8 EC, Endosulfan 35 EC, Lindane-20 EC, Febbrucarb-50 EC, Ethifenphos-10 EC, Fosalone 35 EC, Fenthion 50 EC, Fenthoat 50 EC, Malathion-50 EC, Monocrotofos-36 SL, Methyl parathion 50 EC, Oxydemiton Methyl 25 EC, Phosfamidon 40 S.L., Phosalone 35 EC, Profenfos 50 EC, Triazophos-20 EC, Quinalphos 25 EC, Neem Oil, etc.

(*v*) **Antibiotic**

Example : Aureomycin, Streptomycin, Validamycin 3% etc.

(*vi*) **Acaricides :** Chemicals used for controlling mites.

Example : Ethion 50 EC, Dicofol 18.5 EC, Binapacryl 48% WP and 38.8 EC, etc.

(*vii*) **Nematocides :** Chemicals used for controlling nematodes.

Examples : DBBP, DD, Fenisulphothion 10% and 25 WP, etc.

(*viii*) **Fumigant :** Chemicals used for fumigation.

Example : Aluminium phosphide, Ethylene, Ethylene dichloride etc.

(*ix*) **Rodenticide :** Chemicals used for controlling the rodents.

Example : Zinc phosphide 80%, Koumatatrayl 0.75% and 0.0375%, Bromadiolone 0.005%, (RB)-26 etc.

(*x*) **Weedicides :** Chemicals used for controlling weeds.

Examples : 2,4-D, Alachlor 48% EC and 15%, Butachlor 50 EC, Propanil 35 EC, Nitrophen 25 EC, Paraquate-25 WSC, Simazine-50 WP, Fluchloranin-95 EC, Pendimethanil-30 EC, Atrazine-50 WP, Thirancarb-50 EC, Isoproturon 75 WP, Oxadiazone 25 EC, Oxyflurofen 24 EC, Neuron 50 WP, Trifluranim-48 EC, Diuron-40 WP, Glysophet 41 SL. Metriburine etc.

5. Agriculture and Physics : Mechanisation of agriculture is very much essential for successful crop production. The physicist have developed various agricultural implements to make the agricultural operation easy and economical. Tractors, power tiller, seed drill, potato planter, harvesting and threshing implements, jute ribboner, sprayers, dusters etc. are important implements used for various agricultural operations. Agricultural engineering was mainly concerned with improving farm machinery and implements which will bring the dynamic change in Agriculture.

6. Agriculture and Mathematics : Agriculture is closely related with different branches of mathematics, particularly statistics, statistics is employed for analysis of different experimental results. Confirmation about the experimental result is not possible without the statistical analysis.

7. Agriculture and Economics : Economics has a key role in Agriculture. The production is increased when productive resources (*i.e.* land, labour and capital) are possible to be utilized economically. The study of economics is essential to have knowledge about sales proceedings, demand, and supply, price of produce etc. Besides this, economics provide knowledge about the capital and labour, the most important input in agriculture.

2

CLIMATOLOGY

2.1 INTRODUCTION

Climate, one of the major environmental factors, largely determines the productivity of crops. Climate may be defined as an average condition of weather in a particular region during a comparatively long period (25-30 years or more). The climate is a generalised weather or summoration of weather condition over a given region during comparatively target period. The condition of the atmosphere in a place at a certain time or over a certain period is called weather. It is the result of the composite effect of temperature, rainfall, humidity, frost, snows, sunshine and atmospheric pressure which may vary from moment to moment and time to time. The study of climate and weather is called climatology and meteorology respectively. Climatology may be defined as the science of the average conditions of the weather of any particular place and their effect on plants and animals. Meteorology is the study of physical phenomena which takes place in the atmosphere. The climate deals with atmosphere, radiation, air and water. The sun is the primary cause of climate and its variation. The sun provides heat energy (*i.e.* about 99.97 per cent for all physical processes taking place in earth atmosphere. All life depends on Khitti (earth, soil), appa (water), teja (energy, radiation), byoma (sky, atmosphere) and maruta (air, wind).

2.2 ELEMENTS OF CLIMATE

The important elements that make the climate are as follows :

1. Temperature.
2. Atmosphere.
3. Wind.
4. Humidity.
5. Cloud.
6. Rainfall.

1. Temperature

It is a measure of intensity of heat, sun is the primary source of heat for earth and its atmosphere.

(*a*) ***Air Temperature*** **:** It has great effect on the growth of the crop plants. Different crops grows at different temperature as follows :

Crop	Temperature	Crop	Temperature
Paddy	: 21°C - 35°C	Brinjal	: 21°C. - 29°C
Wheat	: 10°C - 15°C	Tomato	: 21°C. - 24°C.
Potato	: 15.5°C - 22°C	Carrot	: 16°C. - 18°C.
Jute	: 24°C - 37°C	Cabbage	: 10°C.- 25°C.
Sugarcane	: 21°C.- 26°C.	Cauliflower	: 17°C - 29°C.
Mustard	: 21°C - 25.7°C.	Lettuce	: 15.5°C - 20°C.
Seasum	: 21°C.- 27°C.	Knolyhol	: 18°C - 25°C.
Ground nut	: 21°C - 26.6°C	Cucumber	: 18°C - 25°C.
Sunflower	: 28°C - 32°C	Watermelon	: 21°C - 29°C.
Sweet potato	: 21°C - 29°C	Pumpkin	: 18°C - 25°C.
		Lady's finger	: 21°C - 30°C.

(b) Air Temperature and Crops **:** Air temperature has a influence on the growth of the crop to a certain extent. Generally the plant will not grow at below 4.4°C and above 48.4°C. Generally 21.2°C – 26.6°C temperature is required for the proper growth and development of the plants.

(c) Soil Temperature **:** It is one of the important environmental factors for the growth and development of a plant. The exchange of heat between the soil and its surface has pronouncial effect on diurnal surface temperature variation.

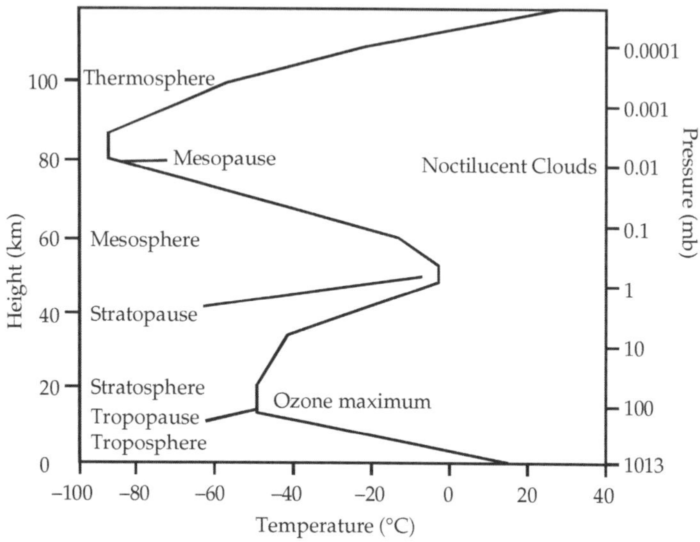

Fig. 2.1 : Generalized vertical distribution of temperature and pressure up to about 10 km.

Soil Temperature and Crops : The soil temperature has an influence on the growth of the plant by affecting some processes as follows :

(*i*) ***Absorption of Water*** : Water absorption by plants is slow both at low and high temperature of the soil. At higher soil temperature water absorption decreases due to faster maturation one root injury.

(*ii*) ***Uptake of Nutrients*** : Nutrient absorption increases with temperature upto certain level, especially in case of anionic forms. Nitrification and ammonification occur at a optimum temperature range of 25°C-30°C and 15°C-40°C respectively. The microbial activity is severaly cusfalied outside the range of 10°C-40°C. The nodule growth in leguminous plants occurs at an optimum soil temperature of 24°C.

(*iii*) ***Germination of Seed*** : Soil temperature has influence on the germination of seed.

(*iv*) ***Microbial Activity*** : Soil temperature affects the microbial activity. The optimum temperature of nitrification is 25°C-30°C and of ammonification is 15°C–40°C.

Table 2.1 : Minimum and optimum temperature for the germination of seed for different crops. (Venkataraman and Krishhan, 1992)

Crops	*Minimum Temperature (0°C)*	*Optimum Temperature (0°C)*
Wheat, Barley and Oat	4.5	20
Rice	10-15	31
Maize	8-10	20
Sorghum.	12-13	25
Soyabean	8-9	29
Tobacco	13-14	29
Cotton	14-15	33
Citrus	15	30
Tomato	11-18	18-25
Cucumber	11-18	25-30
Forage Crop	15	25-30

There are three principle climate on the basis of temperature as follows :

1. Tropical Climate : The lands which lie near the Equator and these tropics are known as Tropical lands and they usually have very hot and wet climate. They are said to have a tropical climate which is found in the Cargo Basin of Africa, the Amaron Basin of South America and Malaysia and Indonesia in Asia. In tropical region, there are other types of climate also as follows:

(*i*) Monsoon type (*i.e.,* hot and humid type).

(*ii*) Hot desert type (*i.e.,* hot and dry type).

(*iii*) Sub-tropical type (climate is not so hot as tropical type)

2. Temperate Climate : The region of tropical climate is found on both sides of the tropical region. In the Northern and Southern Hemisphere, it lies between the Tropic of cancer and

Arctic circle and Tropic of Capricon and the Antarctic circle respectively.

3. Polar Climate : The lands which lie to the north of the Arctic circle *i.e.,* the extreme northern land masses of Eurasia and North America are known as the polar land, They are also known as 'Tunla'. This region remains frozen for most part of the year. The entire vegetation of the earth can be divided into four categories on the basis of temperature requirement as follows :

(*a*) ***Megatherm* :** Plants which require more or less constant high temperature throughout the year for their optimal growth and development are called Megatherm. The plants of tropical region can be included in this class.

(*b*) ***Mesotherm* :** Plants that are capable of enduring considerably lower temperature during some period of the year such as winter month are called Mesotherm. Many plants of tropical and sub-tropical region of the world can be included in this class.

(*c*) ***Microtherm* :** Plants that demand much lower temperature for their growth and development are called Microtherm. Plants of temperate region of the world can be included in this class.

(*d*) ***Hekistotherm* :** Plants that are restricted only to arctic and to alpine regions above 4800 metres in tropics and about 3600 metres in temperate zone of world are called Hekisbotherm. They have the lowest temperature demand of all plants and they are also adopted to short summer.

2. Atmosphere

Atmosphere is a mixture of gases. This gaseous envelop of invisible film of air surrounding the earth is called atmosphere. Air is a mechanical mixture of gases and it can not be felt except when it moves as wind. The atmosphere extends to a height of about 1600 kilometres. However, 9.9 per cent of total mass of atmosphere is within 40 kilometres from the earth.

Table 2.2 : Composition of Air

Gases	*Volume %*	*Weight (%)*
Nitrogen (N)	70.088	75.527
Oxygen (O)	20.949	23.134
Argon (Ar)	0.930	1.282
Carbondioxide (CO_2)	0.030	0.045
Other	Traces	Traces

Nitrogen, oxygen, argon and carbon dioxide account for more than 99 per cent of dry air, and they are mixed in constant proportion upto about 80 kilometres. Besides these gases, hydrogen, methane, neon, hilium, krypton, xenon, nitrous oxide, carbon monoxide, ozone, ammonia, nitrogen oxide, sulphur dioxide, hydrogen sulphide etc. are present in small quantities. In addition to these gases, atmosphere contains variable quantities of water vapour. It is the most important constituent of atmosphere in relation to weather and climates. Atmospheric composition varies with height, latitude and time. Solid particles of dust, smoke, salts etc. are also found in the atmosphere.

Structure of Atmosphere : Based on vertical temperature differences, the atmosphere may be divided into four major layers of strata. Troposphere, stratosphere, mesosphere and thermosphere.

1. Troposphere : It is the lower layer of the atmosphere extending upto a height of 8 to 18 kilometers from the earth depending on latitude and densest part of the atmosphere. Troposphere is thicker at equator than at poles and contains 85 per cent of the atmosphere mass. All weather phenomena like clouds, fog, mist, rain etc. occur in this layer. In this layer, temperature decreases with increase in attitude. The temperature at the boundary of troposphere is roughly –60°C. There is a general decrease in temperature with height at a mean rate of about 6.5°C km^{-1} (3.6°F per 1000 ft). The Zone marking the end of this temperature decrease is called the tropopause.

2. Stratosphere : It is a layer of atmosphere which lies above the tropopause and it contains virtually 15 per cent of the atmosphere's mass. This layer extends upward from tropopause (*i.e.* 8-18 km.) to about 50 kilometres depending on the latitude. It is the seat of most of the photochemical reactions in the air and dust free, cloudless and warmest layer. The air in the stratosphere is thin and highly rarefied compared with air at ground level. This layer absorbs heat from the sun in the form of ultraviolet radiation. Stratosphere contains greatest concentration of ozone. Thus in the process of formation of ozone layer, the harmful ultraviolet radiation is filtered which otherwise makes life impossible on the earth. The term ozone layer or ozonosphere is synonymous with stratosphere.

3. Mesosphere : It is the layer of atmosphere which lies above the stratosphere. The stratosphere and mesosphere are separated by a narrow transitional layer called stratopause. In mesosphere, temperature decreases with increases in altitude upto about 80 kilometres. Above 80 kilometres, temperature again begins rising with height and this inversion is referred to as the mesopause. Pressure is very low in the mesosphere, decreasing from about 1.0 mb at 50 kilometres to 0.01 mb at 90 kilometres.

4. Thermosphere : It is the outermost layer extending from the top of mesosphere where atmosphere densities are very low. The lower layer of thermosphere is called ionosphere. Long distance radio-communication is made possible through this ionised layer. The lower portion of this layer is composed mainly of Nitrogen (N_2) and oxygen in molecular (O_2) and atomic (O) form. The thermosphere is characterised by a steadily increasing temperature with altitude (1200° k at 350 km) owing to absorption of ultraviolet radiation by atomic oxygen.

3. Wind

Air motion in a stream along the surface of the earth is called wind. It is caused due to pressure gradient in the atmosphere. The air moves from high pressure areas to low pressure areas. Hence, temperature variation is the main cause of wind. Wind correct the pressure inequalities on the surface of the earth. A number of winds of local importances develop as a result of local differences in temperature. There are three types of wind as follows :

(*i*) ***Permanent or Prevailing Winds :*** This type of winds blow in the same directions all the year round. The trade wind is permanent wind and it blow from subtropical high pressure belt towards the equatorial low. Pressure belt covering mainly the tropical zone in each hemisphere. Wind blows from south east in southern hemisphere and toward equator in northern hemisphere. They are called south East Trade Winds and North East Trade Winds respectively.

(*ii*) ***Periodic Winds :*** This winds blow during a particular period in definite direction, which are called monsoon wind. This winds are largely the result of unequal heating of continents and oceans.

In several parts of the world, prevailing winds known as monsoons. The word monsoon has been derived from the Arabic word "Mausim", meaning season. Monsoons are well developed in India and adjacent South Eastern Asia, where there is persistent flow from Indian ocean on the land in summer and winds blow out from continent in winter. Pedalaborde (1963) described monsoon as a form of planetary wind affecting the upper layer of the atmosphere over the whole globe. Sometimes, it also refers to regional winds blowing in the lower layer of atmosphere over well defined territory. Average rainfall of India is 120 centimetres per annum.

In India two types of monsoons are recognised as follows:

1. South West Monsoons : The South West monsoon winds start blowing in India from the month of June and lasts upto the first week of October. Arrival of this monsoon brings a great change in weather condition in the subcontinent. The atmosphere becomes saturated with moisture and relative humidity goes with (above 90 per cent). It marks disappearance of dry heat, scorching, '100' winds and appearance of wet and cold air, thunder storms, cumulus clouds hanging over head. The agricultured activities for kharif crops in kharif season begins due to onset of south west monsoon. The South-West monsoon period is called grand period rain fall in India, becomes week by second to third weak of September.

2. North East Monsoon : The North-East monsoon winds start blowing during winter and cold wind blows from the North-East direction. The North-East monsoons while blowing over Bay of Bengal picks of moisture and causes rain fall in the eastern zone of the Peninsular. The North-East monsoon is associated with rainy weather over southern parts of Peninsular India particularly over Southern Rayalaseema and Tamil Nadu from November to January.

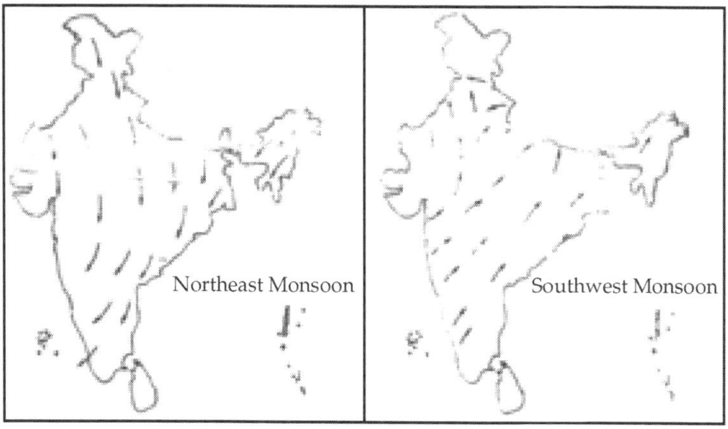

Fig. 2.2 : Monsoon winds over India

3. Variable Winds : Cyclone are the example of variable wind. More severe cyclonic storms usually form in Bay of Bengal and Arabian sea in the transition period, April to June and October to November and causes considerable precipitation in coastal tracts by entering in land.

Wind and Crops : Wind affect the crops in various ways as follows :

(*i*) Wind affects the rate of transpiration in plants.

(*ii*) Normal winds help in the pollination of flower in anemophilly flower such as paddy, wheat, maize, sugarcane etc. and dispersal of fruits, seeds etc.

(*iii*) High wind velocity causes damage to some crops such as paddy, wheat, banana etc.

(*iv*) Dry wind increases the transpiration rate of plants. As a result, wilting of plants occurs.

(*v*) High wind causes mechanical damage to crops such as lodging. Heavy wind accompanied with rainfall causes uprooting of crops mainly in the rainy season.

(*vi*) Hot dry winds may, however, adversely affect photosynthesis and hence productivity by causing closure of stomata even when soil moisture is adequate (Morachan, 1986).

(*viii*) Wind have a direct effect on the dispersal of diseases of crop and seeds of weed.

4. Humidity

Humidity refers to the water vapour content of the atmosphere. Water is always present in the atmosphere in the form of individual water vapour. Liquid water is converted into water vapour by evaporation for which necessary energy is provided by solar radiation in the form of temperature. The amount of water vapour present in the atmosphere depends upon wind and temperature. Warm air can hold more moisture than cold air. Wind distributes the water vapour in the atmosphere. The air is said to be saturated when it holds maximum amount of water vapour at a particular temperature.

Measurement of Humidity : The quantity of water vapour hold in the atmosphere at any particular time can be expressed as absolute, specific or relative humidity.

(*i*) *Absolute Humidity* : It is the absolute or actual quantity of water vapour present in a given volume of air which is measured in gm^3.

$$\text{Absolute humidity } \left(g/m^3\right) = \frac{\text{Weight of water vapour}}{\text{Volume of air}}.$$

Example : Calculate the absolute humidity if 3 cubic metres of air contains 15 g of water vapour.

$$\text{Absolute humidity } \left(gm/m^3\right) = \frac{15}{3} = 5.$$

(ii) Specific Humidity : It is the ratio of mass of water vapour actually in the air to the unit mass of air including its moisture. It is expressed as grams of water vapour present in a kilogram of air.

$$\text{Specific humity} = \frac{\text{Weight of water vapour}}{\text{Weight of air including water vapour}}.$$

Example : Calculate the specific humidity, when the weight of air sample including water vapour is 5 kg and weight of water vapour is 25 gm.

$$\text{Specific humidity} = \frac{25}{5} = 5\,\text{g/kg}.$$

Absolute and specific humidity have lesser effect on crop growth. So relative humidity is widely used in agriculture compared to absolute or specific humidity.

(iii) Relative humidity : It is the ratio between the amount of water vapour present in the air and the amount of water required for saturation at a particular temperature and pressure. It can be expressed as percentage or ratio.

$$\text{Relative humidity} = \frac{\text{Water vapour present in the air}}{\text{Water vapour required for saturation}} \times 100$$

Example : Calculate relative humidity, when water vapour required for saturation of a parcel of air is 50 g and actual amount of water vapour is 40 g.

$$\text{Relative humidity} = \frac{40}{50} \times 100 = 80\%.$$

When RH is 80 per cent, it means that there is deficit of 20, per cent water vapour for saturation.

Humidity and Crops : Humidity affects the growth of crops in various ways as follows :

(*i*) The humidity directly influences the water relation of plants and indirectly affects leaf growth, photosynthesis, pollination and occurrence of insectpests and diseases and finally economic yield.

(*ii*) Humidity has considerable influence on evapotranspiration and hence the water requirement of the crop.

(*iii*) At constant temperature, change in atmospheric humidity affects transpiration by modifying the vapour pressure gradient from leaf to air.

(*v*) In dry region, humidity tends to be low which causes severe water deficit in plants and reduces leaf water potential.

(*vi*) Moist air favours the growth of many fungi and bacteria and these affect the crop seriously. The blight disease of potato and tea spread more rapidly under humid conditions. Similarly many kinds of insect parasites, such as aphids and jassids thrive best under humid condition.

5. Clouds

Clouds is an aggregate of minute drop of water suspended in the air at higher altitudes.

The evapotranspiration supplies water vapour to air.

The word evapotranspiration consists of two words; evaporation and transpiration. Evaporation means change from liquid stage to vapour and it is physical process. Transpiration may be defined as the evaporation of water in the form of water vapour from all aerial organs of the plant, especially from leaves and it is the biological process.

Clouds are classified primarily in the basis of their height, shape, colour, transmission or reflection of light. In accordance to the 1956 International Cloud Atlas of World Meteorological Department, clouds are classified into ten characteristic form as follows :

(*i*) ***Cirrus*** **:** It is the highest of all clouds. It is delicate, detached, fibrous, feather like cloud of silky appearance without shading which appear as bright red or orange before sunrise and after sunset.

(*ii*) ***Cirro-stratus*** **:** It is thin, whitish sheet or cloud and covers all or a good portion of sky giving the sky a slight milky white appearance.

(*iii*) ***Cirro-cumulus*** : It develops from cirro-stratus and appears in patches of small white flaky or sealy globular masses covering small or large portion of the sky having no shading.

(*iv*) ***Alto-stratus :*** This cloud is uniform bluish or greyish white cloud sheet covering all or large portion of the sky. This is generally rain cloud and gives good amount of rainfall in the middle and high altitudes.

(*v*) ***Alto-cumulus :*** This cloud form as elliptical globular units occurring individually or in group.

(*vi*) ***Stratus :*** It is a uniform grey cloud sheet or layer having no particular form or structure and usually completely cover the sky. This cloud usually occur in the vicinity of monsoon depression.

(*vii*) ***Nimbo-stratus*** : This cloud is dense, dark grey and shapeless cloud sheet. It causes steady and continuous rain during monsoon.

(*viii*) ***Strato-cumulus :*** It is a cloud layer with large lumpy masses or dull gray colour with bright intensities, which usually cover all or most of the sky.

(*ix*) ***Cumulus :*** It is a very thick cloud having dome shaped upper surface or cauliflower shaped which may occur at any season.

(*x*) ***Cumulo-nimbus :*** It develops from cumulus. It is known as thunder clouds rise like mountain towers.

6. Rainfall

The word rainfall is often used as a synonym with precipitation. Rainfall is the precipitation in the form of liquid drop larger than 0.5 mm in diameter falling on earth. The falling of any type of condensed moisture to the ground surface is called precipitation. There are two distinct monsoon winds in India, namely South-West and North-East monsoon wind and they are responsible for rainfall in different seasons in the different places. Rainfall is an important element which determines the climate of a place. According to the total amount of annual rainfall, climate may be referred as follows :

(*i*) ***Arid Climate*** **:** The annual rainfall not exceeding 50 cm is called *arid climate.*

Example : Parts of Rajasthan.

(*ii*) ***Semi-Arid Climate :*** The annual rainfall blowing 125 cm. is called *semi-arid climate.*

Example : The climate of the plain of the Punjab, Delhi, Uttar Pradesh, Madhya Pradesh, Hyderabad etc.

(*iii*) ***Humid Climate :*** The annual rainfall exceeding 125 cm is called *humid climates.*

Example : The climate of Assam, West Bengal, Orissa, Tamil Nadu, Mumbai etc.

Rainfall and Crops : The amount and distribution of rainfall have a great influence in the production of a crop as follows :

(*i*) Rainfall provides sufficient moisture in soil for germination of seed, planting of seedings etc.

(*ii*) Plant can absorbs nutrient from the soil when the nutrients in the soil become soluble in rainwater as it acts as solvent and nutrient carrier.

(*iii*) Floods occur due to heavy rainfall and it is harmful to the growing crop in the field.

(*iv*) Drought causes failure of crops totally. Drought occurs due to failure of rainfall.

Climate and Crop Distribution : The cultivation of crop depends on the prevailing climatic condition of a locality. Likewise distribution of crops also depends on the climatic condition. According to Dr. Randhawa, the following region can be defined :

1. Temperate Himalayan Region : This region can be divided into two sub division as follows :

(*a*) *Eastern Himalayan Region :* This include Mishmi Hills, in upper Assam, Sikkim and Bhutan. Rainfall is over 250 cms per annum. There are thick forest of sal. This is a tea growing area. In some places, paddy is grown.

(b) Western Himalayan Region : This region includes Kumaon, Garwal, Shimla Hills, Kulu and Kangra Valley, Himachal Pradesh and Jammu and Kashmir State. The climate is dry, but in northern parts, there is more winter rainfall. Cultivation of maize, potato and paddy is done in some places. Fruit crops particularly apples, cherries, peaches, pears, almond and plums are grown widely.

2. Northern Dry (Wheat) Region : This comprises the Punjab, Haryana, Delhi, North Gujarat, Western Uttar Pardesh, parts of West Madhya Pradesh and Rajasthan. Annual rainfall is between 20 cms and 75 cms. The soil is alluvial and sandy. The chief crops are wheat, barley, gram, maize, jowar, bajra and cotton.

3. Eastern Wet (Rice) Region : This region includes Assam, Meghalaya, Arunachal Pradesh, Mizoram, Tripura, Manipur, West Bengal, Bihar, Orissa, Eastern Uttar Pradesh, Andhra Pradesh, Eastern Tamil Nadu, Kerala and Eastern Madhya Pradesh. The soil is mainly alluvial. This region receives more than 150 cms of rainfall annually. The main crops are rice, jute, sugarcane and tea.

4. Western Wet region : It includes Kerala, Western Coastal strips, Karnataka, and adjoining areas. The soil is lateritic and annual rainfall is over 250 cms. Plantation crop is predominant in this region. The main crops are coconut, rubber, cashewnuts, arecanuts, black pepper and cardamon. Rice is main food crop.

5. Southern Medium Rainfall (Millet) Region : It comprises Southern Uttar Pradesh, South Gujarat, Madhya Pradesh, Western Andhra Pradesh, Western Tamil Nadu, Eastern Maharashtra and parts of Karnataka. Annual rainfall ranges from 50 cms to 100 cms. The soil is partly black cotton and lateritic. The main crops are jowar, bajra, groundnut, castor and cotton.

Climate is, therefore, largely responsible for geographical distribution of crops. Climate is also responsible for the seasonal distribution of crops particularly in Northern India. (V.N. Sahai, 1992).

2.3 METROLOGICAL INSTRUMENTS

1. Rain Gauge : Raingauge is most common instrument used for measuring the amount of rainfall. It is placed on the ground in a open place to collect the rainfall. The depth of water which is collected in the raingauge is measured.

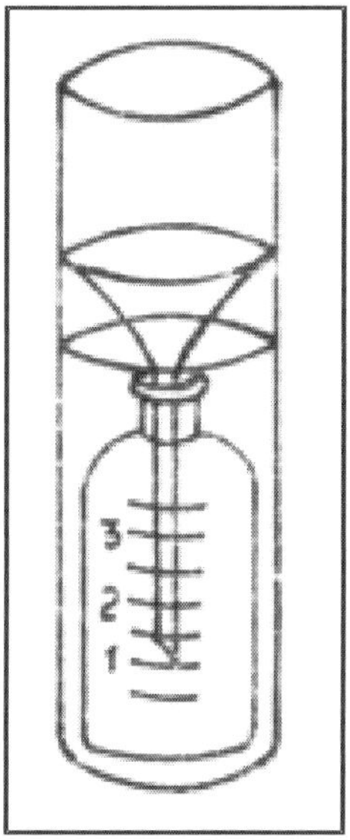

Fig. 2.3 : Rain Gauge

2. Maximum and Minimum Thermometer : Six's maximum and minimum Thermometer is used for recording the maximum and minimum temperature of a day in a particular place.

Temperature is measured either in Fahrenheit or Celsius or in Kelvin units. The melting and boiling point in different scales are as follows :

Scale	*Melting*	*Boiling Point*
Fahrenheit	32°F.	210°F.
Celsius	0°E.	100°C.
Kelvin	273°K.	–

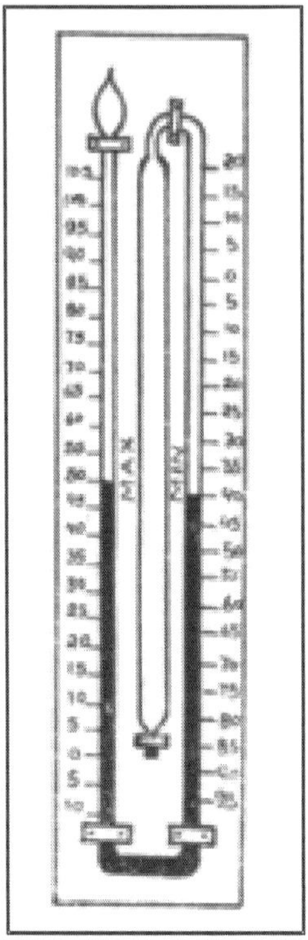

Fig. 2.4 : Six's Maximum and Minimum Thermometers

Conversion formulae of temperature

$$\frac{C}{5} = \frac{F - 32}{9}$$

$$C = (F - 32)\frac{5}{9}$$

$$F = C\frac{(9)}{5} + 32$$

$$K = C + 273.$$

3. Hygrometer : Hygrometer is used for measuring the humidity of surrounding air. The dry and wet bulb thermometer is the most common hygrometer. In this instrument, two thermometers are placed side by side and are graduated in absolute scale. The bulb of one thermometer (*i.e.* wet bulb thermometer) is dipped in distil water and other one is exposed to air when air is not saturated, evaporation takes place from the wet built thermometer and reaching of this thermometer lower down. The humidity of air is calculated from reading of this two thermometer.

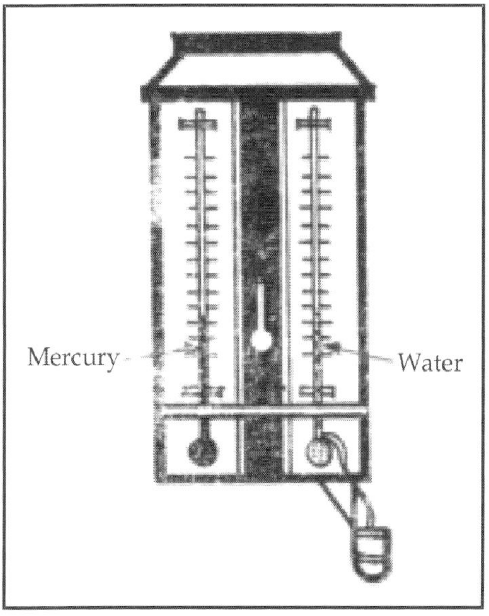

Fig. 2.5 : Hygrometer

4. Barometer : Barometer is used to record the pressure of the atmosphere. There are two types of barometer, such as Fortin's barometer and Aneroid barometer. Out of these, Fortin's barometer is most common type of barometer. The instrument consist of tube in which mercury is poured. The tube is made to stand in an inverted position in a leather cup. The column of the mercury in the tube is supported by the weight of the atmospheric air. The column of mercury rises when pressure of air increases and falls down when pressure of air decreases.

5. Wind Vane : Wind vane is used to indicate the direction of wind. A arrow is mounted on ball bearing in the horizontal plane to enable the arrow to more freely. Four cross arms are fixed rigidly below. The arrow indicating the East, West, North and South. The arrows swings horizontally with slight movement of air and adjusts itself in a position showing the direction of wind.

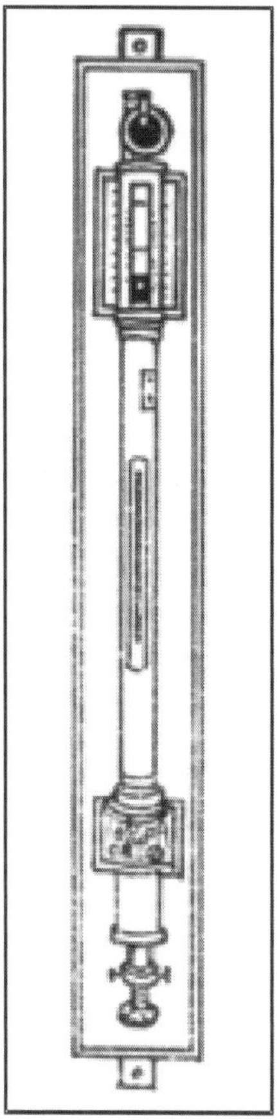

Fig. 2.6 : Fortin's Barometer

6. Anemometer : Anemometer is an instrument used to measure the velocity of wind. It has four cups fixed to the arms to a vertical shaft which moves which the cups are set in motion by wind. The shaft is connected to a dial on which the velocity of wind is recorded.

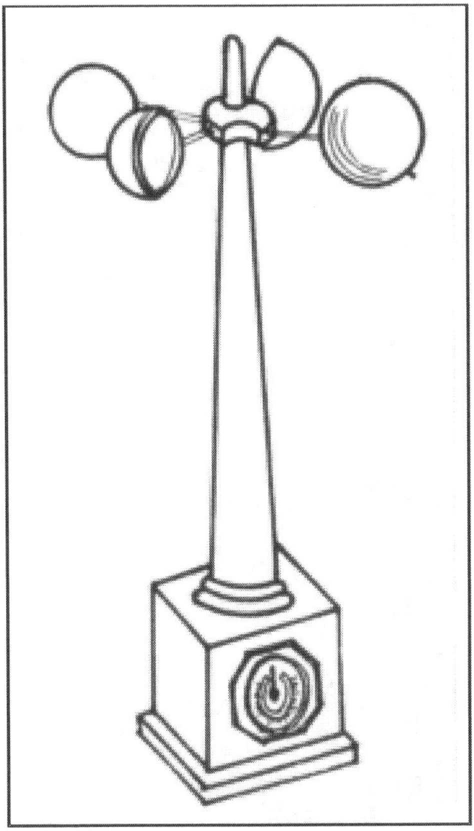

Fig. 2.7 : Wind Vane

2.4 WEATHER HAZARDS AND THEIR MITIGATION

The agricultural activities are dependable on the weather. The unpredictable long and short term fluctuation of weather condition may cause considerable damage to the crops resulting in poor yield. The extreme weather conditions that can cause damage to the crops are called weather hazard. Major abnormalities of weather affecting the crop production are as follows :

1. Flood : Flood occurs due to heavy rainfall. Rainfall of high intensities seals soil pores resulting in surface run off and run off starts when the intensity of rainfall is more than the

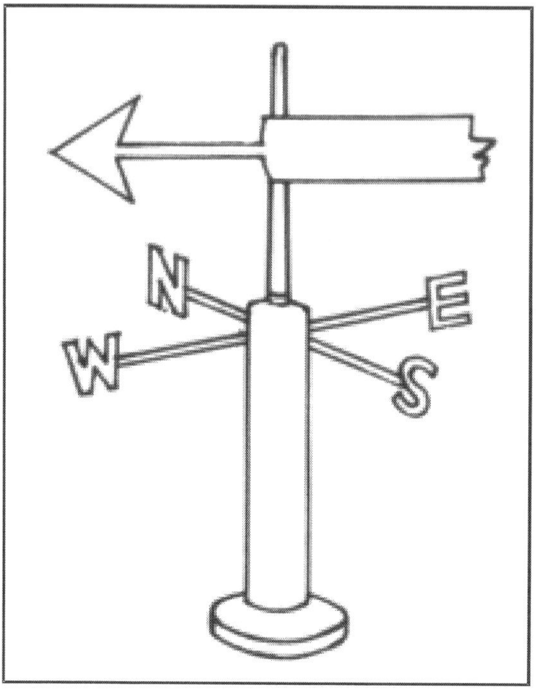

Fig. 2.8 : Anemometer

infiltration capacity of soil, continuous rainfall for a longer times causes flood due to increasing of run off. Floods cause considerable damage to the growing crops in field, live stock and human life.

The methods need to be adopted to overcome the bad effect of flood are as follows :

(*i*) Construction of multipurpose dams on the river.

(*ii*) Afforestation in the catchment areas.

(*iii*) Adoption of watershed development programme.

(*iv*) Provision of adequate drainage.

(*v*) Cultivation of flood resistant variety of paddy. (*e.g.* Jaladhi-1, Jaladhi-2, Jalaprava, FR-13A, F.R. 43B, etc.).

2. Drought : Drought is adverse weather condition. It is occurred mainly due to delay in onset of monsoon, prolonged dry spells during the crop period and early cessation of monsoon. Drought causes crop failure due to deficiency of moisture in the soil.

The management practices that are recommended to mitigate the adverse effect of drought are as follows :

(*i*) Adoption of *in situ* conservation of rain water.

(*ii*) Increasing water holding capacity of soil by application of bulky organic manure.

(*iii*) Adoption of soil and moisture conservation practice to minimise run-off.

(*iv*) Adoption of agronomic practices to overcome moisture stress.

(*v*) Collection of runoff water in reservoir for use in drought period.

(*vi*) Minimising evaporation and transpiration losses by using by mulches, shelterbelts and chemicals.

(*vii*) Cultivation of drought resistant and drought escaping cultivars of crops.

(*viii*) Adoption of intercropping to avoid total crop failure.

3. Untimely Rains : On set and withdrawal of monsoon largely determine the success of dry land agriculture. Timely sowing of crop is not possible if timely rains do not occur. Early withdrawl of monsoon affects the yield of crop due to moisture stress during grain development stage of kharif rice. Rains during pollination, fertilization and harvesting cause damage to the crops.

The management practices that are recommended to mitigate the adverse effect of untimely rains are as follows :

(*i*) Land should be prepared for kharif crop by taking advantage of premonsoon showers.

(*ii*) Short duration variety especially pulse crop should be cultivated.

(*iii*) Intercropping should be practised to avoid total crop failure.

4. Frost and Cold Waves : Frost and cold waves occur in North India and they cause considerable damage to grain and horticultural crops. The leaves show burning effect, wilts and finally drop-off when frost occurs.

The following measures should be undertaken to avoid the adverse effect of frost and cold waves :

(*i*) The frost tolerant crop such as cabbage, cauliflower etc. should be cultivated.

(*ii*) The crop susceptible to frost such as potato, tomato, sugarcane, brinjal, cotton etc. should be cultivated in a manner which facilitate the harvesting of crop before the appearance of the frost.

(*iii*) Irrigation is to be given to the crops succeptible to frost before its appearance.

(*iv*) Heating the garden by lighting small fire help to rise the temperature and due to rise of temperature, frost do not occur.

(*v*) Trees planted as wind break help to protect the vegetation from frost as those trees do not allow the cooling of vegetation on their ice-side.

(*vi*) Covering the ground and crops during night by mats, paper, cloth or cellophane protects valuable crops from the effect of frost.

5. High Winds : High wind causes damage to the crops. High winds are associated with tornadoes are the most destructive among the weather abnormalities. Generally they occur in middle latitude and cause heavy down pour of rain, hail and terific winds, cyclones are most common in India and cause heavy damage to crops, livestock and loss of human life.

Windbreaks and shelterbelts are found effective for considerable protection of high wind to horticultural crops.

6. Heat Waves : It occur in Deccan and central parts of India during hot weather period (March to May). During this period, midday temperature go beyond 43°C. Due to high temperature and hot winds, most of the water bodies dry up quickly resulting in shortage of water. Heat weaves causes pollen and premature fruit crop. Shading of vegetable crops and wind breaks can reduce the heat wave effect on crops. Wind break is found suitable to reduce heat wave effect by decreasing the velocity of hot winds.

2.5 WEATHER AND CROPS

The growth and development of crops and incidence of insectpests and diseases occur at particular weather condition as follows :

1. Rice (*Oryza sativa*) Growth and Development :

Rain fall	: 100 – 150cm.
Temperature	
Optimum	: 21°C – 35°C.
Vegetative growth phase	: 16°C – 27°C.
Flower initiation	: 27°C – 29°C.
Ripening stage	: 23°C.
(Night temperature)	

INSECT PEST

1. Paddy Stem Borer (*Scripophagus incertullus*) : It is the most destructive insectpest of rice. Late varieties are liable for infestation. The meteorological limit for maximum moth emergence and abundance are as follows :

(*i*) Minimum temperature in the range of 14°C – 27°C.

(*ii*) Maximum temperature in the range of 28°C – 30°C.

(*iii*) Soil temperature (at 5 cm. soil depth) in the range of 19°C–30°C.

They are destroyed due to heavy rainfall.

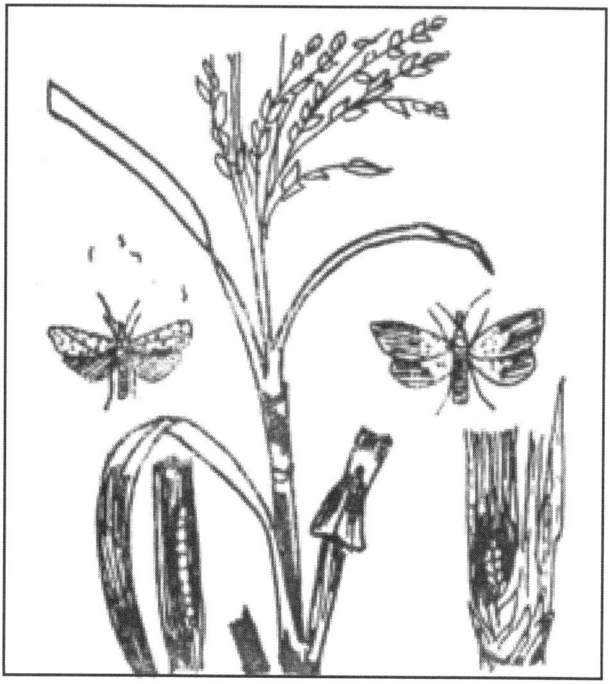

Fig. 2.9 : Paddy Stemborer

2. Gall midge (*Orselia oryzae*) : This pest is most active in the later half of monsoon crop, particularly in tillering stage

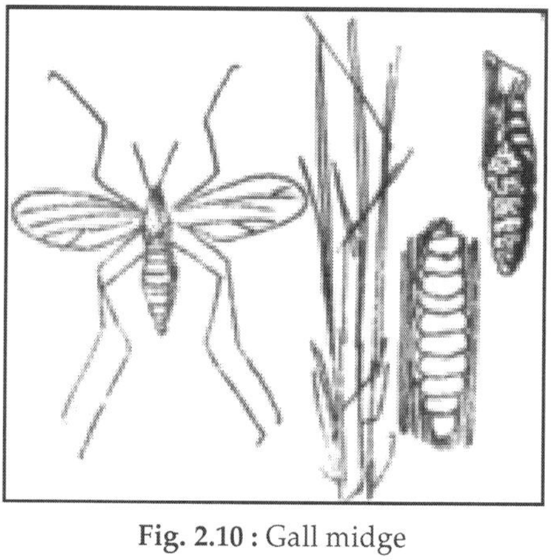

Fig. 2.10 : Gall midge

under condition of land sub-mergence. If rainfalls in May and June are below 250 mm., its first appearance on rice will be delayed. Rice crops in low lying field and in irrigate fields are attached by this pest seriously. Standing water in rice field favour egg laying and hatching of eggs.

3. Rice hispa (*Dicladispa armigera*) : This is serious pest in late planted kharif crop. Heavy rains in July followed by abnormal low rainfall in August-September with lesser rainy days and more bright sunshine, steady temperature with little diurnal average relative humidity (more than 85 per cent) causes outbreak of hispa epidemic (Reddy, 1999).

Fig. 2.11 : Rice Hispa

4. Brown Plant Hopper (*Nilaparvata lugens*) : This is serious pest of high yielding, kharif and boropaddy due to high tillering capacity and due to humid micro environment (RH about 85 per cent) caused by frequent rainfall/irrigations. Low atmos- pheric RH in summer (around 50–60 per cent) is congenial for growth and development of this pest.

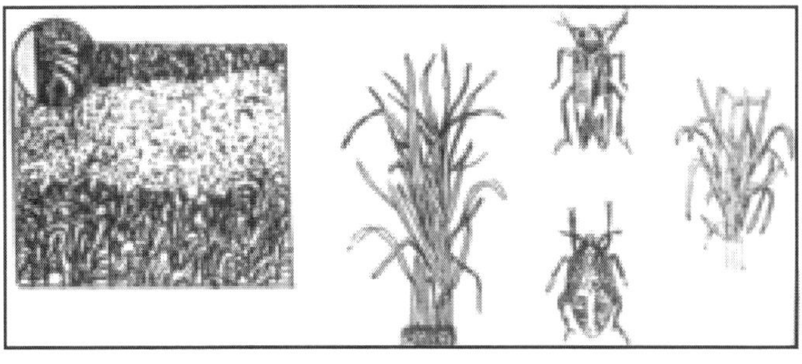

Fig. 2.12 : Brown plant hopper

DISEASES

1. Brown Leaf Spot Disease (C.O. *Helminthosporium oryzae*) : This is a serious disease of rice and it is widely prevalent in all rice growing regions of India. Unusual heavy rains at the time of ear formation followed by flooding and ill drained conditions are good predisposing factors of the disease. Occurrence of either light rainfall of heavy dew continuously for a week is congenial for the spread of the disease. 20°C–30°C temperature is congenial for the spread of the disease.

Fig. 2.13 : Brown leaf spot disease

2. Rice Blast (*Co-piricularia oryzae*) : This is most important fungal disease of rice. It is found in all rice growing region of our country. The temperature around 28°C is the best for the maximum growth of this fungus. High rainfall and more rainy days associated with high humidity causes out break of the disease. The disease incidence is closely related with the temperature and humidity regime. Blast occurs when the minimum temperature remains at 24°C–26°C with RH more than 90 per cent during the seeding, tillering and ear emergence stage of the crop. The severity of the disease increases if temperature runs below this for longer number of days calm to gentle wind intermittent rainfall, rainy days followed by rainless days with bright sunshine and again rainy days with high humidity help in rapid spread of the disease (Lenka, 1998).

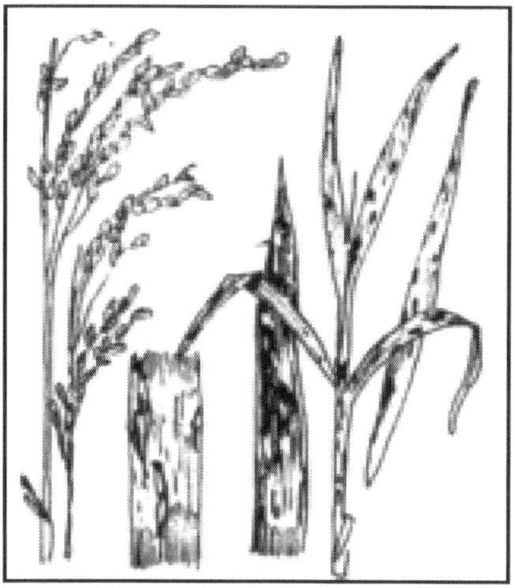

Fig. 2.14 : Rice blast

2. Wheat (*Triticum* sp) : Growth and Development

Rainfall : 25.4–112.5 cm per annum.

Temperature

Seed Germination : 20°C–22°C.

Grain Development :

Minimum temperature : 12°C.

Maximum temperature : 25°C.

Growth of plant : 20°C – 25°C.

DISEASES

1. Brown rust (*Puccinia triticina = P. recondita*)

Mean temperature : 9°C –13°C.

Relative humidity : 70 per cent.

Intermittent clouds from late January to first week of February favour the disease incidence.

2. Yellow rust (*Puccinia glumarium = P. striliformis*)

Mean temperature : 9°C – 13°C.

Relative humidity : 70 per cent.

Partly cloudy condition favour the disease incidence.

3. Sugarcane (*Saccharum officinarum*) : Growth and Development

Rainfall : 150 – 175 Cm per annum.

Temperature :

Germination of selt : 20°C – 23°C.

Plant growth : 21°C – 26.2°C.

Flowering : 27°C – 31°C.

Insect Pest

1. Top Shoot Borer (*Scirpophaga nivella* Fabr.) : It is a serious pest of sugarcane.

Incubation of egg

Temperature : 32°C – 33°5°C.

Relative humidity : 80 – 90 per cent.

Heavy monsoon rains help in profuse multiplication of the pest.

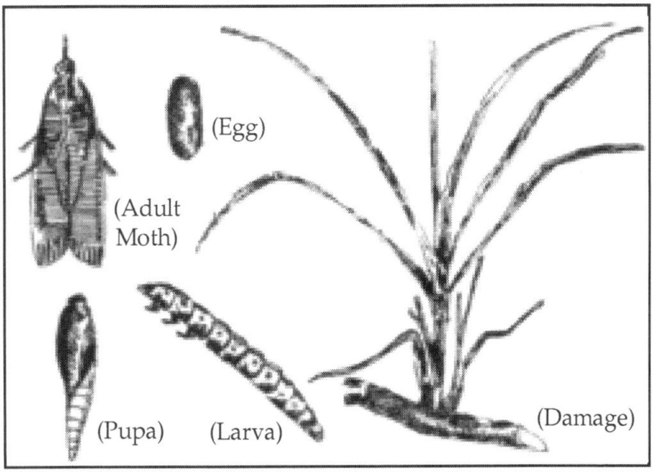

Fig. 2.15 : Sugarcane top Shoot borer

2. Sugarcane Leaf Hopper (*Pyrilla perpusilla* Wlk.): It is a serious pest of sugarcane and widely prevalent in all sugarcane growing regions of our country. This pest multiplies quickly during warm weather having higher relative humidity.

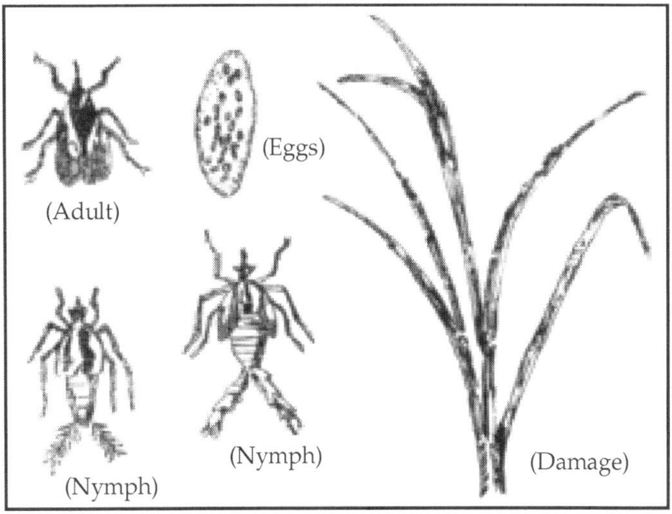

Fig. 2.16 : Sugarcane leaf hopper

3. Potato (*Solanum tuberesum*)

Rainfall : Nil

Temperature

(*a*) Growth of Plant : 15.5°C – 18.3°C.

(*b*) Tuber formation : 17.7°C.

Soil Temperature : 15°C – 18°C.

Soil moisture stress results the lower yield of potato.

Insect Pests

1. Potato Aphid (*Myzus persicae*) : Myzus persicae appears in potato field in the eastern plains (Bihar and West Bengal) in early January and reaches the critical level *i.e.* 20 aphids per 100 leaves by the end of January and in the North Western plains of India from early December and reaches critical level by the end of December. Weekly mean temperature and maximum temperature of 6.3°C to 9.5°C and 25.3°C to 25.3°C have been found as most favourable for the multiplication of *M. persicae*. Temperature above 27°C check their growth. The lower threshold temperature for its development is about 6°C. High humidity does not play direct role either on growth and reproduction, as aphids live in a moist layer. Heavy rains may dislodge many aphids from the plants (Misra *et. al.*, 1995).

Diseases

1. Early Blight (C.O. *Alternaria solani*) : It is one of the most common and widespread diseases of potato and it is observed wherever potatoes are cultivated. The incidence of this disease is mainly influenced by moisture, temperature, variety of potato and age and vigour of the plant.

Optimum temperature : 26°C – 28°C.

Humidity : 91–97 per cent.

Frequent rains followed by warm and dry weather favour development and spread of this disease. Disease occurs from December to January in plains.

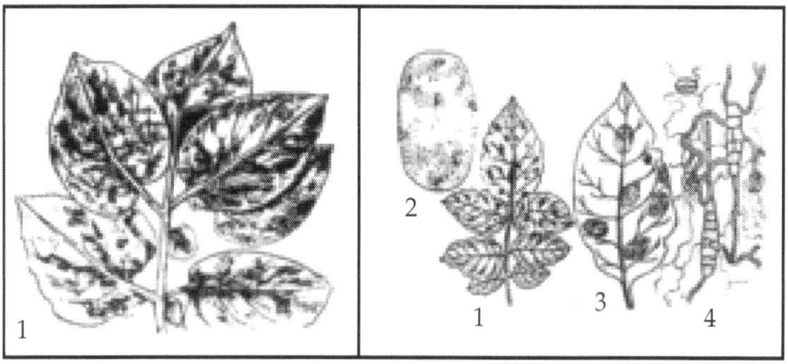

Fig. 2.17 : Early blight of potato

(*i*) 1. Early blight affected leaves. 2. Early blight affected tuber. 3. Early blight symptom in single leaf. 4. Conida.

(*ii*) *Late Blight* (*C.O. Phytopthora infestans*) : Late blight is one of the most serious disease of potato, particularly in temperate region. Moisture and temperature are the main factors limiting the development of late blight and these weather conditions are responsible for geographical distribution of the disease.

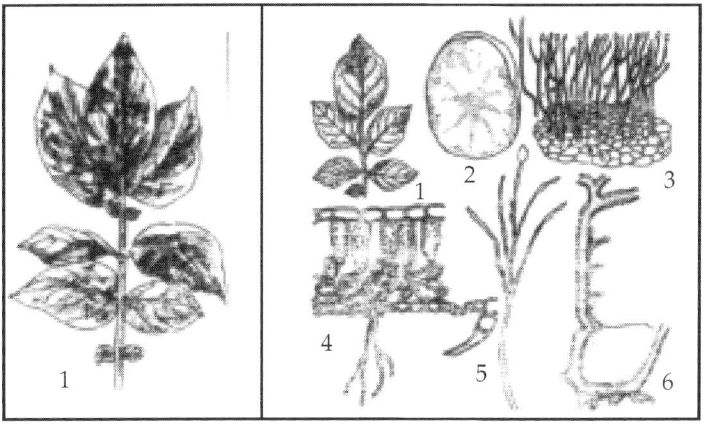

Fig. 2.18 : Late blight of potato

1. Late blight affected leaves 2. Late blight affected tuber. 3. Condiosphore. 4. Position of conidia in affected leaves. 5. Conidia. 6. Penetrating hypae in host plant.

Temperature (Optimum) : 18°C – 22°C.

Humidity : 91–97 per cent.

The cool (12°C – 15°C) and humid weather (Relative humidity above 91 per cent) with heavy dews or rains alternating with warm (20°C) moist period favour the rapid development of this disease.

2.6 CROP SEASON AND CLASSIFICATION OF CROPS

(A) Crop Season : Crop is grown according to the season. Some crop grow well in winter while some other crops grow in summer. Different crops are grown in the following three season such as :

1. Kharif Season

2. Rabi Season
3. Pre-Kharif of Zaid Season

1. Kharif Season : The time between the middle of June to middle of October is called Kharif Season and the crops grown in this season are known as Kharif crop. Kharif crop is generally grown as rainfed crop and high rainfall, high temperature, high relative humidity etc., is essential for this crop.

Examples : Paddy (*Oryza sativa*), Aus and Aman ; Jute (*Corchorus olitorius L., C. capsularis*) Sannhump (*Crotolaria juncea*), Jowar (*Sorghum vulgare*), Bajra (*Pennisetum typhodeum L)* cowpea (*Vigna catjang Walp/V. sinensis Savi*) Castor (*Ricinus communis* L), Ground nut (*Arachis hypogaea*) etc.

2. Rabi Season : The time between the middle of October to middle of February is called *Rabi Season* and the crops grown in this season are known as *Rabi crops*. Rabi crop is generally grown as an irrigated crop and low temperature and short days are essential for this crop.

Examples : Potato *(Solanum tuberosum)*, Cabbage *(Brassica oleracca* Var. *capitata)*, Cauliflower *(Brassica oleracea* Var. *Botrytis)*, Kholkhol *(Brassica culorapa)*, Wheat *(Triticum Sp.)* Barley *(Hordeum vulgare) Mustard (Brassica* Sp*)*, Gram *(Cicer arietinum)*, pea *(Pisum arvense)* Lentil *(Lens esculentus)*, Beet *(Beta vulgaris)*, Turnip *(Brassica campestris Var. rape* L.*) (Brassica rapa)*, Carrot *(Daucus carota)* Radish *(Raphanus sativus)*, Spinach *(Spinacia oleracea)* Safflower *(Carthamus tinctorius)*, Tomato *(Lycoprescon esculentum)* etc.

3. Zaid Season or Pre-Kharif Season : The time between middle of February to middle of June is known as Zaid season and the crops grown in this season are known as Zaid crop. High temperature, long day etc. is essential for the Zaid crop.

Example : Blackgram (*Phaseolus mungo* Var. *radiatus*). Green gram (*Phaseolus aureus*), Water melon (*Citrulus vulgaris*) etc.

Crop season has remarkable effect on crop production. For this reason, all crop do not grow at all times of the year.

The Kharif crop cultivated in rabi season or rabi crop cultivated in Kharif season do not grow well or fails to grow. For this crop planning should be made by keeping notion on the crop season. There are some crops that can grow in all times of year as follows :

Brinjal (*Solanum nigrum*), pumpkin (*Cucurbita pepo, Cucurbita moschata duch*), H.Y.V. paddy (*Oryza sativa*), Maize (*Zea mays*) Radish (*Raphanus sativus*), Chilli (*Capsicum frutescens*) etc.

But it is necessary to cultivate those crop by selecting variety suitable for the particular season. Otherwise, desirable yield is not possible. As for example, chaitali pumpkin grown in rainy season or rainy season pumpkin grown in winter, do not yield satisfactorily. Sweet potato (*Ipomoea batatus*) etc.

2.7 CLASSIFICATION OF CROPS

Crop plants are classified in a number of ways. From agronomic point of view, crop plants are classified as follows:

1. Cereals : Sources of carbohydrate : Paddy (*Oryza sativa*), Wheat (*Triticum* sp), Oat (*Avena sativa*), Barley (*Hordeum vulgare*).

2. Millets (Sources of Carbohydrate) : Millet are small grained cereals. It is used as stable food in drier regions of developing countries.

Example : Bajra (*Pennisetum typhoides*), Jowar (*Andropogon sorghium*) *Boot/Sorghum vulgare per*), Maize (*Zea mays*), common millet (*Panicum millaceum*), finger millet (*Eleusine coracana*), little millet (*Panicum miliare*) etc.

3. Oil Seed (Sources of Oils) : Rapeseed and mustard (*Brassica sp*), Sesamum (*Sesamum indicum/S. orientale*) Linseed (*Linum usitatissinum*), Castor (*Ricinus communis*) Sunflower (*Helianthus annuus*), Safflower (*Carthamus tinctorius*), Groundnut (*Arachis hypogaea*) etc.

4. Pulses (Sources of Protein) : Gram (*Cicer arietinum*), Pea (*Pisum sp*), Red gram (*Cajanus cajan*), Lentil *(Lens esculenta/* L. *culinaris*), Gram (*Phaseolus aureus*), Black gram (*Phaseolus* *mungo*). Cowpea (*Vigna sinensis*), Soyabean (*Glycine max*) etc.

5. Sugar Crops (Sources of Sugar) : Sugarcane (*Saccharum officinarum*), Sugar beet (*Beta vulgaris*) etc.

6. Fibre Crops (Sources of Fibre) : Jute *(Chorchorus* sp.*)*, Cotton (*Gossypium* sp.), Mesta (*Hibiscus sp.*), Sannhemp (*Crotalaria juncca*) etc.

7. Vegetables (Sources of Vitamins and Minerals) :

(*a*) ***Country Vegetables :*** Pumpkin (*Cucurbita pepo, C. moschala Duch*), Karela *(Momordica charantia*), Cucumber (*Cucumis sativus*), Lady's finger (*Hibiscus esculentus/ Abelmoschus esculentus Mocnch),* Indian bean or Deshi Shim (*Dolichos lablab*), Ridge gourd *(Luffa acutangula*), Pointed gourd (*Trichosanthes dioica*), Snake gourd *(Trichosanthes anguina*), Smooth gourd *(Luffa cylindrica*) etc.

(*b*) ***English or Exotic Vegetables :*** Potato (*Solanum tuberosum*), Cabbage *(Brassica oleracea var capitata),* Cauliflower *(Brassica oleracea* var. *botytis),* Kholkhol *(Brassica caulorapa/B.oleracea* Var. *Caulocarpa),* Tomato *(Lycopersicon esculentum),* Lettuce *(Lactuca sativa)* , Radish *(Raphanus sativus),* Carrot *(Daucus carota)* Turnip *(Brassica rapa),* Beetroot *(Beta vulgaris),* Spinach *(Spinacia oleracea)* etc.

8. Spices and Condiments (Sources of spices products): Anise (*Pimpinella anisum*), Cumin (*Cuminum cyminum*), Ginger (*Zingiber officinale Rosc),* Coriander (*Coriandrium sativum*), Black pepper (*Piper nigrum*), Chilli (*Capsicum frutescens*), Methi (*Trigonella foenumgraecum*), Onion (*Allium cepa*), Garlic (*Allium sativum*), Cardamon (*Elettaria cardamomum*), Nutmeg (*Myristica laurifolia/M. fragrans Hautt.*) etc.

9. Pot Herb (Source of Leafy Vegetables) : Spinach (*Spinacia oleracea*), Bottle gourd (*Lagenaria leucantha* Rusby/L. *siceraria* Stanll. Amaranth (*Amaranihus* sp).

10. Narcotics (Source of Narcotic Products) : Tobacco

(*Nicotiana tabacum* L). Indian hemp. Or ganja (*Cannabis sativa* L) Poppy (*Papaver somniferum* L), Betel vine (*Piper betle*).

11. Root Crops (Sources of Vegetables and Spices) : Sweet potato (*Ipomoea batatus*), Radish (*Raphanus sativus*), Carrot (*Daucus carota* L), Beet (*Beta vulgaris*).

12. Stem or Tuber Crop : (Sources of vegetable and spices) potato (*Solanum tuberosum*), ginger (*Zingiber officinale*) Rosc, Turmeric (*Curcuma longa*).

13. Forage crops (Sources of Fodder for Cattle) : Egyptian clover or Berseem (*Trifolium alexandrium* L), Alfalfa or Lucem (*Medico sativa L*), Guinea grass (*Panicum maximum* Jacq.), Napier grass (Pennisetum purpureum Schum), Indian clover (*Melilotus parviflora* desv.), Johnson grass (*Sorghum halepense*), Rhodes grass (*Chloris gayana* Kunth), Sudan grass (*Sorghum sudanense* stapf).

14. Dyes (Sources of Dye Products : Safflower (*Carthamus tinctorius*), Mehedi (*Lawsonia alba* Lamk/L. *inermis* L), Indigo (*Indigofera sativa* L). etc.

15. Plantation Crop (Sources of Beverage and Fruit): Areca palm (*Areca catechu* L), Cacao (*Theobroma cacao* L.) Cinchona (*Cinchona officinalis* L), Coffee (*Coffea arabica* L), Para rubber (*Hevea graziliensis; Hevea brasiliensis* Mullang) Tea (*Camellia thea, Camellia sinensis* O. Ktze.) etc.

16. Medicinal Plants (Sources of Medicine) : Penicillium (*Penicillium notatum*), Sarpagandha (*Rauwolfia serpentina*), Vasaka (*Adhatoda vasica*), Tulsi (*Ocimum sanctum*), Kalmegh (*Andrographis paniculata*), Cinchona (*Cinchona succirubra*), Nux vomica, Belladona, Ipecac etc.

17. Fruits Crops (Sources of Vitamins and Minerals) Mango (*Mangifera indica*), Jackfruit (*Artocarpus integrifolia* L: *Artocarpus heterophyllus* lam), Apple (*Pyrus malus* L), Banana (*Musa sp*), Bullock's heart (*Anona reticulata* L), Cape groseberry (*Physalis peruviana* L) Cashew (*Anacardium occidentale* L), Cherry (*Prunus avium* L), Custard apple (*Annona squamosa* L), Date palm (*Phoenix dactylifera* L), Fig (*Ficus carica* L), Grape fruit (*Vitis* *vinifera* L), Guava *(Psidium* guajava L), Jujube *(Zizyphus jujuba* L*), Zizyphus mauritiana* Lam), Lemon (*Citrus limonia* Ozbeck, *Citrus limon* Burm f.) Lime Acid one sour (*Citrus aurantifolia* swingle), Litchi (*Litchi chinensis* sonn.), Loquat (*Eriobotrya japonica* Lindl), Mangosteen (*Garcinia mangostana* L), Mulbery (*Mulbery morusals* L*)*, Papaya (*Carica papaya* L*)*, Peach (*Prunus persica* Batsch). Pear (*Pyrus communis* L), Pine apple (*Ananas sativa* Schult, *Ananas comosus* Merr.) Plum *(Prunus domestica* L), Pomegranate (*Punica granatum* L), Sapota (*Achras sapota* L), Pomelo (*Citrus decumana* Murr. *Citrus maxima* Merril, *Citrus grandis* Osbeck), Santra orange, Mandarim orange (*Citrus reticulata* Blanco), Strawberry (*Fragaria vesca* L), Sweet orange mosambi (*Citrus sinensis* Osbeck) etc.

18. Timber Crop : (Sources for fuel and timber for furniture making). Sal (*Shorea robusta*), Segoon or Teak (*Tectona grandis*), Akashmoni (*Acacia auriculiformis*), *Gamar* (*Gmelina arborea*), Sishuo (*Dalbergia sisso*), Sirish (*Albizia lebbeck*), Mehogini (*Swietenia macrophylla*) etc.

19. Green Manuring Crops : (Crops used for green manuring) Dhaincha (*Sesbania aculeata*), Sannhemp (*Crotalaria juncea*), Cowpea (*Vigna catjang*), Lentil (*Lathyrus sativus*), Soybean (*Glycine max*), Maize (*Zea mays*) etc.

20. Perfume Producing Crops : (Sources of perfume) Sandal (White sandal - *Santalum album*, Red Sandal- *Pterocarpus santalinus*), Neem (*Azadirachta indica*) Mutha (*Cyperus rotuntus*).

21. Beverage Crops (Sources of Beverage) : Tea (*Camellia thea*), Coffee (*Coffea* sp), Cocoa (*Theobroma cacao*) etc.

22. Essential Oil Producing Crops : (source of essential oil). Eucalyptus (*Eucalyptus* sp), Lemongrass etc.

REFERENCES

Anandeshwari, Awasthi, Indian Climatology (1995), A.P.H. Publishing Corporation, New Delhi.

Balasubramaniy P. and Palaniappan, Principles and Practices of Agronomy

(2004), Agrobios (India), Jodhpur.

Chritechfield. H.J. General Climatology (1968), Prentice Hall of India Pvt. Ltd., New Delhi.

Gopalaswamy W. Agricultural Meteorology (1994), Rawat Publication, Jaipur.

Kakde, J.R. Agricultural Climatology (1985), Metropolitian Book Co. Pvt. Ltd., New Delhi.

Mavi. H.S. Introduction to Agrometerology (1986), Oxford and IBH Publishing House, New Delhi.

Reddy S.R. Principles of Agronomy (1999), Kalyani Publisher, Ludhiana.

Sankaran S. and Mudalian V.T.S. Principles of Agronomy (1991). The Bangalore Printing and Publishing Co. Ltd., Bangalore.

3

GROWTH AND DEVELOPMENT OF CROPS

3.1 GROWTH OF CROPS

Definition : Growth may be defined as permanent and irreversible change in volume or size of an organism accompanied by 'an increase in dry weight' (Ganguly *et. al.*, 1972).

Growth is one of most important characteristics of plant and it is associated with photosynthesis.

3.2 PHOTOSYNTHESIS

Photosynthesis (photo = light, synthesis = putting together) is the synthesis of energy rich carbohydrates from water and carbon dioxide in illuminated chloroplastids. This conversion of carbon dioxide and water into carbohydrates is an energy consuming process. The chlorophyllus cells of the plant are able to absorb light (*i.e.* active or kinetic energy) and to transform it into chemical energy (*i.e.* potential energy) which is stored in energy rich carbohydrates.

The chemical equation of photosynthesis is as follows:

$$6CO_2 + 12H_2O \xrightarrow[\text{Chlorophyll}]{\text{Sunlight}} C_6H_{12}O_6 + 6H_2O + 6O_2$$

Photosynthesis provides the raw materials for plant growth and for the development of storage organs such as leaves, stems, tubers and fruits. Growth in plants has various changes and this phases starts with activation of embryo and ends with maturation of seed.

(*a*) Gemination of Seed : The process by which the embryo awakens from its dormant condition and begins to grow is known as germination.

The seeds, when sown in moist soil, absorbs water and as a result of absorption of water, seeds germinate gradually. The cotyledon of a seed provide food material during the process of germination. The embryo contains the miniature shoots are roots.

Favourable conditions are necessary for the germination of seed and the production of normal and healthy seedlings. Seeds are grouped into three categories on the basis of the requirement of light for germination as follows :

(*i*) ***Positive Photoblastic*** : Seeds that require light for inducing germination.

(*ii*) ***Negative Photoblastic*** : Seed that do not germinate when exposed to light.

(*iii*) ***Non-Photoblastic*** : Seeds that germinate either in light or in dark. Seeds of many cultivated crop belong to this group.

(*b*) Emergence of Seedlings : The cotyledons swells greatly due to absorption of water and the testa busts at some point near the micropyle causing an opening through which radicle and plumule comes out. The radicle will always go downward and form the root. The plumule is pushed through the soil and comes out of soil due to subsequent growth. This is known as emergence of seedling. The emergence of seedling is critical for good stand establishment and it is influenced by soil crust, high salt concentration, water lodging condition etc.

(*c*) Root Growth : The primary roots begins its development in the radicle. At this stage the radicle is a mass of undifferentiated meristematic tissue and is usually referred to as promeristem. The root of monocotyledonous and dicotyledonous plant is called fibrous root and tap root respectively.

(*d*) Shoot Growth : The primary shoot begins its development in the plumoule. The shoot consists of stem, branches, leaves and flower. Leaves are always arise from nodes in stem or branches and invariably buds are found in their axil.

(*e*) Inflorescence : Reproduction is the important function of a plant which comes in the plant that racfully developed of their vegetative organ (*i.e.*, root, stem, leaf etc.). The reproductive organ in flowering plant are developed in specially modified shoot called flower. The branch system bearing a collection of flower or cluster of flower is named as inflorescence.

(*f*) Seed and Fruit Growth : The seed and fruit growth starts just after fertilization. The union of male gamets with egg or female gamets is known as Fertilization. Pollination follows fertilization. The transfer of pollen grain from the anther to stigma is known as pollination. As a result of fertilization, the ovary and ovule increase greatly in size and transformed into fruit and seed respectively. Photosynthates are supplied to seed and fruit from other parts of the plant.

Sometimes seeds and fruits maturation varies from plant to plant.

3.3 DEVELOPMENT OF CROP

The change over from vegetative stage to reproductive stage leading to the differentiation of floral parts and its subsequent growth is referred to as development. It denotes phasic change. The term development is need to encompass the activities resulting from growth are differentiation (Noggle and Fritz, 1950). Developmental process relate to varying degree of expansion or growth in weight of an organ. The developmental schedule in relation to the growth of early varieties of rice (duration : 110-120 days), wheat or any dicotyledonous plant is as follows :

(*i*)	Germination of seed	:	2-3 days
(*ii*)	Emergence of seedling	:	1-2 days
(*iii*)	Vegetative growth (*i.e.*, filtering, crown, root initiation or production of branches)	:	38-40 days

(iv) Growth in reproductive state :

Flower initiation	:	20-21 days.
Pollination	:	12-14 days.
Fertilization	:	3 days.
Development of seed	:	28-30 days.
Dormancy of embryo in seed	:	6-7 days.

3.4 PLANT GROWTH SUBSTANCES

Hormone is called plant growth substance. It is a biochemical product produced in plants or synthetic substances, applied to plants (exogenous) which causes modification in plant growth and development. Hormones are organic substances which regulates physiological process in living organism.

Table 3.1 : Specific Function of Hormones

Name of hormone	*Function*
Auxin (IAA, IBA etc.)	Stimulate elongation of cells of stems and roots.
Cytokinins	Stimulate cell division, promote orderly development of embryos in seed, break dormancy of seeds and buds and delay senescence.
Gibberellins	Increase cell division and cell elongation, breaking of dormancy of seed and inducing flowering.
Florigen	Flower initiation.
Abscisic Acid	Induces leaf and fruit abscission, dormancy of seed and bud etc.
Ethylene	Stimulates the swelling or isodiametric growth of stem and roots. Regulation of ripening of fruits or delay in ripening process.

Table 3.2 : Regulatory Activities of Major Growth Substances (*i.e.* Hormones)

Nature of growth	*Names of hormones associated with the growth*
Rooting	Auxin, gibberellins, cytokinins and ethylene
Growth	Auxin, gibberellins, cytokinins, ethylene and abscisic acid.

Nature of growth	*Names of hormones associated with the growth*
Flower initiation	Auxin, gibberellins, cytokinins, ethylene and abscisic acid.
Abscission	"
Senescence	"
Fruit growth	"
Fruit set and Fruit ripening	All except abscisic acid.

Table 3.3 : Care, cultural operations and nutrition that are essential in different stages of growth of some crops

Name of Crops	*Stages of Growth*	*Requirement of care, cultural operation and nutrition in different growth stages*
Aman Paddy	Vegetative growth phase	This stage is consigned in the stage of growth of seedling filtering and vegetative large phase after the germination of seed. 2-3 inches (5-7.5 cm) water is required in the filtering phase. Growth phase continues from seven week to flowering stage and the growth of plant is more in this phase. 9-12 inches (22.5-30 cm) water is required in this stages. Water management and top dressing of fertilizers are the important cultural operation in this stage. Weeding should be done in this stage.
	Reproductive growth phase	This exists from panicle initiation stage to flowering stage. Top dressing should be done before the panicle initiation stage. It is necessary to maintain optimum water in this stage.
	Maturity phase.	This phase exists from flowering to ripening state of this crop. No cultural operations are required in this stage. Drainage of excess water is to be done after ripening of grains. Proper plant protection measure should be undertaken at each stage is necessary.

Name of Crops	*Stages of Growth*	*Requirement of care, cultural operation and nutrition in different growth stages*
Wheat	Vegetative growth phase	This phase exists from germination of seed upto before the flowering crown. Root initiation stage appears within 18-22 days of sowing seed. This is the most important stage of wheat first irrigation and top dressing should be done in this stage.
		Second irrigation should be given at late jointing stage (*i.e.* 65 days)
	Reproductive growth phase	Tillering and flowering stages appear at 40-45 and 60-70 days (Days after sowing) respectively.
		Irrigation schedule should be maintained on the availability of water as follows.
		For two irrigation : First at crown root initiation stage (*i.e.* 18-22 days) and second at late jointing stage (*i.e.* 65 days).
		For three irrigations : First at crown root initiation stage, second at late jointing stage and third at flowering stage.
		For four irrigations : First at crown root initiation stage, second at late fillering stage (*i.e.* 45 days), third at flowering stage and five at milk stage of the grain.
		For five irrigations : First at crown root initiation stage, second at late tillering stage, third at late jointing stage, fourth at flowering stage and fifth at milk stage of the grain.
		For six irrigations. : First at crown root initiation stage, second at tillering stage, third at jointing stage, fourth at flowering stage, fifth at milk stage and six at dough stage of the grain.
	Maturity state	Grain formation starts at 90-95 days. Last irrigation should be given at this stage. The wheat is harvested at full ripe stage.

Name of Crops	*Stages of Growth*	*Requirement of care, cultural operation and nutrition in different growth stages*
Potato	Germination phase	Germination commences with the planting of seed tubers and ends with the emergence of the plant with the ground. Proper moisture in the soil is essential for the germination. Irrigation should be given according to mixture content in the soil.
	Vegetative growth phase	With the emergence of the plant above the ground, the plant phases into a stage of intense vegetative activity. The growth is a continuous process and proceeds till near maturity of the plant. The soil should have sufficient moisture for the good growth of the plant. Irrigation should be given to maintain proper moisture in the soil for the uniform growth of plant. Hoeing and weeding are done for breaking the hard soil court which is formed due to rains or irrigation after planting but before germination is completed and for the control of weeds respectively. Nitrogen is essential for the growth of plant. Top dressing with nitrogenous fertilizer should be done before earthing up of the plant.
	Tuberization Phase	Potato shoot emerges out of the soil in about three week and emergence is completed in five weeks. The tubers are usually formed near the surface and many of them become unfit for taste purposes unless haulms are properly earthen up. The first earthing up should be done when the plants are about 10-15cm. high. A second earthing up may also be required to cover up the fusers properly to prevent greening of tubers. The furrow is made due to this operation in which water is applied. Irrigation should be given at an internal of 5-7 days depending on the moisture content of

Names of Crops	*Stages of Growth*	*Requirement of care, cultural operation and nutrition in different growth stages*
		the soil. Suberization stage appears after completion of tuberization and this stage appears during the last stages of growth of plant. The soil should have sufficient moisture in this time and for this irrigation should be given in regular way.
	Maturity stage	Tuberization takes place after 45 days of planting and the development of tuber continues till the vinces dic. The leaves turn yellow and are shed in course of time and haulms dry up and die which may be taken to be an initiation of maturity. The haulms could be cut to enforce maturity, particularly on late varieties. During the time of harvesting the soil must be moist and easily workable. It harvesting is to be delayed it is best to leave the soil dry and light irrigation is given at about the time of harvest, just a couple of days before to make the harvesting crop.

REFERENCES

Das, P.C. Crops and their Production Technology under different condition (2000), Kalyani Publishers, Ludhiana.

Ganguli, H.C. K.S. Das and C. Dutta. College Botany. Vol-1 (1972), New Central Book Agency, Kolkata.

Singh, S.S. Crop Management under Irrigated and Rainfed condition (1990), Kalyani Publishers, Ludhiana.

4

SOIL

4.1 INTRODUCTION

The word soil is derived from latin word "solum" meaning floor. Soil is a complex system made up of mineral matter, organic matter, soil moisture and soil air. Soil constitutes the natural medium which under suitable condition can support plant growth. Soil serves as a reservoir of food material and water needed by the plant and provides mechanical anchorage to them. The approximate composition of soil is as follows:

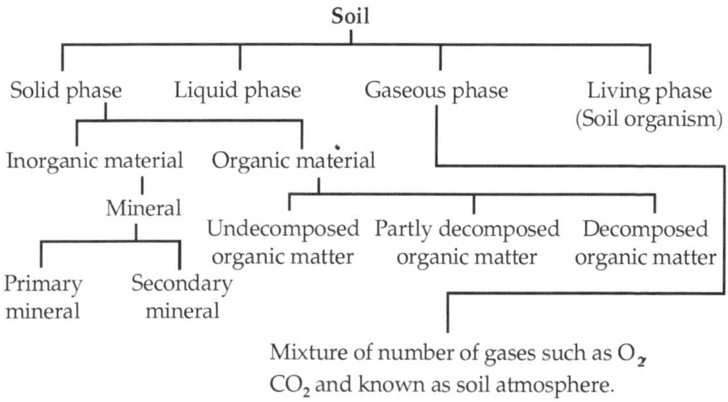

4.2 DEFINITION OF SOIL

1. Soil is the upper weathering layer of earth crust.

   –Ramann

2. Soil is more or less loose and friable material in which by means of their root plant may or do find foot hold

and nourishment as well as other condition of growth.
–Hilgard

3. Soil is the upper part of a earth crust in which plants are anchored. *–Daubenimire*

4. Soil is top most layer of earth crust capping the rock. *–Walia*

4.3 COMPOSITION OF SOIL

Soil is most important constituents of earth's surface. Soil is a solid, liquid and gaseous phase of earth surface which under suitable condition can support plant growth. It is a complex system made up of mineral matter, organic matter, soil moisture and soil air. Soil constitutes the natural medium which support the growth of plant on earth's surface. Soil serves as a reservoir of food material and water needed by the plants and provide a mechanical anchorage to them. Soil is a product of parent rocks and minerals by the process of weathering *i.e.* physical, chemical and biological. Weathering is the initial stage and the soil forming process or pedogenic stages comes afterwards. Weathering process leads to disintigation and decomposition of complex mineral substances that are locked up in the rocks and eventually simple compounds are formed. Soil is a mixture of different constituents as follows :

(*i*) *Solid Phase :* Mineral matter and organic matter.

(*ii*) *Liquid Phase :* Water, known as soil solution.

(*iii*) *Gaseous Phase :* Air.

(*iv*) *Living Phase :* Soil organism.

4.4 FORMATION OF SOIL

Soils are formed by disintegration and decomposition of rocks and of vegetables and animal fossils embedded in them. The conversion of present material into soil is known as formation of soil. There are three types of weathering for the formation of soil.

1. Physical or Mechanical weathering.

2. Chemical weathering.
3. Biological weathering.

1. Physical Weathering : The principal agent of physical weathering are as follows :

(*i*) ***Heat and Cold :*** The general rule of expansion and contraction due to heat and cold (*i.e.,* variation of temperature) is true to rocks also. Rocks are composed of various mineral crystals which posses different co-efficient of expansion and contraction of adjacent unlike minerals due to temperature changes of a day and night loosen the crystals causing the rocks to crumble. In this way, rocks are weathered and finally reduced to soil material by other agencies. This type of weathering is most common in dry climate.

(*ii*) ***Freezing and Thawing :*** When water freezes, its volume increase by about nine per cent and the force exerted is 150 tons per square feet. In the crevices of rock, water frequently freezes and the crevices enlarged by breaking of fragments and when thawing occurs they may roll down the slope.

(*iii*) ***Glaciers :*** Glaciers are the drainage system of the region of perpetual snow. The moving or ice obeys the same laws as streams and does the same kind of work. As the ice is the solid body, it has a great grinding power.

(*iv*) ***Water :*** Torrential rain dislocates solid particles from rock and puddle the surface of some parent materials. The stream of water with their loads of various material has a grinding capacity. The debris is then transported to a long distance by flood water of streams and river and deposited as a soil in low lying areas, mouth of river and under the sea.

(*v*) ***Wind :*** Sand storms in deserts, Semi-arid region and dry river beds and high winds on see shore have both erosive and transporting action.

The soil formed by physical weathering is known as "skeletal soil".

2. Chemical Weathering : In the chemical weathering, the minerals in the parent rocks are decomposed and as a result the new substances are originated. The chemical weathering of rocks is brought about by the following way:

(*i*) ***Water :*** Most of the minerals in the parent rocks are more or less soluble in water. But the solvent capacity of pure water is less. The decomposing action of water is increased by the presence of carbon dioxide and organic acid formed from the oxidation of sulphur compounds or iron sulphides also play an important role in the chemical weathering of rocks. Gypsum and calcarous rocks which are easily soluble in water makes the soil after dissolving in water.

(*ii*) ***Hydration :*** Hydration is the chemical combination of water with other chemical substances. It occurs more commonly in humid region. The union of water with minerals softens the rocks which then result in the weathering of rocks and ultimately brings about their decomposition. Hydration is also sometimes accompanied with an increase in volume, which may bring physical disintegration of rocks. Example of hydration is as follows :

$$\underset{\substack{\text{Hematite} \\ \text{(red)}}}{2Fe_2O_3} + \underset{\text{Water}}{3H_2O} = \underset{\substack{\text{Limonite} \\ \text{(Yellow)}}}{2Fe_2O_3\ 3H_2O}$$

(*iii*) ***Hydrolysis :*** Hydrolysis is essential an exchange of constituent part between water and mineral. The minerals which are affected by hydrolysis are the complex silicates of Calcium, Magnesium, Potassium, Sodium, Aluminium and Iron. Examples of hydrolysis is as follows :

$$\underset{\text{Orthoclase}}{KAl\ Si_3O_8} + \underset{\text{Water}}{HOH} \rightarrow \underset{\substack{\text{Acid silicate} \\ \text{clay}}}{HAl\ Si_3O_8}\ \underset{\substack{\text{Potasium} \\ \text{hydroxide}}}{KOH}$$

$$\underset{\text{Potash felspar}}{K_2O, Al_2O_3, 6SiO_2} + \underset{\text{Water}}{2H_2O} \rightarrow \underset{\text{Intermediate compound}}{H_2O, Al_2O_3, 6SiO_2} + \underset{\text{Potassium hydroxide}}{2KOH}$$

$$\underset{\text{Intermediate compound}}{H_2O, Al_2O_3, 6SiO_2} + \underset{\text{Water}}{H_2O} \rightarrow \underset{\text{Kaolinite}}{Al_2O_3, 2SiO_2, 2H_2O} + \underset{\text{Silicon dioxide}}{4SiO_2}$$

(*iv*) ***Carbonation :*** Carbonation is the chemical combination of CO_2 which is present in atmosphere or released by the decomposition of soil organic matter and respiration of plant roots and micro-organism, with other chemical compounds. When CO_2 combines with water, it forms carbonic acid which increases the solvent power of water. Carbon dioxide or carbonic acid may also react with other chemical compounds resulting in soluble carbonates which dissolve in water and weakens the rock and help its decomposition.

$$\underset{\text{Feldspar}}{2KAl\ Si_3O_8} + \underset{\text{Water}}{2H_2O} + \underset{\text{Carbon dioxide}}{CO_2} \rightarrow \underset{\text{Hydrated}}{H_4Al_2Si_2O_9} + \underset{\text{Potassium carbonate}}{K_2CO_3} + \underset{\text{Silicon dioxide}}{4SiO_2}$$

$$\underset{\text{Calcium hydroxide}}{Ca(OH)_2} + \underset{\text{Carbon dioxide}}{2CO_2} \rightarrow \underset{\text{Calcium bicarbonate}}{Ca(HCO_3)_2}$$

(*v*) ***Oxidation :*** Oxygen combining with various minerals produces soluble oxides which on dissolving in water weaken the rocks and help its disintegration.

$$\underset{\text{Ferrous silicate}}{2FeSiO_3} + \underset{\text{Water}}{3H_2O} + \underset{\text{Oxygen}}{O} \rightarrow \underset{\text{Ferric hydroxide}}{2Fe(OH)_3} + \underset{\text{Silicon dioxide}}{2SiO_2}$$

$$\underset{\text{Iron pyrite}}{2FeS_2} + \underset{\text{Water}}{7H_2O} + \underset{\text{Oxygen}}{15O} \rightarrow \underset{\text{Ferric hydroxide}}{2Fe(OH)_3} + \underset{\text{Sulphuric acid}}{4H_2SO_4}$$

$$\underset{\text{Ferrous oxide}}{4FeCO_3} + \underset{\text{Oxygen}}{O_2} \rightarrow \underset{\text{Ferric oxide}}{2Fe_2O_3} + \underset{\text{Carbon dioxide}}{4CO_2}$$

(*vi*) ***Reduction :*** Under aerobic conditions, when soil pores are saturated with water, oxygen may be absent and reduction may occur.

$$\underset{\text{Ferric oxide}}{Fe_2O_3} + \underset{\text{Hydrogen}}{H_2} = \underset{\text{Ferrous oxide}}{2FeO} + \underset{\text{Water}}{H_2O}$$

3. Biological Weathering : Physical and chemical weathering brings about physical disintegration and chemical decomposition of rocks. Both the plants as well as animals are also largely responsible for further reduction of rock minerals particles into soil in which Crop Plants can grow. Lower plants such as mosses and lichens can grow on almost bare rocks causes gradual disintegration. Grasses, shrubs and trees growing in rock crevices help in extending rock cracks by their root growth. When the plants die, they leave organic matter. The soil organisms convert the organic matter into humus which is very suitable medium for the growth of more plant. Animal like earthworm, termites, ants, moles, rats and rodents burrow underground and loosen hard soil or change its character by mixing upper and lower layer of soil.

4.5 FACTORS AFFECTING SOIL CHARACTER

There are a number of factors which is largely control the kind of soil that finally develops. The factors are shown by the following equation :

$$S = f\,(Cl, o, r, P, t\text{----------})$$

where, Cl = Climate (mainly temperature and precipitation)

O = Living organism (Specially native vegetation)

r = Topography of area

P = Parent material

t = Time.

(*i*) ***Climate :*** It is most influential of all factors and determine the nature of weathering that occurs. Climate control vegetation of a particular area. Thus in humid region, heavy rainfall not only encourages rapid chemical changes but also provides an environment favourable for the growth of plants.

(*ii*) ***Living Organism :*** Mosses and Lichens can grow on bare rocks. After the death of these plants, the organic matter content of rock increases and thus helps to grow the higher plant. Animals like earthworms, ants, termites etc. help to change the character of soil.

(*iii*) ***Parent Materials :*** The nature of parent materials may influence in the characteristics of a well matured soil. Soil develop from sand stone contain more sand and its texture become coarse. Black cotton soil of South India developed from Basalt rock. The texture of this soil is fine and it is very fertile soil.

(*iv*) ***Topography :*** The character of soil depends on the topography. In the smooth rock, water stands on the rock which help to grow the plant and rock is modified in various forms to make the soil.

(*v*) ***Time :*** The actual length of time that the materials are subjected to weathering plays an important role in soil formation.

4.6 TYPES OF SOIL

According to origin and texture, the soils are of various types *e.g.*

1. Types of Soil According to Their Origin : According to origin, the soils are of two types :

(*i*) ***Local or Residual or Indigenous Soil :*** This types of soil is found to exist on the parent rock from which the soil is originated. Red soil of Santhal Pargana of Bihar and Black cotton soil of South India is the example of local soil.

(*ii*) ***Transported Soil :*** This type of soil is not so stable and is synthesized during weathering. This soil is transported from original place by air or by other means. Sometimes on primary soil, secondary soil is deposited. The Primary soil in this case is termed as 'Fossile'.

2. Types of Soil According to Their Texture : Soil texture refers to the relative size of soil particles. According to the texture, the soil particles are of three types :

(*i*) ***Sand Particles :*** The diameter of sand particles is 2-0.2 mm. The sand particles whose diameter is 2-0.2 mm, 0.2-0.02 mm and more than 2 mm is known as coarse sand, fine sand and gravel respectively. The size of sand particles is big. There is a large number of big pore space in sand particles. As a result of which, the water holding and nutrient retention capacity of sand particles is very poor. But the aeration and drainage capacity of sand particle is very good.

(*ii*) ***Silt Particles :*** The diameter of silt particles is 0.02 – 0.002 mm. The silt particles are most friable and of fine texture. The silt particles is very plastic and it has a good cohesion, nutrient and water holding capacity. It contain large amount of organic matter.

(*iii*) ***Clay Particles :*** The diameter of clay particles is 0.002 mm or less. The clay particles have water holding and nutrient holding capacity. But the aeration and drainage is not good in the clay particles. Clay particles are of three types *e.g.* montmorillonite, Kaolinite and illite.

Table 4.1 : Important Physical Properties of Different Soil Fraction

Properties	*Degree to which the different soil fraction exhibit physical properties*			
	Coarse sand	*Fine sand*	*Silt*	*Clay*
1	2	3	4	5
1. Water holding capacity	Very low	Medium	High	Very high
2. Capillarity	Do	Good	Do	Do

1	2	3	4	5
3. Rate of water movement	Very fast downward	Fast downward	Slow up and downward	Very Slow
4. Nutrient holding capacity	Very slight	Slight	High	Very high
5. Cohesion and plasticity	None	Very slight	High	Very high
6. Aeration	Very good	Fairly good	Poor	Very poor
7. Temperature	Very warm	Fairly warm	Cold	Very cold
8. Drainage	Very good	Fairly good	Poor	Very poor.

4.7 PHYSICAL PROPERTIES OF SOIL

1. Soil Texture : The relative size of soil particles is expressed by the term textures, more specially the texture is the relative proportions of different size, groups or separates.

The soil separates are classified into sand, silt and clay according to their size.

(i) Sand **:** Sand particles are classified into two categories according to their diameter as follows :

Coarse sand : 2.0 – 0.20 mm.

Fine sand : 0.02 – 0.002 mm.

Macropores exist in sand particles due to their large size, which facilitate free movement of air and water. The water holding capacity of sand particles are very poor. The sand particles are very friable. The sand particles have no cohesion and plasticity if the clay and silt are not mixed with sand, sandy soil possess good drainage capacity and aeration. Sand particles are of coarse texture.

(ii) Silt : The size of silt particles is 0.02 – 0.002 mm. It is predominant in river belts area. Silt particles exhibit some plasticity, cohesion, adhesion, and adsorption. The silt particles are most friable and of fine texture. The silt particles become sticky and plastic on wetting. The water holding capacity, nutrient holding capacity, cohesion are more in silt particles than sand. Drainage capacity of silt particles is very low. The silt particles have more adsorption surfaces for nutrients.

(iii) Clay : The diameter of clay particles is 0.002 mm or less. Clay particles are of three types such as Kaolinite, montmorillonite and illite. Clay particles are generally very fine and they have more cohesion and adhesive capacity. The clay particles have more plasticity, capillarity, aggregation capacity, water holding and nutrient holding capacity. The clay particles swells on wetting and becomes more soft and sticky and shrink on drying. Clay particles are more reactive in nature but poor in aeration and drainage.

1. Soil Texture and Plant Growth : Soil texture is grouped into coarse texture, medium texture and fine texture. The growth of plant depends on texture of soil as follows :

(i) Coarse Texture Soil : Sandy soil belongs to this group. Soil aeration and drainage capacity are very good in sandy soil. The sandy soil can not retain enough water and nutrients for the growth of the crop and it needs frequent application of fertilizers and light irrigation to meet the demand of the cultivated crops. Sandy soil have good drainage capacity. The plants which require good drainage will thrive well in sandy soil, sandy soils are usually loose and friable and hence easy to handle in tillage operation.

(ii) Medium Texture Soil : Loamy soil belongs to this group. This soil are good for cultivation of most crops because they are well drained and aerated and can retain nutrients and enough water for the growth of most plants. Loamy soils are considered best for crop production because they retain more water and nutrients than sandy soils and have better drainage, aeration etc. than clay soils.

(iii) Fine Texture Soil : Clay soil belongs to this group. Clay soils have the capacity to retain plant nutrients and water, but clay is not permeable to air and water. The plants that are not sensitive to water logged condition can grow well in clay soil. Paddy can grow well in clay soil as it requires stagnation of water through its life cycle. Clay soils are difficult to handle in tillage operation.

2. Soil Structure : The arrangement of particles in the soil mass is called *soil structure.* The primary soil particles such as sand, silt and clay are usually grouped together in the form of aggregates. Soil structure is an important soil property which influence the soil environment through its effect on the amount and size of pore space, water holding capacity, availability of plant nutrients and growth of microorganisms.

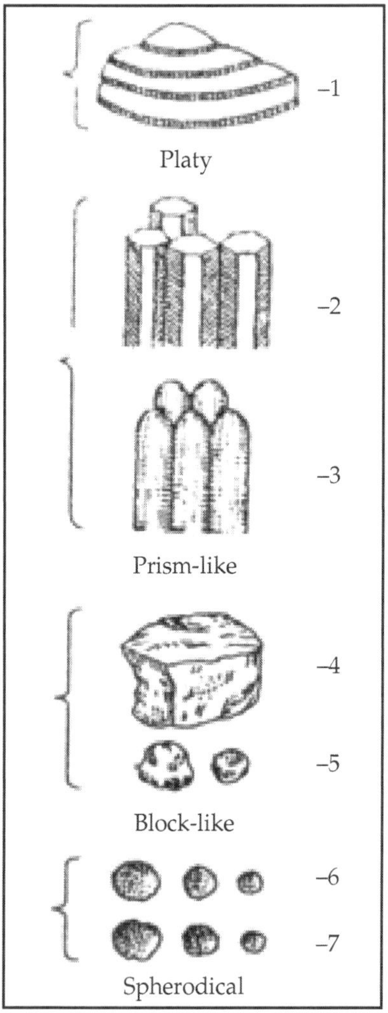

Fig. 4.1 : Soil structure 1. Platy

1. Platy,
2. Prismatic (Level tops),
3. Columnar (Round tops),
4. Blocky (Cube like),
5. Blocky (Sub-angular),
6. Granular (Porous),
7. Crumb (Very porous).

The shapes of aggregates observable in the field can be grouped as follows :

1. Platy
2. **Prism like structure** : (*a*) Columner and (*b*) Prismatic.
3. **Block like structure** : (*a*) Angular blocky and (*b*) Sub-angular block.
4. **Spheroidal :** (*a*) Granular and (*b*) Crumb.

The single grains, granular are crumb structure, are generally found in surface. The other three types are generally found in sub soil.

Soil Structure and Plant Growth : Soil structure influences plant growth rather indirectly. Crumb structure is desirable for agricultural soil so as to retention of nutrient and water holding capacity are concerned. In crumb structure pore spaces are more and this structure is favourable for germination of seed and emergence of seedling. In crumb structure, peds are soft and porous. As a result, the roots of seedling can penetrate in the soil easily.

3. Density of Soil : Density represents the mass (or weight) of a unit volume of soil. The soil density is expressed in two manner such as particle density and bulk density. The weight per unit volume of the soild portion or soil is called *Particle density*. Bulk density is the mass (weight) of a unit volumes of dry soil in its natural structure. Bulk density, therefore, considers the entire soil volume, where as particle density is concerned only with the volume of soil particles.

4. Pore Space : The pore space of a soil is that portion which is occupied by air and water. Pore space is necessary for retention of soil moisture, aeration of plant root and for proper drainage. In general, pore space are of two types such as Macropores or non-capillary pore and Micropores or capillary pore. Water and air movement can take place easily in macropore. But in micropores, the movement of air is difficult and movement of water is restricted largely to slow the capillary movement. Sandy soils have macropores whereas clay soil have micropore. Macropore is necessary for proper drainage and aeration of plant root and micropore is necessary for retention of moisture which is necessary for plant.

5. Soil Consistence : Soil consistence is defined as a term to designate the manifestation of physical forces of concession and adhesion acting within the various level of moisture content.

Consistence is important in determining the practical utilization of soil. The soil will be puddled when ploughing is given in a soil which is in a high degree of sticky consistence at moisture content and soil structure will break down. On the other hand, clods will form if the soil is hard and this condition is not suitable for sowing of seed.

6. Soil Colour : The colour of soil is probably the first soil property for human perception and it is one of its obvious property. Colour indicates, approximately the organic matter content of the soil. Black coloured soil are rich in organic matter and therefore fertile. Soil colour influences greatly the soil temperature. The dark coloured soil absorbs heat more readily than light coloured soil. Red coloured soils are rich in ferric oxide. Highly weathered and of poor quality. When superphosphate is applied in red soil (pH 5.3 – 6.0), the phosphorus is converted to iron and aluminium phosphate. The phenomenon is called *'Phosphorus fixation'*.

7. Soil Temperature : The main source of heat energy for soil is solar radiation. The temperature of soil is very important, because plant growth as well as chemical and biological weathering are greatly influenced by the soil temperature.

Soil temperature has a profound influence on germination of seed, plant growth, absorption of soil water by plant root, availability of nutrients to plants, decomposition of soil organic matter, soil formation and physical properties of soil. Soil microorganism shows maximum growth and activity at optimum soil moisture range. Soil temperature controls the rate of microbiological process involved in the mineralization of nitrogen, nitrogen fixation etc. The optimum soil temperature range for microbiological activity is between 25°C to 35°C.

4.8 PLANT NUTRIENTS

The green plants are autotrophic or self-nourishing *i.e.* they are able to manufacture carbohydrate food from the raw or inorganic materials by the process of photosynthesis (photo-light, synthesis = putting together) and nourish themselves. The conception of elements as plant nutrients was originated by Liebig in 1840 when he put forward his revolutionary theory of mineral nutrition.

Classification of Essential Nutrients : Plants contain small amounts of 90 or more elements but 16 elements are known to be essential and they are divided into 2 groups on the basis of their quantity required by plants as followed :

1. Major or Macronutrients : The nutrients which are used by plants in relatively large quantity (*i.e.* concentration exceeding 1 ppm) are called *Major nutrients* or *Macronutrients. They* are further classified into two categories as follows :

(*i*) ***Primary Nutrients :*** Carbon (C), Hydrogen (H), Oxygen (O), Nitrogen (N), Phosphorus (P), and Potassium (K) are the primary nutrients and they are essential right for the germination of seed. Out of these Carbon (C), Hydrogen (H), and Oxygen (O) are provided by air and water and the rest are either obtained from soil or supplied through chemical fertilizers.

(*ii*) ***Secondary Nutrients :*** Calcium (Ca), Magnesium (Mg) and Sulphur (S) are the secondary nutrients. They are secondary in the sense that they are needed only when the plant have started growing or they have grown for some periods.

2. Minor or Micronutrients : The nutrients which are used by the plants in very small quantity (*i.e.* less than 1 ppm) are called *Minor or Micronutrients.* Manganese, (Mn), Iron (Fe), Molybdenum (Mo), Copper (Cu), Boron (B), Zinc (Zn) and Chlorine (Cl) are the micronutrients.

Table 4.2 : Essential plant nutrients, forms of uptake and discovery (Fageris, Baligar and Junes, 1991)

Name of nutrient	*Principal form for uptake by plants*	*Year of discovery*	*Discovered essential by*
Macronutrients			
Carbon (C)	CO_2	1882	J. Sachs
Hydrogen (H)	H_2O	1882	J. Sachs
Oxygen (O)	H_2O , O_2	1804	T. De Saussure
Nitrogen (N)	NH_4^+ , NO_3^-	1872	G.K. Rutherford
Phosphorus (P)	$H_2PO_4^-$, HPO_4^-	1903	Postermak
Potassium (K)	K^+	1890	AFZ Schimper
Calcium (Ca)	Ca^{++}	1856	F. Salm-Horstmar
Magnesium (Mg)	Mg^{++}	1906	Willstatter
Sulphur (S)	$SO_4^=$ SO_2^-	1911	Peterson
Micronutrients			
Iron (Fe)	Fe^{++}, Fe^{+++}	1860	J. Sachs
Manganese (Mn)	Mn^{++}	1922	JSMC Hargue
Boron (B)	H_3BO_3	1923	K. Warington
Zinc (Zn)	Zn^{++}	1926	AL Sommer and CB Lipman
Copper (Cu)	Cu^{++}	1931	C, Blipman and G. Macc Kinnery
Molybdenum (Mo)	MoO_4 =	1938	DL. Arnon and P.R. Stout
Chlorine (Cl)	Cl^-	1954	T.C. Broyer *et. al.*

Sodium (Na), Silicon (Sl), Cobalt (Co) and Vanadium (Vn) are beneficial for some plants. But they have not been established as essential elements for higher plants. Nicholas (1963) used "Ultra micronutrient") for Molybdenum (Mo) and Cobalt (Co) and they are required as one part per billion by most of the plants.

Mengel and Kirby (1978) classified plant nutrients on the basis of their bio-chemical behaviour and physiological functions into four groups as follows :

Group 1 : Carbon (C), Hydrogen (H) , Oxygen (O), Nitrogen (N) and Sulphur (S). These are major constituents of organic material involved in enzymatic processes and oxidation-reduction reactions.

Group 2 : Phosphorus (P) and Boron (B). These are involved in energy transfer reactions and esterification with native alcohol groups in plants.

Group 3 : Potassium (K), Calcium (Ca), Magnesium (Mg), Manganese (Mn) and Chlorine (Cl). This group plays osmotic and ions balance role plus more specific function in enzyme conformation and catalysis.

Group 4 : Iron (Fe), Copper (Cu), Zinc (Zn) and Molybdenum (Mo). They present as structural chelates of metallo proteins. These elements enable electron transport by valency change.

Table 4.3 : Elements Required by a Particular Species of Plant

Name of elements	*Plant Species*	*Year of Discovery*	*Discovered by*
Aluminium	Ferns	1942	K. Tanbock
Gallium	*Aspergillus niger*	1938	R.A. Stienberg.
Iodine	Polysphonia	1966	I. Foics.
Selenium	*Astragalus sp.*	1938	S.F. Trelease and H.M. Trelease.
Silicon	Diatoms	1962	J.C. Lewin.
Vanadium	Scene desmus obliquns.	1953	D.I. Amon and G. Wessd.

4.8.1 Functions of Essential Nutrients

Carbon : Carbon in the form of carbondioxide is used by

green plant in the process of photosynthesis. It is also required for cell formation in plants.

Hydrogen : It is essential for cell and tissues formation in plants and it plays a central role in plant metabolism.

Oxygen : It is most important element required for respiratory activity. It helps in formation of tissue and cells.

Nitrogen : It imparts dark green colour to plants, as it is an integral part of chlorophyll. It is a component of many important organic compounds ranging from protein to nuclear acid. It promotes vegetative growth. It is a regulator that governs to a considerable degrees the utilization of phosphorus, potassium and other constituents.

Phosphorus : It stimulates root development, flowering fruit setting and seed formation and encourages greater growth and shoots. It induces nodule formation in leguminous plants and rhizobial activity.

Potassium : It is an essential elements for the development of chlorophyll and photosynthesis. It increases resistance in plants against drought, heat, frost and various diseases. It plays a key role in production of quality vegetables. It increases the plumpness and boldness of grams and seeds.

Calcium : It is an important constituent of cell wall and required in large amounts for cell division. It promotes root development and growth of plant, induces stiffness of straw and thereby reduces lodging in cereals, enhances nodule development in leguminous plant. It neutralises the organic acid formed within the plant body and eliminates their toxic effects and also increases the crop resistance to certain disease.

Magnesium : It is a constituent part of chlorophyll and chromosomes, and needed by plants for formation of oils and fats. It regulates the uptake of nitrogen and phosphorus from the soil and increases the crop resistance to drought and diseases.

Sulphur : It is involved in the formation of chlorophyll and thereby encourages vegetative growth. It hastens root growth and stimulates seed formation, promotes the nodule formation on the root of leguminous plants and causes pungent colour of onion and garlic. It help in the synthesis of protein and amino acids.

Manganese : It helps in the synthesis of chlorophyll and acts as catalyst in oxidation-reduction reactions within the plant tissue.

Molybdenum : It helps in the synthesis of protein and amino acid, regulates the activities of several enzymes and accelerates the nitrogen fixing efficiency of aerobic bacteria (*i.e.* Azotobacter etc.), anaerobic bacteria (*i.e.* Clostridium etc.), blue green algae, azolla and symbiotic bacteria (*i.e.* Rhizobium etc.).

Copper : It is a constituent of chlorophyll and thus helps in the synthesis of chlorophyll. It helps in oxidation-reduction reaction in plants. It has indirect effect on nodule formation in leguminous plants.

Boron : It plays an important role in the development and differentiation of tissue, carbohydrate metabolism and translocation in sugar in plants, helps in root development and flower and pollen grain formation and normal growth of plant and absorption of nitrogen in the soil. It makes up the calcium deficiencies to some extent.

Zinc : It takes part in the synthesis of chlorophyll and biosynthesis of plant growth hormone (IAA) and in reproduction process of certain plants. It has positive role in photosynthesis and nitrogen metabolism.

Iron : It takes part in the synthesis of chlorophyll and impart dark green colour of plants. It regulates respiration, photosynthesis and reduction of nitrates and sulphates. It is essential for the synthesis of protein content in chloroplast.

Chlorine : It is essential for photosynthesis and as an activator in spliting of water. It is thought to be associated with the evolution of oxygen during photosynthesis.

4.8.2 Nutrient Deficiency Symptoms in Plant

Nitrogen : Plant remains stunted with yellowish or light green colour of leaves, the yellowing always starts from apex or outer margins of leaf blades of older leaves and gradually spreads towards the base of the leaves which slowly spread towards younger leaves.

Phosphorus : Root and shoot is restricted, leaves may shed prematurely. Flowering and fruiting may be delayed considerably.

Potassium : Plants remain's stunted with bushy appearance. Chlorosis and dying backs of tip of shoot. The older leaves show the deficiency symptoms earlier.

Calcium : The root may become short, stubby and brown. The normal growth of the plant is arrested and young leaves of terminal buds die back at the tips and margins.

Magnesium : Premature defoliation and chlorosis are the main deficiency symptom, which are usually observed on older leaves and the veins of cholorotic leaves remains green.

Sulphur : The older leaves may develop orange or reddish tints and they may shed prematurely. The stem and petioles become brittle and they may collapse. Sulphur deficiency symptoms usually appear first on the youngest leaves.

Manganese : Chlorosis which is first visible on youngest leaves. Manganese deficiency may result in disorders such as grey speck of oats, speckled yellow of sugarbeet, marsh spots of peas and pahala blight of sugarcane (De, 1990).

Molybdenum : Chlorotic mottling between the veins in older and middle leaves. The leaves show scorching and withering which starts from margins and spread over entire lamina except petiole and it gradually extend to younger leaves. The affected younger leaves fails to expand, Molybdenum deficiency produces 'whip tail' in cauliflower, broccoli and brassica crop, 'downward cupping' in raddish and (scald) in beans.

Copper : Chlorosis, withering and often distortion of the terminal leaves, multiple bud formation in the leaf axil and malformation of leaves, gum pocket under the bark and die-back of shoots in citrus.

Boron : The leaves turn yellow or red and thereby plant growth is retarded, malformation of reproductive organs and seeds also fail to mature. Boron deficiency causes corking and pilting of fruits in tomatoes, hollow stem and bronzing of curd in cauliflower, brown heart disease in table beets and turnip (Dutt, 1967).

Zinc : Chlorosis of the foliage particularly in lower leaves with the reduction in the size of young leaves, shortening of internodes and stunted growth of plants, delay in flowering, fruiting and maturity, shoots may die off and leaves fall prematurely, Zinc deficiency is responsible for white bud in maize, khaira disease in paddy etc.

Iron : Chlorosis of young leaves and the veins remain green, leaves become dry and papery and may turn brown and necrotic, complete leaf fall and die of shoots.

Chlorine : Chlorosis and overall wilting of younger leaves.

4.8.3 Usefulness of Some Non-essential Elements

Stile (1961) reports the following non-essential elements for certain species.

Vanadium : It is found very useful for algae. It helps in the growth of fungi and it is involved in the process of photosynthesis.

Sodium : It effects water absorption and metabolism in plant. It gives resistance against drought especially in sugarbeet.

Silicon : It is a major constituent of soil and present in many plant species. It releases fixed phosphorus from deficient soil and promotes the growth of all kinds of crop plants. It increases the resistance against some insectpests and diseases of the crop of gramineae family especially rice.

Strontium : Tensho (1959) reported the absorption of strontium by rice plants. It is similar to calcium in chemical behaviour.

Cobalt : It increases the photosynthesis and growth of plant. It helps in the fixation of nitrogen by bacteria. It activates many enzymes (Sahai, 1992).

4.8.4. Sources of Plant Nutrients

Soil is the main store house of plant nutrients. Crop depletes the nutrient from the soil and lowers its fertility. Soil must be provided with appropriate nutrients in suitable proportion to get the higher yields from the cultivated crop, through some materials as follows :

1. **Manure :** Compost, F.Y.M. etc.
2. **Green Manures:** Sannhemp, Dhaincha etc.
3. **Fertilizers :** Urea, Superphosphate, Muriate of Potash, Gromor 28:28:0, IFFCO Gromor/Paras 10:26:26, DAP. Suphala 20:20:0, Suphala 15:15:15 etc.
4. **Soil Amendments:** Lime, Gypsum etc.
5. **Fungicides :** Copper oxychloride 50% WP, Mancozeb 75% WP etc.

4.8.5 Integrated Nutrient Management (INM)

Fertilizers are used extensively to increase the nutrient status of soil for getting higher yield. Out use of fertilizers has some demerits. For this, nutrient supply both from fertilizers and manures is most important for getting good quality yields. The INM envisages the use of organic manures, green manures, and biofertilizers alongwith fertilizers. The INM system improves the physical condition of soil and this includes the followings :

(*i*) ***Recycling of Organic Wastes* :** The organic wastes (*i.e.* crop residues, dung and urine farm wastes etc.) have good effects on soil. The organic wastes can be better applied by composting and vermi composting.

(*ii*) ***Green Manuring* :** It is an age old practice for supplying part of nutrient requirement of crop and improving soil fertility. Biological nitrogen fixation is possible through some organism as follows :

Rice Field : Blue green algae (BGA), Azolla etc.

Leguminous Plant : Rhizobium etc.

Others : Azotobacter, azospirillum etc.

There are some fungus (*i.e. Bacillus subtiles, B. circulans, Aspergillus niger* etc.) Who can solubilize the unavailable phosphorus and increase its availability to the plant. There are some factors of importance in nutrient management as follows:

1. Physical Condition of Soil : Aeration, aggregation and drainage are the good physical condition of soil and under such condition, nutrients remain available to plants. In poor aerated soil, absorption of nutrients such as nitrogen, becomes low. Under water logged condition due to poor aeration nitrification is decreased and denitrification is stimulated.

2. Inhibitory Factors : Acidity, salinty and alkalinity of the soil are most important inhibitory factors affecting nutrient availability. Soil that is about neutral (pH 6.0–7.0) is suitable for the availability of all most all nutrients. In very acid soil, Iron (Fe), Aluminium (Al), and Manganese (Mn) may become toxic to plant. In alkali soil, high percentage of Sodium inhibit the absorption of Calcium (Ca) and Magnesium (Mg) by plant roots.

3. Moisture in the Soil : Plant absorbs nutrient in soluble form. So sufficient moisture is therefore required for the uptake of nutrients from the soil.

4. Manures and Fertilizers : Manures and fertilizers are the good sources of plant nutrients. Manures should be applied in addition to the fertilizer. Fertilizers influence the reaction of the soil. But manures improves physical property (*i.e.* water boiling capacity, permeability, granulation etc.) and physico–chemical properties (*i.e.* Buffering capacity, cation exchange capacity etc.) of the soil.

5. Bio-fertilizer : It is the culture of microorganisms used for inoculating seed or soil or both under ideal conditions to increase the availability of plant nutrients, mainly nitrogen. Bacteria (*i.e.* Azotobacter, Rhizobium, Clostridium etc.), Blue-green algae and Azolla are the important organisms capable of fixing atmospheric nitrogen. Azotobacter and clostridium are non-symbiotic and Rhizobium is symbiotic nitrogen fixing bacteria. Blue green algae and azolla fix atmospheric nitrogen in the flooded rice field.

It was observed that combined use of manures and fertilizers gave higher yield without any deterioration in the soil. So, INM is only the way to maintain and improve the nutrient status of Indian Soil.

4.9 SOIL FERTILITY

Soil fertility and soil productivity are the two terms which are used in agriculture in determining the quality of the soil. Soil fertility may be defined as the inherent capacity of the soil to supply plant nutrients in adequate amounts and suitable proportion and free from toxic substances. It indicates plant growth in relation to nutrients available in soil. Soil productivity is the ability of the soil to produce crop per unit area under specified system of management. It is expressed in terms of crop yield. A fertile soil may or may not be productive depending upon crops, marketing condition and several other factors (*i.e.* excessive acidity or alkalinity, the presence of toxic substances, poor physical properties, or an excess or deficiency of water). But every productive soil has to be fertile (Singh, 1990). Soil productivity mainly depends on fertility of soil to a great extent.

The following things must be properly followed for the maintenance of soil fertility.

1. Proper use of land.
2. Good tillage.
3. Crop rotation.
4. Control of weeds.
5. Maintenance of optimum moisture in the soil.

6. Control of soil erosion.
7. Cultivation of green manuring crops.
8. Application of manures.
9. Cultivation of cover crops.
10. Removal of excess water.
11. Application of fertilizers.
12. Maintenance of proper acidity and alkalinity of soil.

The fertilizers are applied to supply the plant nutrients in the soil. But it is necessary to apply balanced fertilizers and to avoid the ill effects of unbalanced use of fertilizers timely as follows :

1. Excess of elements may be toxic to living materials (*i.e.* protoplasm) of the plant which is very common in case of manganese, copper and zinc.
2. Excess of one plant food element may cause deficiency of another such as :
   (*a*) Excess of nitrogen cause potassium starvation in certain crops like potato.
   (*b*) Excess of potassium causes deficiency of Manganese in tomato.
   (*c*) Excess of phosphorus causes deficiency of zinc in most of the crop (Singh, 2002).

4.9.1. Soil Fertility Evaluation Methods

1. Biological Test

(*a*) *Field Test :* It is most direct way of diagnosing nutrient status of the soil. In this method, the effect of additional amount of nutrient in the form of fertilizers applied to the soil on the growth of the crop is measured. It is subjected to statistical analysis.

2. Laboratory and Green House Test

(*a*) *Mitscherlich's Post-Culture Technique* : This technique helps in finding out the extent of increase in yield of the crops as influenced by growth factors. This technique consist of growing of oats to maturity in a pots containing 2.5 kg. Soil taken from the field in which growth factors Nitrogen (N), Phosphorus (P) and Potassium (K) are to be studied. The yields of N.P. and K treatment are expressed as percentage of yield from complex N.P.K. treatment.

The level of treatment application is as follows :

(*i*) Nitrogen as Ammonium Nitrate @ 1.0 gm. in 20 ml. of water.

(*ii*) Phosphorus as Super phosphate @ 1.1 gm. in 50 ml. of water.

(*iii*) Potassium as Potassium sulphate @ 1.5 gm. in 50 ml. of water.

Thus, the limiting factor can be detected and the extent of reduction in yield can be determined after comparing the yield of each treatment with the treatments which has received all nutrients. The equation given by Mitscherlich helps in finding out the percent increase in yield corresponding to added nutrients is as follows :

$$Y = A(1 - 10 - CK)$$

Where, Y = yield produced by a given quantity of growth factor.

A = Maximum possible yield.

C = Constant depending on the nature of growth factor.

(*b*) Neubauer Seedling Method : This method is used for detecting available plant nutrients percentage in soil. It is based on the uptake of nutrients by large number of plants grown on a small amount of soil. The plant nutrients which are removed by plants are determined by chemical analysis of plant tissue.

Neubauer has mentioned the utilization co-efficients for different crops as follows :

For Potassium (K_2O) : Barley – 12%

Wheat – 15%

Potato – 25%

For Phosphorus (P_2O_5) : Barley – 20%

All other common plant – 33%.

3. Microbiological Method : In this method, certain microorganisms are used instead of higher plants.

(i) Azotobacter Plaque Test : This test is conducted to determine the deficiency of lime, phosphorus and potassium. Azotobacter is very sensitive to acidity and low phosphorus availability.

5 g. of starch is added to 100 g. of soil and it is divided into four equal parts and treatment is done as follows :

First Part : Control to which no nutrients are added.

Second Part : 2.5 ml. of 3 per cent potassium sulphate solution (K_2SO_4) is added for testing the deficiency of potassium.

Third Part : 2.5 ml. of 6 per cent Na_2HPO_4 solution is added for testing the deficiency of potassium.

Fourth Part : 2.5 ml. of 3 per cent solution of K_2HPO_4 solution in added for testing the deficiency of phosphorus and potassium.

Result : The abundance of Azotobacter calories proves the proper amount of phosphorus (P_2O_5) and potassium (K_2O), Because the growth of Azotobacter will be poor under acidic reaction and poor content of phosphorus and potassium.

(ii) Aspergillus niger Test : In this test, nutrient solution is added in a small quantity of soil. It is inoculated with the suspension of *Aspergillus niger* and then incubated for 4 days at 35° C. After four days, the weight of mycelial pad is taken as a measure of nutrient deficiency. The amount of nutrient absorbed by mycelial is used to measure nutrient deficiency.

(iii) Use of Indicator Plant : There are some plants which are very sensitive to particular elements in the soil and they start showing the deficiency symptom in scarcity of that elements in the soil. They are called indicator plants as follows:

Plant nutrient	*Indicator plants*
Nitrogen and calcium	Cabbage and Cauliflower
Phosphorus	Rape
Potassium and Magnesium	Potato
Sodium	Sugar beet
Magnesium	Oats and Sugar beet
Iron	Cabbage, Cauliflower, Potato and Oats
Boron	Sunflower

3. Visual Symptoms : It is most cheapest method for identifying nutrient disorder in crop plants. But it is sometimes confused with diseases, insectpests of soil moisture stress. There are three steps for identifying nutrient disorder by visual symptoms as follows :

(*i*) ***Growth and Development of Plant (Stunted Growth):*** It may be due to deficiency or toxicity of all elements. But the deficiency of nitrogen and phosphorus have more influence on the reduction in the growth of plant.

(*ii*) ***Plant Part Affected :*** Deficiency symptoms appear on lower leaves or older leaves.

(*iii*) ***Nature of Symptoms :*** Chorotic, necrotic or deformed.

The leaves showing the symptom indicate the deficiency of plant nutrients as follows :

1. **Lower Leaves :** It may be due to deficiency of mobile nutrients such as nitrogen, phosphorus, potassium and magnesium. Mobile nutrients are those which trans-located with the plants.

2. **Upper Younger Leaves :** It may be due to the deficiency of immobile nutrient such as calcium, copper, sulphur,

boron, manganese and molybdenum. Immobile nutrients are those which are not translocated to growing region of the plant, but remain in older leaves where they were originally deposited after absorption from the soil.

4. Chemical Test for Evaluating Soil Fertility : The chemical tests are employed to determine the soil fertility status and it includes the analysis of soil and plant tissue.

4.10 SOIL MOISTURE

Water is indispensable for the growth and development of plant. It is essential part of protoplasm of living being and an important ingredient in the photosynthesis, the food manufacturing process in green plant. It is also required for translocation of nutrient in the plant.

Soil moisture is depleted by various method as follows :

(*i*) Run off losses of water.

(*ii*) Deep percolation into soil beyond root zone.

(*iii*) Evaporation from soil surface.

(*iv*) Transpiration through the plant.

(*v*) Weed depletes soil moisture by absorbing the same from the soil.

The availability of water is reduced to crops and the plants are subjected to water stress. It is possible to conserve soil moisture by reducing the losses of soil moisture. For this some measures should be followed as follows:

(*i*) Water losses can be checked by providing bund around the land. Terracing and contour bund are found effective in controlling the water loss in slope land.

(*ii*) The evaporation losses of water from the soil can be checked by adoption of mulching. Mulching is the covering on the soil surface to conserve soil moisture by materials like polythene film, straws and dry grasses, old gunny bags etc. which are employed to reduce the evaporation or loss of moisture from the soil surface. A practical mulch should be easily available, cheap and easy to apply. Mulching materials may be found in yards, garden centres, sawmills, dairy firm etc. Mulches should be applied prior to active weed growth ard summer droughts.

Mulching has several advantages and disadvantages as follows :

Advantages

1. Mulches save on the labour of cultivation since the emerging small weeds are killed under dark, barrier.
2. Mulches reduces the intercultivation and the use of weed control chemicals.
3. Water is conserved, because mulches reduce the evaporation of soil moisture by lowering soil temperature.
4. Mulches are excellent conservation agent of moisture and soil.
5. The insulating property of mulch avoids drastic fluctuation of soil temperature, keeping the soil warmer in winter and cooler in summer. Summer mulches improves both not growth and nutrient availability. Winter mulches reduce the risk of root damage due to cold.
6. Organic mulches improves the fertility of soil as lower layers decompose and becomes incorporated into the soil. At the end of growing season, organic mulches can be tilled into the soil to further increase the organic matter content of the soil.

Disadvantages

1. The application of mulch requires labour of varying degrees depending on the materials selected.

2. Mulch and moisture around newly emerging seedling or perennials can provide an ideal environment for diseases such as damping off or crown rot.
3. Mulches provide attractive habitat for insectpests and rodents. They can cause damage to the growing plants.
4. Dry mulches that are combustible may be a fire hazard near building or in public places.
5. Woody mulches can be vector for termites near building.

4.11 RECLAMATION OF SOIL

Timely application of adequate amount of fertilizer is necessary for getting higher yield. The success of fertilizer application depends on the physical, chemical and biological condition of Soil. The plant nutrient remains available to plants at a pH 6.5 – 7.5. If the pH is below 7.0 and above 7.0 the soil becomes acidic and alkaline respectively.

1. Acid Soil : When the hydrogen (H^+) ion is more than Hydroxyl ion (OH^-) in the soil solution, the soil becomes acidic. The following equation represents the mode of increasing acidity in the soil :

Ca^{++} (Clay colloid) + 2HOH $\rightarrow$ H^+ (Clay colloid + Ca^{++} + $2OH^-$

The soil acidity are of two types :

(*a*) ***Active Acidity :*** Dissociated Hydrogen ion in the soil solution determine active acidity and it can be measured by pH.

(*b*) ***Reserved or Potential Acidity:*** The absorbed or exchangeable Hydrogen ion concentration of soil colloid is designed as the reserved or Potential acidity.

Formation of Acid Soil : The acid soils are generally found in humid region. The factors which will help in the release and removal of bases from the soil will help in the development of acidity. If H^+ ion become more than OH^- ion in soil solution, the soil becomes acidic. There are various causes of development of soil acidity. Such as :

(*i*) Leaching of some nutrient such as Calcium, Magnesium etc. from the soil solution helps in the formation of acid soil.

(*ii*) Soil develops from acid rock (e.g. granite etc.) becomes naturally acidic in nature.

(*iii*) Decomposition of organic matter in the soil increases the soil acidity. During the decomposition, large number of organic and inorganic acids are produced. Out of those, most spectacular one is the formation of H_2CO_3.

(*iv*) Application of fertilizer like Ammonium Sulphate, Urea etc. is also an additional causes of developing soil acidity. On nitrification of ammonium ions forms nitrates which may produce nitric acid, may be an additional causes of development of acid soil. Similarly application of sulphur or its compound exert a similar action by producing sulphuric acid.

Soil Acidity and Plant Growth : The acid soil may effect the plant in the following way or manner.

(*i*) In acid soil, concentration of Al^{+++} and Mg^{++} is generally increased due to their high solubility in acid medium. This high concentration of Al^{+++} and Mg^{++} may become toxic to plant and this adversly affect the plant growth.

(*ii*) In acid soil, basic substance is very low. As a result, the plant may suffer due to the deficiency of calcium in acid soil which is one of the major essential plant nutrients.

(*iii*) In acid soil, the fixation of Phosphorus is increased to a great extent due to high solubility of Aluminium and Iron. Therefore, the plant may suffer from the deficiency of Phosphorus, though the soil may contain high percentage of phosphorus.

(*iv*) The large number of organism which is responsible for mineralization process becomes inactive in very acid soil. As a result of which, the process of mineralization becomes slow and plant do not get their required quantity of nutrient from soil.

Reclamation of Acid Soil : Losses of Calcium and Magnesium from the soil increases the acidity of soil. Addition of these metal whose removal causes the development of acid soil decreases the acidity of soil. The metallic cations best suited for reclamation of soil acidity are calcium and magnesium, specially the calcium fits the requirement rather perfectly. They are cheap, plentiful and leave of favourable effect on Physical condition of Soil. The action of Calcium hydroxide, Calcium bicarbonate, Calcium oxide with acid soil may be shown as follows :

$$\begin{matrix} H^+ \\ H^+ \end{matrix}\boxed{\text{Soil}} + Ca(HCO_3)_2 \rightleftharpoons Ca^{++}\boxed{\text{Soil}} + 2CO_2\uparrow + 2H_2O$$

$$\begin{matrix} H^+ \\ H^+ \end{matrix}\boxed{\text{Soil}} + CaCO_3 \rightleftharpoons Ca^{++}\boxed{\text{Soil}} + CO_2\uparrow + H_2O$$

$$\begin{matrix} H^+ \\ H^+ \end{matrix}\boxed{\text{Soil}} + CaO \rightleftharpoons Ca^{++}\boxed{\text{Soil}} + H_2O$$

$$\begin{matrix} H^+ \\ H^+ \end{matrix}\boxed{\text{Soil}} + Ca(OH)_2 \rightleftharpoons Ca^{++}\boxed{\text{Soil}} + 2H_2O$$

Thus the amount of active calcium has been increased and as a result, the acidity is decreased and the density of minor elements *e.g.* Fe^{++}, Mg^{++}, Cu^{+}, Zn^{+} etc. decreases that were toxic in acid soil. As a result, the activity of soil organism and the structure of soil is improved. Amount of lime required for reclamation of acid soil depend on the soil reaction, soil texture, type of clay particles, amount of organic matter in the soil and crops to be grown in future.

Large amount of lime should not be applied at a time to reclaim the acid soil in a short time. Because the application of lime in a large quantity at a time hamper the availability of some plant nutrient *e.g.* Fe^{--}, Mn^{++}, Zn^{+} etc. Besides this the large amount of Calcium decreases the intake capacity of K^{+} and Mg^{++} by plant root. So, it is better to apply lime on the basis of soil testing report. Amount of lime required to apply in different soil to reclaim acidity is given below :

Different Types of Substances	*Light Soil (Sand and Sandy Loam)*	*Medium Soil (Loam, Silt Loam)*	*Heavy Soil (Clay, Clay Loam)*
Calcarous rock	2.500 ton	5.00 ton	7.500 ton
Basic slag	1.250 ton	2.500 ton	3.750 ton
Dust lime	1.250 ton	2.500 ton	3.750 ton
Wood ash	1.250 ton	2.500 ton	3.750 ton

Source : Saar O Mati, Director of Agriculture, Govt. of W.B.

Effect of Liming on Soil and Plants

(*i*) Lime neutralizes soil acidity.

(*ii*) Lime stimulate the activity or soil micro-organism. As a result, the decomposition or organic matter and fixation of nitrogen by symbiotic and free living organisms is enhanced and the availability of nitrogen is also increased.

(*iii*) Lime makes Phosphorus more available by reducing the solubility of Iron and Aluminium. Otherwise the Phosphoric acid is fixed as Iron and Aluminium Phosphate in acid Soil.

(*iv*) Limining helps the availability of Potassium and Molybdenum.

(*v*) Lime supply two essential elements namely calcium and magnesium (If lime is dolomite) for plant nutrition.

(*vi*) Liming improves the physical condition of soil and increases the effectiveness of fertilizers.

(*vii*) Lime decreases the plant diseases favoured by acid soil.

(*viii*) Lime converts the injurious aluminium and manganese into available compound.

1. Time of Limining : Lime should be supplied at least one month before the sowing and transplanting of crops. The lime should be mixed with soil by ploughing. It should be remember that lime and nitrogen and water soluble P_2O_5 should not be applied at a time, because the application of lime and nitrogenous fertilizer at a time causes the loss of nitrogen in the form of Ammonia and also decreases the availability of phosphorus.

2. Saline and Alkali Soil : The soil develop under imperfect drainage and arid region, possess the properties of high percentage of soluble salt and high base status. During rainy season, the salt may be move downward to the lower soil layer, but after rainy season the intense evaporation brings the salt back to the surface and makes the soil saline. Such soil is known as halomorphic soil and have been classified under three heads:

(*i*) Saline or White alkali soil.

(*ii*) Saline-alkali soil.

(*iii*) Non-saline alkali soil.

Land under Saline and Alkali Soil in India is 6 million hectare and in the West Bengal is 0.8 million hactres. Region of Saline and Alkali Soil *i.e.* abundant in Saline and alkali soils are:

(*i*) Rajasthan and Punjab.

(*ii*) West to 80° E in U.P. and Delhi.

(*iii*) Upland and Deccan Plateau in the area between Narmada, Godavari and Burma region.

(*iv*) The Western region in deltas of Ganga and Cavari river.

(*v*) The Salt flats of Run and Kutch.

Saline Soil : The soil which containing high percentage of soluble salt to interfere the growth of most plant seriously is known as Saline Soil. Chloride and sulphate of Na, Ca and Mg are the dominant salt present in the saline soil. Besides these, smaller amount of Potassium Salt and bicarbonate and nitrate also present in this soil. The Saline Soil contains less than 15 per cent exchangeable Sodium to total base exchange capacity of Soil. Due to presence of neutral Salt, pH of the soil is less than 8.5. Hilgard regarded these soils as white alkaline soil, because sometimes a white incrustation of salts are found on the surface of Saline Soil.

Saline-Alkali Soils : The soil which contains appreciable quantities of neutral soluble salts and enough absorbed sodium ions to effect plant is known as Saline-Alkali soils. The exchangeable sodium of this soil is more then 15 per cent of base exchange capacity of soil. The pH of the soil remains below 8.5.

Non-Saline Alkali Soils : This soils do not contain appreciable quantities of neutral soluble salt but contains exchangeable sodium ions to effect plants seriously. The exchangeable sodium is more than 15 per cent of total base exchange capacity of soil. Due to hydrolysis of sodium, the pH of the soil is above 8.5, sometimes raising as high as 10.0.

Saline-Alkali Soils and Plant Growth

(*A*) *Saline Soils* : The effect of saline soils on plant growth can be discussed under two heads :

(*i*) *Osmotic Effect :* Due to increasing concentration of Soluble Salt in Saline Soil, the osmotic pressure of soil solution is increased. As a result of this, the absorption of water by plant root is inhibited and in extreme cases, plasmolysis of plant cell may occur.

(*ii*) *Ionic Effect :* The ionic effect of the Saline Soil may be exhibited in two ways. Such as (*a*) Toxic effect of some ions and (*b*) Nutritional disorder. Some ions of Soluble salts of sodium and chlorine have been proved by many as toxic to plants. High concentration of Sodium and Sulphate may depress the absorption of Calcium and may cause Calcium deficiency to plants.

(*B*) *Alkali Soils* : The effect of alkali soil on plant growth can be discussed under two heads :

(*i*) *Physical Effect :* High percentage of exchangeable Sodium may help to develop a poor structure. Soil become impervious and tight which is not congenial to proper aeration, drainage and penetration of soil mass by plant root.

(*ii*) *Nutritional Effect :* High percentage of exchangeable Sodium inhibit the absorption of calcium and magnesium by plant root. This may cause calcium deficiency. The active Sodium ions adversely effect the plant metabolism nutrition.

Development of Saline and Alkali Soils : The accumulation or distribution and concentration of Salts are necessary for the formation of the saline and alkaline soil. Initially the salts are derived from weathering of minerals. These salts may be transported by water and accumulated to the soils of lower horizons or regions. The presence of such salts in the soil surface is due to the following factors:

(*i*) Arid and Semi-Arid climate.

(*ii*) Impervious Sub-Soil or hard pan.

(*iii*) Temporary abundance of humidity in the Soil, interspread with dry period.

In the arid climate, loss of water by evaporation from the soil is more than the water gained by soil through precipitation i.e. rainfall. As a result of these, net movement of water and of the salts dissolved in it is occurred from the lower horizons to the surface soil. These phenomenon help in the continuous accumulation of Salts on the soil surface and thus causes the development of Soil salinity.

The amount of exchangeable Sodium in greater quantities in the soil makes soil alkalinity. The Sodium ion easily displaced the Calcium from the Clay Colloid and make the sodium mixed Clay particles. This Sodium is converted into Sodium hydroxide by hydrolysis as per the following reaction.

$$2Na^{+} + CO_3^{-} + 2H_2O \rightleftharpoons 2Na^{+} + 2OH^{-} + Na_2CO_3$$

The OH^- thus formed increases the Soil pH.

Reclamation of Saline Soil : There are three general ways in which Saline Soil may be reclaimed or managed in order to avoid, at least partially, the injurious effect to plants. Such as :

1. Eradication : The most common method used to remove the excess salt partially are (*i*) Under drainage, (*ii*) Leaching, (*iii*) Scraping and (*iv*) Trenching.

A combination of first two, flooding after the tile drains is most effective and satisfactory process. The soils are irrigated heavily and repeatedly. As a result of which the soluble salts are leached out from the soil solum and drained off through the tile. The leaching method is useful in pervious saline soils where soluble salts are mainly neutral and high in Calcium and Magnesium. Sometimes scraping of accumulated Salts on the surface of soil is done. But it is not a satisfactory process as large amount of salts remain in the soil and also this method is very costly.

2. Conversion : Gypsum is used for reclamation of Saline Soil. If Calcium percentage is low, application of Gypsum is helpful for reclamation of Saline Soil. When Gypsum ($CaSO_4$) is added to the irrigation water, the calcium percentage is increased in the soil with the result that Calcium displaced sodium from exchangeable complex and a favourable physical change in the Soil takes place. The Gypsum reacts both with Sodium carbonate and absorbed Sodium in the following manner.

$$Na_2CO_3 + CaSO_4 \rightleftharpoons CaCO_3 + Na_2SO_4 \text{ (Leachable)}$$

$$\begin{matrix} Na^+ \\ Na^+ \end{matrix} \boxed{\text{Soil}} + CaSO_4 \rightleftharpoons Ca^{++} \boxed{\text{Soil}} + Na_2SO_4 \text{ (Leachable)}$$

3. Control : The retardation of evaporation and artificial drainage is an important feature of Saline Soil control. Mulching is an important practice for retardation of evaporation and deep ploughing or deep chiselling the soil with impervious layer in order to open the soil for desired downward movement of water and salts is also an important operation for the control of saline soil. The cultivation of salt tolerant crops is another important features of successful management of Saline Soil. Relative tolerance of Crops to Salt is given below :

High Salt Tolerance	*Medium Salt Tolerance*	*Low Salt Tolerance*
Barley	Wheat, Rice, Sorghum	Bean
Cotton	Maize, Sweet clover, Beet	White Clover
Barseem	Spinach, Tomato, Cabbage	Radish
Lucern	Cauliflower, Lettuce, Sweet Corn, Carrot, Onion, Peas, Cucumber, Squash	Celery

Reclamation of Alkali Soils : There are some method for reclamation of Alkali Soils *e.g.*

1. Eradication : Same as in Saline Soil except conversion is done previous to leaching. Because Sodium does not leach by irrigation water from exchangeable phase.

2. Conversion : The following reaction takes place due to the application of Gypsum. The Gypsum reacts with Sodium carbonate and absorbed sodium and produces sodium sulphate. This sodium sulphate become soluble in water and leached away.

$$\begin{matrix} Na^+ \\ Na^+ \end{matrix}\boxed{\text{Soil}} + CaSO_4 \rightleftharpoons Ca^{++}\boxed{\text{Soil}} + Na_2SO_4 \text{ (Leachable)}$$

$$Na_2CO_3 + CaSO_4 \rightleftharpoons CaCO_3 + Na_2SO_4 \text{ (Leachable)}$$

(i) Application of Sulphur : When Sulphur is applied, it is oxidies and produces SO_3 and formed Sulphuric acid by reacting with water. The Sulphuric acid poor with Sodium carbonate in alkali soil and alkalinity is removed.

$$2S + 3O_2 \longrightarrow 2\ SO_3 \text{ (Sulphur trioxide)}$$

$$SO_3 + H_2O \longrightarrow H_2SO_4 \text{ (Sulphuric acid)}$$

$$H_2SO_4 + Na_2\ CO_3 \longrightarrow CO_2 + H_2O + Na_2SO_4 \text{ (Leachable)}$$

$$\begin{matrix} Na \\ Na \end{matrix}\boxed{\text{Soil}} + H_2SO_4 \longrightarrow \begin{matrix} H \\ H \end{matrix}\boxed{\text{Soil}} + Na_2SO_4 \text{ (Leachable)}$$

(ii) Application of Sulphuric Acid : When the alkalinity is very high, sulphuric acid is applied for reclamation of alkali soil. But it is a costly and risky method.

(iii) Application of Lime or Lime Sulphur : The reaction due to application or lime or lime Sulphur is as follows :

$$\begin{matrix} Na^+ \\ Na^+ \end{matrix}\boxed{\text{Soil}} + CaCO_3 \longrightarrow Ca^+\boxed{\text{Soil}} + Na_2CO_3$$

$$CaS_5 + 8O_2 + 4H_2O \longrightarrow CaSO_4 + 4H_2SO_4$$

(iv) Application of Ferrous Sulphate or Any Other Soluble Salt of Aluminium : The reaction due to application of ferrous sulphate or any other Soluble Salt of Aluminium is as follows:

$$FeSO_4 + H_2O \longrightarrow H_2SO_4 + FeO$$

$$CaCO_3 + H_2SO_4 \longrightarrow CaSO_4 + H_2O + CO_2 \text{ (Leachable)}$$

(v) Application of Mollases : Two ton of mollases, one ton of press mud (by-product of Sugar industry) per acre is required for reclamation of Alkali Soil. This produces organic acid when applied in Soil which neutralises alkalinity, Calcium in it is also helpful for exchange and phosphate help in Nitrogen absorption.

Table : 4.4 : How Much Gypsum and Sulphur is to be Applied

Exchangeable Na^+ (me/100 gm of Soil)	*Gypsum (Ton/acre)*	*Sulphur (Ton/acre)*
1	1.7	0.32
2	3.4	0.64
3	5.2	0.96
4	6.9	1.28
5	8.6	1.60
6	10.3	1.92
7	12.0	2.24
8	13.7	2.56
9	15.5	2.88
10	17.8	3.20

REFERENCES

Agarwal R.R., Jadar, J.S.P. Gupta, R.N. Saline and Alkali Soil of India (1982), I.C.A.R., New Delhi.

Bennett, Hough Hamm, Elements of Soil Conservation 2nd edn. 2009, Biotech Books, Delhi.

Biswas, K. and Mukherjee. S.K., Text Book of Soil Science (1987), Tata McGraw Hill, New Delhi.

Brady Nyle C., The Nature and Properties of Soil (1974), Eurasia Publishing House (P) Ltd., Ramnagar, New Delhi.

Daji, J.A., A Text Book of Soil Science (1968), Media Promoters and Publishing Co., Bombay.

Das D.K. Introductory Soil Science (1977), Kalyani Publishers, Ludhiana.

Das, P.C. Soils in India (2002), Kalyani Publishers, Ludhiana.

De Gopal Chandra, Fundamental of Agronomy (1990), Oxford and IBH Publishing Co. Pvt. Ltd., New Delhi.

FAO, Green Manuring for Soil Productivity Improvement, 2007, Daya Publishing House, Delhi.

Gupta, P.K., A Hand Book of Soil Fertilizer and Manures (2004), Agro-Bios (India), Jodhpur.

Kolay, A.K. Basic Concept of Soil Science (1993), New Age, International (P) Ltd., New Delhi.

Miller, C.E., Soil Fertility (2007) Biotech Books, New Delhi.

Rai, M.M. Principles of Soil Science (1995), MacMillan India Pvt. Ltd., New Delhi.

Reddy S.R. Principles of Agronomy (1999), Kalyani Publishers, Ludhiana.

Reddy T. Yellamanda and Reddi G.H. Sankara, Principles of Agronomy (1992), Kalyani Publishers, Ludhiana.

Russell, E. John and E. Walter Russell, Soil Condition and Plant Growth, 2008, Biotech Books, Delhi.

Sohai, V.N., Fundamentals of Soil (1990), Kalyani Publishers, Ludhiana.

Singh. S.S. (Editor), Hand Book of Agricultural Sciences (1991), Kalyani Publishers, Ludhiana.

5

CULTIVATION

5.1 INTRODUCTION

Crops are generally cultivated on soil. But all soils on earth crust is not suitable for cultivation of crops. Some mechanical operations are required to be done to bring the land under cultivation. The crop production by those mechanical function such as tillage, sowing of seed, inter-cultural operations, harvesting, threshing etc. is collectively known as cultivation.

5.2 LAND SELECTION

India is an agricultural country. More than 75 per cent of its populations depend directly and indirectly on agriculture. Crops differ in their adaptability in different soil. Crop cultivation is mainly controlled by climate. We have no control over climate. The cropping pattern should be adjusted according to the prevailing climatic condition of a locality. Otherwise, there is a every possibility of crop failure. Land should be selected after considering something else *e.g.* nature of soil, soil fertility, irrigation and drainage facility, soil acidity and alkalinity etc. before the cultivation of crops. Aman paddy grow well in clay soil. So Aman paddy should be cultivated on clay soil for getting higher yield. Otherwise, yield upto our expectation is difficult to get. So, we should have sound knowledge about the crop–soil association. Growing of a crop in any soil without proper planning results heavy loss. So the importance on land selection should be given before the crop cultivation.

5.3 TILLAGE

Tillage is the physical manipulation of soil with tools and implements to result in good tilth for better germination of seed and subsequent growth of crops (Ready *et. al.*, 1993). Tillage is the tilling of land for bringing about conditions that are favourable for the cultivation of crops. The word tillage is derived from the Angle-Saxon word *'tilion'* and *'teolian'* meaning to plough and prepare soil for sowing the seed, transplanting the seedlings and raising the crops. Tilth implies to the physical condition of soil in its relation to plant growth. Tilth is brought out by tillage. Tillage is the primary function of cultivation and it is laborious and expensive cultural practice. Tillage helps to replace natural vegetation with useful crops and is necessary to provide a favourable edaphic environment for establishment, growth and yield of crop plants. Tillage helps to improve the physical condition of soil, control of weeds, insect pests and diseases and also bring the nutrient available to plant. The cultivation is not possible without tillage operation. The crops production depends on good tillage operations.

The word of manures is originated from the French word "Manoeuvrer" which refers to 'Work with Soil', that is why the word tillage and manure where synonyms as is clear by the statement of Jethro Jull (1700 B.C.) "Tillage is manure", there must be sufficient moisture in the soil for good tillage. Tillage is quite impossible in fully dry soil. On the other hand, tillage in wetland results in clogging of soil. Tillage in wetland having sufficient water brings puddling condition of soil which is favourable for cultivation of transplanted paddy, onion etc.

5.3.1 Objectives of Tillage

There are several objectives of tillage as follows :

(*i*) The main objectives of tillage is the preparation of land that becomes suitable for crop cultivation. The soil becomes hard and compact after the harvest of crop and this soil is quite unsuitable for crop cultivation. The soil should be loose, and friable with sufficient air and water for germination of seed.

(*ii*) Tillage helps to control weeds by uprooting and smothering the young seedling or weeds. Tillage may induce germination of Weed seeds which can be destroyed by subsequent tillage. The grown up weeds may also be controlled by tillage.

(*iii*) Tillage improves the soil structure. The soil become loose, friable and granular if it is ploughed properly at optimum moisture level.

(*iv*) Tillage helps to incorporate weeds, crop residues, green manures and other organic manures and, fertilisers, soil amendment and other agro-chemicals applied for the control of weeds, insectpests etc.

(*v*) Tillage helps to improve the capacity of soil to receive rain or irrigation water and to retain the moisture for crop growth.

(*vi*) Tillage improves the soil aeration which is beneficial for respiration of plant root and soil microorganism and their multiplication.

(*vii*) Tillage exposes the lower soil to weather and places the surface soil underneath. As a result of which the soil inhabiting organism comes out and destroyed by heat and predatory animals and birds.

(*viii*) Tillage improves the soil permeability and help to conserve soil moisture as a result of soil mulch. It also check the loss of water through percolation in lowland paddy field by making an impervious soil layer.

(*ix*) Tillage provides condition suitable for growth, nutrition and development of beneficial soil organism.

(*x*) Tillage remove the hard pan that increases the soil depth for water absorption.

(*xi*) Tillage helps to modify slightly the thermo capacity of soil.

(*xii*) Proper tillage results in soil and moisture conservation through higher infiltration, reduced run-off and increase depth of soil for moisture storage.

(*xiii*) Organic matter decomposition is hastened resulting in higher nutrient availability if the soil is ploughed properly at optimum moisture level.

(*xiv*) Tillage makes the soil level and flat that are suitable for uniform movement of irrigation and drainage water.

(*xv*) Tillage brings the soil in such a condition that are conductive for root penetration as the root fails to penetrate in hard soil.

(*xvi*) The most important objective of tillage is to prepare good seed bed. Good seed bed is necessary for early seed germination and initial good stand of the crop.

(*xvii*) The stubbles of previous crop which harbour insectpests are removed following tillage resulting in reduced pest attack on the succeeding crop.

5.3.2 Types of Tillage

(A) Preparatory Tillage : Tillage operations that are carried out to prepare the field for raising crops from the harvest of a crop to the sowing of the next crop are known as preparatory tillage. It is divided into primary and secondary tillage operations.

(i) Primary Tillage or Ploughing : The tillage operation that is done after the harvest of crop to bring the land under cultivation is known as primary tillage. Ploughing is opening of the compact soil with the help of different ploughs. Primary tillage is done mainly to open the hard soil and to separate the top soil from lower layers and to uprooting of weeds also. This operations consist of ploughing, discing, harrowing and levelling of the field. Country plough, Disc plough, Mould board plough. Bose plough, Tractors, Power tiller etc. are used for primary tillage.

Depth of Ploughing : Depth of ploughing may ranges from 10 to 30 cm depending on the root system of the crop that is to be cultivated. Deep ploughing is required for the tap rooted crops while the shallow ploughing required for shallow and fibrous rooted crops.

Time of Ploughing : Soil moisture determine the optimum time of ploughing. When the soil, particularly heavy soil is dry, it is difficult to open the soil and great clods are turned up, if ploughing is done by any means which is difficult to work into a good seed bed. On the other hand, the soils, when they are too wet are ploughed, get puddled and becomes impervious to air and water and also structureless. This soil becomes hard on drying and unworkable untill remoistened.

Number of Ploughing : The number of ploughings necessary to obtain a good tilth depends on soil type, weed problem and crop residues on the soil surface. The heavy soil needs more number of ploughings, the range being 3-5 ploughings to obtain a good tilth. On the other hand, the light soils require 1–3 ploughings to obtain good tilth. More number of Ploughings are necessary for a soil having more weed growth and plant residues.

(ii) Secondary Tillage : The tillage operations that are performed on the soil after primary tillage to bring a good soil tilth are known as secondary tillage. Secondary tillage consist of lighter or finer operation which is done to clean the soil, break the clods and incorporate the manures and fertilisers. Harrowing and planking is done to serve those purposes. Planking is done to crush the hard clods, level the soil surface and to compact the soil lightly. Harrows, Cultivator, Spade, Ladder, Khurpi, Nirani etc. are used for secondary tillage.

(B) After Tillage : The tillage operations that are done in the standing crop after the growing or planting and prior to the harvesting of the crop plants are called after tillage. This is also known as intercultivation or post seeding/planting cultivation. It includes harrowing, hoeing, weeding, earthing up, drilling or side dressing of fertilisers etc. spade, nirani, wheel hoe, paddy weeder, etc. are used for intercultivation.

Tillage operations depend on the crops to be cultivated as follows :

(i) Dry Tillage : Dry tillage is generally practised for the crops that are sown or planted in dry land condition having sufficient moisture for germination of seeds. Dry tillage is suitable for the crops such as broad casted Paddy, Jute, Wheat, Oilseeds crop (*i.e.* mustard, sesame, linseed, sunflower etc.), Pulses (*i.e.* Gram, Black gram, Green gram, Lentil, Cowpea etc.), Potato, Radish, Carrot, Beet, Turnip and Vegetable crops. Dry tillage is done in a soil having sufficient moisture (21 – 23 per cent). The soil becomes more porous and soft due to dry tillage. Besides this, the water holding capacity of soil and aeration is also increased. This conditions are more favourable for soil micro-organism, the most important convertor of nutrients into available form to plants.

(ii) Wet or Puddling Tillage : The tillage operation that is done in a land with standing water is called wet or puddling tillage. Puddling operation consist of ploughing repeatedly in standing water untill the soil becomes soft and muddy. Puddling creates an impervious layer below the surface to reduce deep percolation, losses of water and to provide soft seed bed for planting rice. Puddling tillage is also done in both the direction for the incorporation of green manuring crops and weeds. Wet tillage destroys the soil structure and the soil particles that are separated during puddling settle later. Wet tillage is only the means of land preparation for transplanting semi–aquatic crop plant, such as rice etc. Planking after wet tillage makes the soil level and compact. Puddling hastens transplanting operation easily and smoothly as well as the establishment of seedlings. Wet land ploughs or wornout dry land ploughs are generally used for wet or puddling tillage.

5.3.3 Factors Influencing Preparatory Tillage

The factors influencing preparatory tillage are as follows:

(*i*) Crop : It dictates the type and extent of preparatory tillage. The small size seed crop (*i.e.* Mustard, Jute, Sesame, etc.) requires fine seed bed for good germination of seed and stand establishment. Pulse crop (*i.e.* gram, pea, pigeon pea, green gram etc.) does not require fine seed bed. The root crop (*i.e.* Radish, Carrot, Beet, Turnip etc.) and tuber crop (*i.e.* Potato etc.) require deep and fine tilth. Crops like Brinjal, Cabbage, Cauliflower, Knolkhol, Chilli, Onion etc. are transplanted in the main field after raising the seedlings in the seed bed. The seed bed and main field require repeated ploughings followed by plankings for the sowing of seed and transplanting seedlings. Paddy require puddling condition which is achieved by tilling the land in standing water in the field.

(*ii*) Soil : The light soil requires less tillage and heavy soil requires more tillage to bring the soil favourable for sowing seed and transplanting of seedlings.

(*iii*) Climate : The climatic condition affects on the moisture content of the soil which creates condition favourable for tillage.

(*iv*) Cropping System : Cropping system involving different crops need different types of tillage. Crops following pulse crops need lesser tillage than that of following maize, sugarcane etc. The crops following potato require minimum tillage. But the crops following paddy need repeated preparatory tillage for obtaining an ideal seed bed for sowing seed or transplanting seedlings.

(*v*) Type of Farming : Tillage operation varies with the type of farming. Deep ploughing is necessary in dry land agriculture to eradicate perennial weeds as well as to conserve soil moisture. In intensive irrigated farming, tillage is required for each crop throughout the year.

5.3.4 Modern Concept of Tillage

Jethro Jull (1674–1741) gave a concept that 'Tillage is manure and this statement is correct in many sences'. Because tillage makes the soil a better medium for the growth and development of the plant. It improves the physical condition of the soil and makes better air-water temperature relationship in the soil. This condition helps in the better growth and activities of the soil microorganism. Tillage makes the soil suitable for sowing of seed and it also controls weeds in the field. In traditional tillage, too much tillage, is recommended to get fine tilth and it is expensive, time and energy consuming which often destroys the soil structure else. In recent time, the concept of tillage has been changed considerably and new concept have been introduced namely minimum tillage, zero tillage, stubble mulch tillage.

1. Minimum Tillage : Minimum tillage is aimed at reducing tillage to the minimum necessary for ensuring a good seed bed, rapid germination, a satisfactory stand and a favourable growing conditions. Tillage can be reduced into two ways such as (i) by emitting operations which do not give much benefit when compared to the cost and (ii) by combining agricultural operations like seeding and fertilizers application (Reddy and Reddi, 1993).

Minimum tillage can be practiced by different methods as follows :

(*i*) ***Row Zone Tillage :*** In this method, after completing the primary tillage in the entire field, the secondary tillage is done in row zone area only into which seeds are sown.

(*ii*) ***Plough Plant Tillage :*** In this method, after completing primary tillage in the entire field, the crop row area is pulverised and seeds are sown or seedlings are transplanted. In this system, ploughing and seeding or planting may be completed on one operation.

(*iii*) ***Wheel Tract Tillage :*** In this system, after completing the primary tillage as usual, tractor is used for sowing of seed and the wheel of tracted pulverises the soil of the crop row zone only.

Minimum tillage has some advantages and disadvantages as follows :

Advantages

(*i*) Improvement of soil condition due to decomposition of plant residues *in situ*.

(*ii*) Cost and time of soil preparation are reduced.

(*iii*) Higher infiltration caused by the vegetation presentation on the soil and channels formed by decomposition of seed, plant roots.

(*iv*) Less resistance to root growth due to improvement of soil structure.

(*v*) Less soil compaction by the reduced movement of heavy tillage implements.

(*vi*) Less soil erosion compared to conventional.

Disadvantages

(*i*) Seed germination is lower with minimum tillage.

(*ii*) Poor root and nodule development in some leguminous crops like peas and broad beans.

(*iii*) Reduced rate of decomposition of organic matter.

(*iv*) Necessity of herbicidal use to control weeds. Continuous use of herbicides causes pollution and dominance of perennial problematic woods.

Minimal Cultivation of Wheat : Under All India Co-ordinated Wheat Improvement Project, at B.C.K.V., West Bengal, conducted by Dr. N. Das, the following recommendation have been made regarding cultivation of Wheat without any tillage. In silt loam and sandy loam soil a furrow of 5-7cm is made in between two rows of stubbles after the harvest of rice with the help of hand tyne. No pre-sowing irrigation is required. Fertilizers @ 110 Kg of Urea, 250 kg of Single Super Phosphate and 68 kg of Muriate of potash per hectare should be mixed with soil within the furrow. Then seeds are to be sown manually @ 100 kg per hectare. The Sonalika and Janak variety should be sown and then the seeds are to be covered with soil. Urea @ 110kg per hectare will be top dressed, 18-21 days after sowing before giving irrigation. After first irrigation, when the soil will reach field capacity, the stubbles should being corporated with help of hand spade and thereafter weeds and stubbles will decompose in the soil. The other irrigation will be given at late tillering, flowering and milk stage. The minimal cultivation in this way can be practiced in flood affected areas where it becomes sometimes difficult to plough the field for sowing the wheat in times.

2. Zero Tillage : Zero tillage is a tillage system in which mechanical soil manipulation is reduced to traffic and seed bed preparation (Bacumer and Bakerman, 1973). It is an extreme form of minimum tillage. It reduces labour cost, runoff, prevent soil erosion and crust formation. The plant residues remain on the soil surface with zero tillage which provides a clothing on the surface. As a result of which soil erosion and loss of soil moisture by evaporation is controlled.

The zero tillage system was first used successfully in 1950 in pasture renovation in USA with mulch farming practices and till plant system for row crops in order is provide year round protection of the soil form erosion and to minimise planting cost. This system is found to be an alternative to the conventional tillage system which are subjected to wind and water erosion, the timing of tillage operation is too difficult. The requirements and cost of energy and labour are too high and performance is insufficient (De, 1960).

Zero tillage areas often face serious weed infestation and require harbicides application for the control of weeds. Weed infestation causes troubles for smooth sowing of seed and it becomes necessary to apply broad spectrum non-selective herbicides like paraquat, glyphosate etc. Application of selective herbicide after emergence of crop is required to avoid manual weeding in the field. There are some disadvantages of zero tillage as follows :

(*i*) The seedling establishment is 20-30 per cent less than the conventional method of sowing seeds.

(*ii*) The crop needs about 15-20 per cent higher dose nitrogen because of retarded rate of mineralisation of organic matter.

(*iii*) Occurrence of higher infestation of weeds (both annual and perennial).

(*iv*) Infestation of insect pests and diseases.

(*v*) Higher residual toxicity of applied herbicides.

3. Stubble Mulch Tillage : A new approach was developed for keeping the soil protected at all times by leaving crop residues or stubbles on the soil surface during fallow period. It is known as stubble mulch tillage or stubble mulch farming. In this system, the crop is harvested leaving stubbles of 10-15 cm height. Then the tillage is done to loosen the soil and kill the weeds present in the fallow land. The seeds are sown in the field by using a narrow planter shoe to open narrow furrow in which seeds are dropped. This method have some practical problem. The stubbles or residues left on the soil surface interfere with the seed bed preparation and sowing of seed.

5.4 SOWING OF SEED

Tillage is followed in used bed preparation. Seeds of some crops such as Wheat, Paddy, Jute, Sunhemp, Mustard, Sesamum, Linseed, Groundnut, Gram, Pea, Lentil, Soyabean, Spinach etc. are directly sown in the main field. But seeds of some crops such as Paddy, Vegetables, Seasonal flowers etc. are sown in the seed bed or nursery bed. Seedlings are raised in the nursery bed for the following reasons :

(*i*) The seedling of some crops are raised in the nursery bed whose seeds are smaller in size and seedlings need sufficient care during early stage or growth.

(*ii*) The seeds of some crops that do not response well in direct seeding in the main field are sown in the nursery bed and the seedlings are transplanted in the main field.

(*iii*) The seedlings are transplanted just after the harvesting of the crops. On the other band, the crops will be late, if the seeds are sown after the preparation of land.

(*iv*) Seedlings are raised in nursery bed with sufficient care and they are protected from the attack of insect pests and disease.

(*v*) Rogueing of crops become easy.

(*vi*) The crops such as vegetables, season flower etc. yield better if they are transplanted in the main field after raising the seedlings in the nursery' bed. Care should be taken for the following points for raising seedlings in the nursery bed.

1. Selection of Land for Seed Bed : The seed bed is to be prepared in the land which is loamy, slightly elevated above the surrounding areas and is getting bright sun shine throughout the day. The soil should be rich in organic matter and loose and friable with water retention capacity. Drainage facility is must. Low land having no drainage facility is totally unsuitable for the vegetables cultivation.

2. Preparation of Seed Bed : The land is brought to find tilth by repeated ploughing followed by planking. The nursery bed should be raised 15–20 cms from ground level, specially during rainy season. The seed beds are usually four feet wide (120 cm) and having convenient length. But the common length for grain is 25ft. (7.5m) and for vegetable crops is 6–8 ft. (1.8–2.4 m). A space of 1 ft. (30 cms) may be left around each bed for ensuring good drainage of the beds. Each bed must be worked throughly to obtain good tilth. The clods should be broken to a fine pieces. Weeds stones and other undesirable material should be removed. The seed bed at 1/10th of an acre (400 sq.m) should receive 500 kg. F.Y.M., 1kg. K, 1kg. P_2O_5 and 1 kg. K_2O for proper and healthy growth of seedlings.

3. Soil Treatment : The seedlings are usually affected by soil borne organism. The soil of nursery bed must be sterilized by any one of the following methods to prevent the infection of "damping off" disease. Because this disease in many times becomes a troublesome factor in crop production.

Following are the general method of Sterilizing the seed bed.

(*a*) Dry refuges are spread on the bed and burnt. The weeds, insect pests and disease materials which are likely to be present are killed by heat.

(*b*) Dr. Metha (1959) recommends the following methods for Sterilizing the soil. One part of formalin mixed with

100 parts of water should be applied @ 4.55 liters per 0.09 square metres of surface soil as to saturate it upto a depth of 152.4 mm. The Soil mixture is covered with a gunny bags or a thick paper for a day or so. So that fumes may penetrate the Soil and kill all fungi and insect pests. The soil mixture is then uncovered and loosened and left for a few days till the smell has disappeared.

The sterilized soil mixture is then put through a 6 mm mesh screen. The coarser material separated by the screen may be put in the bottom which will improve drainage and the finer mixture may be put at the upper layer pressing firmly especially at the sides and the corners of the boxes.

Basic organic manures such as compost, F.Y.M. etc. should be mixed with the soil of the bed. The surface of the bed should be fairly smooth and friable and free from clods in order to plant at uniform depth and to have a good coverage of seed.

Individual seedlings may also be raised in earthen pots, polythene tubes and paper pots. So that they can be transported and transplanted to any distant place without much injury to their root system.

4. Selection of Seed : Seed is a fertilised ripened ovule. A seed consist of the following three parts.

(*i*) ***Embryo*** **:** It gives rise to new plant in due course under suitable conditions necessary for germination of seed.

(*ii*) ***Endosperm :*** It contain the substances that nourish the embryo during the development prior to and for sometime after germination.

(*iii*) ***Seed-Coat :*** It shields the embryo and endosperm and protect the seeds.

Seeds are the protectors and propagators of their kind. Good seed is the base of any successful farming industry. The success of a crop depends on the quality of seed used for sowing or planting. The seed must possess the following characteristics:

(*i*) It should be true to type (Pure) with good yield potential.

(*ii*) The seeds must be healthy, fresh, pure and free from all the inert materials and weed seeds.

(*iii*) They should be free from seed borne diseases.

(*iv*) The seed should be large (according to the variety), plump, bold, uniform in size, shape, colour and development.

(*v*) The seed must not be affected by any insect attack.

(*vi*) The seed must be viable and should have high germinating capacity.

Plants reproduce sexually by seeds and asexually by vegetative parts. Most crops are usually propagated by seeds. But few are propagated by tuber (*e.g.* potato), stem cutting (*e.g.* sweet potato, pointed gourd etc.), bulbs (*e.g.* onion, garlic etc.) and rhizome (*e.g.* Yam, colocasia etc.).

5.4.1 Types of Seeds

The seeds are evolved, tested and if found to be good they are multiplied and released for commercial production of the crop. Four classes of pure seed have been recognised by the international crop improvement association. The seed of any improved variety which is to be multiplied for distribution to the farmer for commercial cultivation has to pass through these classes as follows :

(*i*) Breeder Seed or Nucleus Seed : It is the initial amount of pure seed of the improved variety available with the breeder. This seeds are produced by hybridization, Selection and mutation under the strict supervision of the plant breeder at the main breeding station. It is always cent percent pure genetically as well as physically. This seeds are available in little quantity and are often costly.

(*ii*) Foundation Seed : Any seed which is produced on government multiplication farm is usually called foundation seed. And it is second grade seed order its genetic purity because there may be slight degeneration during the process of multiplication of nucleus seeds. Foundation seed is the source of all other certified seed classes, either directly or through registered seed. And it is relatively less pure compared to breeder's or nucleus seed. It is also called as mother seed and is produced from the breeder's seed under the supervision of trained seed scientists, by certain organisation viz. National Seed Corporation, Tarai Development Corporation and State Seed Corporation.

(*iii*) Registered Seed : This is the name given to the seed multiplied from either the breeder seed or the foundation seed. These seeds are produced by the farmers and other growers under the special contract with the seed certifying agency and they are provided with the initial amount of pure seed alongwith complete technical advice. The seeds thus produced are purchased by the government at fixed price and handed over to the distribution agencies.

(*iv*) Certified Seed : This seed is actually the progeny of registered seed. It is the name given to any seed that has been approved and certified by seed certifying agency as a good seed for sale and distribution to the farmers for commercial cultivation. This seeds are produced at the seed farm of the national seed corporation of state seed farm and also by certified farmers under grower programme. The certified seed is relatively less pure compared to the other classes of seed.

5.4.2 Dormancy of Seed

Dormancy is the state of rest period of a seed in which it does not germinate even they are placed in a conditions suitable for germination. Dormancy is survival mechanism of a seed. Dormant Seeds can be carried for a distance place and stored for longer time.

The dormancy of the seed may be classified according to their occurrence and it may be classified as follows :

(*i*) Primary or Natural Dormancy : The seeds that are incapable of germination after ripening if they provide suitable condition of germination. The seeds belong this category are said to have primary dormancy. ***Example :*** Potato etc.

***(ii)* Secondary or Induced Dormancy :** The seeds that are capable of germination under favourable condition just after ripening become incapable of germination when they are stored under unfavourable condition even for a few days. This type of dormancy is said to be secondary dormancy. Such dormancy lasts untill the seed experiences the condition necessary to break it (Fryer and Mukepeace, 1977).

Causes of Dormancy : There are some seeds that do not germinate even though they are exposed to a favourable condition considered adequate for germination. This is due to the dormancy of seed which may be due to any single cause or a combination of more than one of the following causes.

(*i*) ***Physical Impermeability of Seed Coat to Water :*** The seeds of legumes and cotton have a seed coat that are impermeable to water and they fail to germinate even though they are exposed to a favourable condition considered adequate for germination.

(*ii*) ***Hard Seed Coat :*** The seeds of mustard, amaranthus contain hard and strong seed coat. They fail to germinate if the seed coat do not burst.

(*iii*) ***Immature Embryo :*** The seeds of many orchids that are apparently ripen, contain a rudimentary or immature embryo. The germination of such seeds naturally gets delayed untill the embryo develops properly.

(*iv*) ***Seed Coat Being Impermeable to Oxygen :*** The seed coat of *Xanthium* sp are impermeable to oxygen and they fail to sprout if the seed coat do not rupture at due time.

(*v*) ***Dormant Embryo :*** There are some seeds such as apple, peach etc. that do not germinate even though the embryos are completely developed and the seeds are provided all favourable conditions for germination. In such cases some physiological changes called 'after ripening' take place during the period of dormance which enables the seeds for germination (Singh, 1990).

(*vi*) ***Synthesis and Accumulation of Germination Inhibitors within the Seeds :*** The chemical compounds that are synthesised by plant organs are accumulated in the seed at maturity and they inhibit the germination of seed. As for example, the tomato seeds fail to germinate if they are treated with their fruit juices.

Method of Breaking Dormancy of Seed

(*i*) ***Scarification :*** The dormancy due to hard or impermeable seed coat can be broken by scarification. Chemical scarification with Hydrochloric acid (HCl), Sulphuric acid (H_2SO_4) etc. or mechanical scarification with sand paper etc. may be helpful for breaking the dormancy of seed. But this can be done in such a way that the embryo is not injured at all.

(*ii*) ***Storage :*** The dormancy due to physiological change may be broken by storing the seeds under dry storage condition.

(*iii*) ***Exposure of Seeds to Light :*** The light has an effect in breaking the dormancy of seed. The seeds should be exposed to bright sun for some hours before sowing them in the field.

(*iv*) ***Application of Pressure to Seeds :*** Germination of *Medicago sativa* is found to be increased from 50 to 200 per cent when a hydraulic pressure of 2000 atmosphere at 18°C is applied. The probable reasons for higher germination may be an increase is permeability of seed coat to water and Oxygen (Singh, 1990).

(*v*) ***Chemical Treatment :*** There are some chemicals that are employed for breaking dormancy of seed. Denny (1962) shortened the rest period of potato by treating them with ethylene chlorohydrazine, thio–urea or Potassium thiocynate.

5.4.3 Seed Treatment

Seeds act as a carrier of disease producing organism and

insect pest. Seed treatment is a process of application of chemicals to the seed that prevent the transmission of disease producing organism through seeds.

Seeds before sowing either in nursery or in main field should be treated by any one of the following method :

(*i*) **Hot Water Treatment :** Soaking of seeds for 15 – 20 minutes at hot water at 122° F (50°C) help to control some disease *e.g.* Black rot of cabbage.

(*ii*) **Seed Treatment with Organic-Mercuric Compound** : Most of cereals and vegetables seeds are treated with seed treatment chemicals (vide: Table 6.1).

For treating seeds before sowing, seeds are dressed with disinfectants in dry, Wet or slurry conditions.

(*i*) **Dry Treatment :** The seed dressing chemicals are mixed with water soaked seed in a seed dressing drums and the seeds are treated quickly and uniformly by this method. This method reduces the inhalation hazard.

(*ii*) **Wet Treatment :** The water soluble chemicals are dissolved in water in earthen pot and the seeds are dipped in the solution for some time. The seeds are taken from the solution and dried under shed for some time. The treated seeds are then sown in the field or stored in the suitable container for future use. The seeds should be properly dried if they are stored. Otherwise they are likely to be damage during storage.

(*iii*) **Slurry Treatment :** The wettable chemicals are mixed with small quantity of water and this mixture is applied then to the seed in automatic or semiautomatic machine suitable for large scale seed treatment. This is most suitable method for applying wettable powder formulation to seeds.

Table 5.1 : List of Seed Treatment Chemicals

Chemical Name	*Trade Name*	*Amount of Seed Treatment Chemical*
1. MEMC. (Methoxy ethyl Mercury chloride)	Emisan-6, Baga 1101–6, Areton, Cerason, Agnosan-G.N. etc.	(*i*) Dry seed treatment : Paddy, Wheat and other crop : 3.0 g per kg. of seed. (*ii*) Dry seed treatment: Vegetable seeds : 2.0g per kg. of seed. (*iii*) Wetseed treatment: Paddy, Wheat etc. : 1.5 g Per litre of water for treating one kg. of seed. (*iv*) Wet seed treatment: potato etc. : 1.5 kg per litre of water (100 g chemical is required for treating 100 kg. of seed. (*v*) Wet seed treatment: Sugarcane etc.: 2.5 g. per litre of water (100 g chemicals is required for 150 kg of sugarcane.
2. Carbendazim-25% & 50% WP	Bavistin. Derosal, Bengad, Arrest, Kar-zim, Dom, Instin, Sevistin, Volgm, Fungigard, Sevistin. Palstin etc.	(*i*) Dry seed treatment : Paddy, Wheat, Maize etc.: 2.5 – 3.0 g per kg. of Seed (*ii*) Wet seed treatment : Paddy, Wheat etc.: 1.5 per litre of water for treating one kg of seed.
3. Thiram–75% W.P.	Thiram, Thiram – D.S. etc.	Dry seed treatment: Paddy, Wheat, Maize etc.: 2.5 –3.0 gm per kg of seed. (*i*) Dry seed treatment : Jute: 5.0gm per kg of seed. Paddy:30.g per kg of seed. Vegetable : 2.5 g per kg seed.
4. Mancozeb–75% WP.	Dithane M–45, Indo-fil M–45, Uthane, Savier, Shield–45	(*i*) Dry seed treatment : Jute: 5.0 g per kg of seed. Paddy: 3.0g per kg. of seed. Vege-table : 2.5 g per kg of seed.
	Dhanuka–M–45, Anu M–45 etc.	(*ii*) Wet seed treatment : Potato, sugarcane etc.: 2.5

Chemical Name	*Trade Name*	*Amount of Seed Treatment Chemical*
		gm per litre of water (100 g chemical is required for treating 100 kg of seed/ SCH).
5. Captan 50% WP	Captan, Dhanustan etc.	Dry seed treatment: Jute : 5.0 g per kg of seed. Vegetable : 3.0 g per kg of seed.
		Dry seed treatment : Paddy : 4.0 g per kg of seed.
6. Trichoderma viridi (bacterial fungicide	Si–guard, Stanes Biocure (F),	Dry seed treatment : Paddy, Wheat, Potato, Pulse crops, Sugarcane & Vegetable crops: 5.0 gm per kg. of seed.
7. Pseudomonas Fluorescens (Bacterial and bacteriacide)	Sajag	Dry seed treatment : Paddy, Pulse crops, oilseeds and Vegetable crops: 10.0gm per kg of seed.
8. Biofertilizer (Bacterial culture)	(*i*) Azospirillum (*ii*) Rhizobium	(*i*) Cereals, millet, sesame, cotton etc.: 600gm for seed required for a hectare.
		(*ii*) Leguminous crops 600gm for seed required for a hectare.

The seeds of pulse crop should be treated with Rhizobium culture when this crops are grown for the first time in any field as the soil have no particular Rhizobium species.

5.4.4 Method of Sowing

Sowing is an art of placing seed in the soil to have good germination and stand of crop in the field. There are some common methods used for sowing crops as follows :

(*a*) Broadcasting : Broadcasting is a very common method of sowing seed. Broadcasting is the process of random scattering of seeds on the surface of seed bed. Broadcasting of seeds can be done both manually or mechanically. Seeds are broadcast on the soil having sufficient moisture for the germination of seed. Seeds may also be broadcast in the wet field. The germinated paddy seeds are generally sown by this method in the wet land. When broadcasting is done manually, the uniformity of scattering seeds depends on the skill of the man. The seeds are to be covered with soil by planking or some other devices, just after the sowing is completed. When broadcasting is done mechanically, a machine is used for scattering the seeds on the surface of one seed bed at controlled rates. Broadcasting have some advantages and disadvantages as follows :

Advantages

(*i*) It is an easy, quick and cheap method of sowing seed.

(*ii*) More land can be covered within a short time.

(*iii*) No need of sowing implement. So, cost of sowing becomes less.

(*iv*) Crop plants not requiring special types of cultural practices, such as earthing up etc. may be sown by broad casting.

(*v*) Under the situation where the number of plants per unit area (Per M^2) is more important than that of definite spacing from plant to plant, broadcasting is the usual method of sowing.

(*vi*) Broadcasting is the usual method of sowing seeds for mixed cropping (De, 1990).

(*vii*) Broadcasting needs less labour.

Disadvantages

(*i*) The requirement of seed per acre is more.

(*ii*) There are difficulties in uniform distribution and placing of seeds in optimum and uniform depth of soil. So the germination is not uniform.

(*iii*) The cost of weeding and thinning in broadcast crop is more.

(*iv*) The intercultural operations such as earthing up, manuring, irrigation etc. cannot be carried out successfully.

(*v*) Broadcasting needs planking for covering the seed with soil. On the other hand, drilling needs no planking.

(*vi*) Broadcast crops do not grow uniformly and desired yield is not possible.

(*b*) Dibbling : Dibbling is the process of placing the seed in the holes made by any convenient implement such as spade, khurip, nirani etc. or by hand in the seed bed and covering them with soil by hand. In this method, the seeds are placed in holes made at definite depths at fixed spacing. The equipment used for dibbling the seeds is called a dibbler.

Dibbling have some advantages and disadvantages as follows :

Advantages

(*i*) It is the most suitable method for wider spaced planted crops, such as cucurbitaceous vegetables, fruit crops etc.

(*ii*) Sowing of seed at definite spacing is possible.

(*iii*) Dibbling gives more or less rapid and uniform germination with good seedling vigour.

(*iv*) The cultural practices such as weeding, thinning, manuring, irrigation, earthing up, roughing, spraying, harvesting etc. can be done successfully.

(*v*) The entire field need not be prepared for sowing seed by this method. Preparation of only the seeding zone may be helpful.

Disadvantages

(*i*) Uniform germination is not possible if all seeds are not placed at uniform depth.

(*ii*) Dibbling is a more laborious, time consuming and expensive process compared with broadcasting.

(*iii*) It is difficult to maintain uniform seed rate per acre.

(*c*) Drilling : Drilling is a practice of dropping seeds in furrow lines in continuous flow and covering them with soil. Sowing implements such as seed drill or seed-cum-fertiliser drills are used for sowing seeds. During, drilling seeds may be sown continuously or at regular intervals in rows.

Drilling have some advantages and disadvantages as follows :

Advantages

(*i*) The seed rate per acre becomes less.

(*ii*) Drilling maintains uniform depth and spacing of seed and also uniform plant population per unit area.

(*iii*) Seeds are placed at uniform depth and covered and also compacted uniformly in this method.

(*iv*) Drilling facilitates thinning and roughing of weak and disease affected plant.

(*v*) Weeding can be done profitably within a short time by wheel hoe, Japanese Paddy Weeder etc.

(*vi*) The intercultural operation such as earthing up, manuring, irrigation, spraying etc. can be done successfully in the drilled crops.

(*vii*) The drilled crops get light, air, nutrients equally as they are spaced at uniform distance.

(*viii*) Harvesting of drilled crops is easier and advantageous. So harvesting cost becomes less.

(*ix*) Drilling may be adopted for both pure cropping and intercropping situations.

(*x*) The cost of cultivation in drilled crop becomes less. The per acre yield of drilled crop increases.

Disadvantages

(*i*) Drilling requires an implement such as a seed-drill which increases the cost of cultivation.

(*ii*) Drilling requires more time, energy and cost.

(*iii*) An expert technical person is required for running of a seed dill.

(*iv*) Drilling needs more time in comparison to broadcasting.

(*v*) Drilling is not feasible in clay and stony soils.

(*d*) Sowing of Seed Behind the Plough : In this method, some crops such as Paddy, Wheat, Maize, Jute etc. are sown in the well prepared field. A man drops seed in the furrow behind the plough. The sowing is done by hand behind the plough, Kera in the irrigated area or by pora in the rainfed condition. Pora consist of a bamboo tube provided with a funnel shaped mouth, which is attached to the handle of the plough. The ploughman drops the seeds through the funnel. This is a slow and laborious method.

(*e*) Sowing of Seeds in Deep and Shallow Furrow :

(*i*) A deep furrow (15–20 cm) is made at a required distance by spade and sugarcane setts are planted in furrow. Then the setts are covered with soil afterward.

(*ii*) A shallow furrow (8–10 cm) is made at a required distance by spade or small country plough and then the potato tubers are planted in furrow at a distance of 15–20 cm. The tubers are covered with soil afterwards.

More time is needed for sowing seed by this method. This is a slow and laborous method.

(*f*) Sowing of Seed in Ridges : The seeds are sown in ridges which is made at required distance to avoid spoiling of seed in rainy season. The seeds are sown on the ridges by dibbling method. Some crops such as lady's finger, maize, sunflower, etc are sown by this method. This is slow, laborious and costly method.

5.5 RAISING OF SEEDLING

1. Paddy : Paddy seedlings required for transplanting are raised in specially prepared seed bed and they may be of three

Table 5.2 : Seed Rate and Spacing for Important Field Crops

Sl. No.	*Crops*	*Seedrate (kg/ha)*	*Time of Sowing*	*Spacing (cm)* Row to row	Plant to plant
1.	Paddy				
	Aus	62.5	April-May	20	10
	Aman	37.5-62.5	Mid. June to Mid. August	15-20	10-15
	Boro	50.0-62.5	December to Jan.	15	10
2.	Wheat	100-125	Oct. to Nov.	25-30	10
3.	Maize	37.5-50.0	Kharif : June-Oct.	60	20
			Rabi : Oct.-Feb.		
4.	Barley	75.0-80.0	November	20-25	10
5.	Pulses				
	Gram	Desi : 60-75 Kabuli : 95-100	Mid. Oct. Ist week of November	30	8-10
	Pea	62.5-75.0	Oct. to Mid. of Nov.	30-60	5.75
	Lentil	25-35	Oct. to Nov.	22.5-30.0	10
	Cowpea	30.0	Kharif : May-June	Kharif : 30-45	Kharif : 8-10
			Rabi : Oct. to Nov.	Rabi :25-30	Rabi 8-10
			Spring : Feb. to Mar.		
	Pigeon pea	10-25 (Drilling)	Last week of June to Ist week of July	90-120	30
	Green gram	15-20 (Drilling)	Kharif : Mid. June-Mid. July	45	10
			Rabi : Oct. to Nov.	30	10
			Summer : Mid. Mar. to Mid. June	30	10
	Soyabean	75-100	Kharif : June or July	40-60	5-10
			Rabi : Oct. to Nov.		
6.	Oil seed crop				
	Rapeseed and Mustard	6.25-7.5	Sep. to Nov.	30-45	10-15
	Sesame	6.5-7.5	Mid. January to Mid March	30-45	10-15

Sl. No.	*Crops*	*Seedrate (kg/ha)*	*Time of Sowing*	*Spacing (cm)*	
				Row to row	*Plant to plant*
	Ground-nut	100-125	Kharif : Mid. April-Mid. July Rabi : Oct. to Dec.	30-45	8-15
	Linseed	10 (Drilling)	October	25-30	4-6
	Sun-flower		Kharif : June-July Rabi : Oct. - Nov.	60-80	20-30
7.	Fibre Crop Jute	Capsularis : 6.25-7.5 Olitorious : 3.75-5.0	Mid. February to Mid. May	Capsularis: 30 Olitorius : 22.5-25.0	4-7

types, namely (*i*) Dry seed bed, (*ii*) Wet seed bed and depog method of nursery.

(i) Dry Seed Bed or Dry Nursery : The aus and aman seedlings can be raised in dry nursery. Light loamy soil having adequate drainage facility should be selected. The land is brought to fine tilth by repeated ploughings followed by plankings.

The area and measurement of dry seed bed are as follows:

Area : 1/10 the hectare of land (*i.e.* 1000 sq. meters) for seedbed. The seedlings raised in this area will be sufficient for transplanting in one hectare main field.

(*a*) *Measurement*

Breadth : 4 feet (1.2 metre)

Length : Convenient length. But the common length is 25 feet (7.5 metres).

Height : 6 inches (15 cm) high from the ground level.

There should be a space of one foot (30 cm) between the two bed. And it is used as road, irrigation and drainage channel.

Manuring

(a) Basal Dressing

Compost or F.Y.M. : 500 kg.

Nitrogen (N) : 1 kg.

Phosphorus (P_2O_5) : 1.5 kg.

Potassium (K_2O) : 1.5 kg.

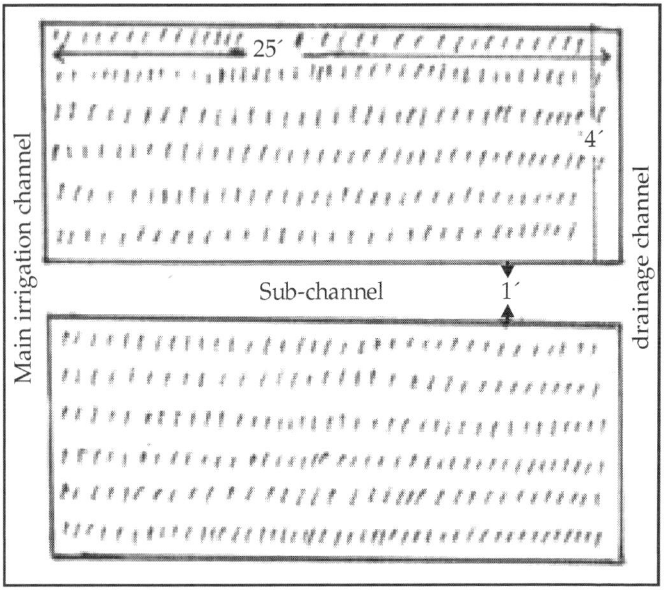

Fig. 5.1 : Seed bed (Paddy)

The manures and fertilizers should be evenly spread and incorporated in the soil.

(b) Top Dressing : Top dressing of 5.5 kg. of Urea is to be done for the proper growth of seedlings after 15 days from date of sowing seeds.

The seeds are sown in the prepared seed bed and the sown seeds are covered with a layer of fine compost and earth. Dry nursery should be irrigated by sprinkling water. The seedlings get ready for transplanting in 25-45 days depending upon the duration of the variety and rate of growth of seedlings in the nursery. The nursery period is roughly one week for every month of the maturation period of the variety. The seed bed is irrigated before lifting the seedlings to reduce injury to the root system of young seedlings.

(*ii*) Wet Seedbed or Wet Nursery : Mainly Boro and sometimes Aman seedlings can be raised in wet nursery. Clay and clay loam soil should be selected for wet nursery. The soil is brought to tilth by repeated ploughings followed by ladderings. Then the field is puddled by giving three or four ploughings in standing water with an interval of about 5 or 6 days. The field is then levelled properly and divided into small raised beds, 4 feet (1.20 metre) wide and of any convenient length. The beds are throughly levelled and one or two inches (1.25 cm or 5 cm) of water is allowed to stand in the field.

Area : 1/6th hactare of land *i.e.* 1666 sq.meters) for seedbed of raising boro seedlings.

Manuring :

(*a*) *Basal Dressing* :

Compost or F.Y.M. : 750 kg.

Nitrogen (N) : 2.kg.

Phosphorus (P_2O_5) : 2 kg.

Potassium (K_2O) : 2 kg.

(*b*) *Top Dressing* : Top dressing of 8.25 kg of urea in to be done for the proper growth of seedlings after 15 days of seed sowing.

The sprouted seeds are broadcasted evenly on the soft mud in the seedbed. Next day, the water of the seedbed is completely drained off. After 3 or 4 days, depending upon the physical condition of the soil, when fine cracks are forming, the nursery is irrigated lightly. This is continued until the seedlings are about two inches (5 cm) tall and thereafter 2 or 3 cm of water is maintained.

(*iii*) Dapog Method of Nursery : Dapog method of raising paddy nursery has been introduced in India from Philippines. The seedlings raised by this method get ready for transplanting within 10–14 days of sowing regardless of the duration of the variety. 25 square meters of area is enough to raise seedlings sufficient for planting one hectare of land. Dapog method of raising nursery consist of growing seedlings on a concrete floor or raised seedbed covered with polythene sheet or banana leaves or cement or fertilizers bag. Pregerminated seeds are shown on the top of cover 1.0 kg per square meter of nursery and covered with a moist gunny bag. When the germination is completed, the gunny bag is removed. The seedlings should be pressed with a slight pressure twice a day either with hand or a light wooden plank at 8–10 days after sowing. This operation keeps the roots of the seedlings in contact and prevents the roots being exposed to air, water should be sprinkled in morning and evening before pressing the seedlings. After three or four days, the seedbed should be irrigated continuously to a depth of 1.0 to 2.0 cm of water. The polythene sheet containing seedlings is rolled like a mat with roots outward and carried to the field where the seedlings are easily separated and transplanted. The seedlings should be transplanted in the puddle field containing about 2.0 cm deep water.

Modified 'Dapog' Method of Raising Paddy Seedlings: A modified Dapog method has been described by Reddi and Rao (1969) which is given below :

(*i*) Separate beds are prepared with channels all around. The beds are one metre wide, a metres long and 10 cm. high.

(*ii*) The surface of the bed is packed tightly and kept at uniform level.

(*iii*) The beds are covered with polythene sheets. Sands or ash upto a thickness of 2.5 cm is spread on the bed.

(*iv*) The seeds are soaked in water for 12 hours and kept for sprouting for 24 hours.

(*v*) The pre-germinated seeds sown in sand or ash media uniformly are one grain thickness. Seed rate is one kilogram of seed for square metre of nursery area.

(*vi*) The seeds are covered with gunny bags for the first three days and kept moist by watering with rosecane thrice a day. Both in morning and evening, the germinated seeds are gently pressed with bare hand so that the root system may not exposed.

(*vii*) The channels around the beds are filled with water upto the brim.

(*viii*) After three days, when the critical period of germination is over, the gunny bags covered on the seeds are removed and one centimetre of water is maintained in the nursery for the rest of the period by raising the edges of nursery beds under ploythene sheet with mud. Watering with rose cane both in evening and morning is given in the nursery bed.

(*ix*) From the eight day after sowing, urea dissolved in water at the rate of 150 gm per litre of water is applied to the seedlings on alternate day upto 16th day.

(*x*) By applying urea in liquid form to the seedlings, the seedlings grow tall, healthy and robust.

(*xi*) No further manuring is given to the seedlings after 16th day.

The advantages of the modified Dapog method *i.e.* spreading of 2.5 cm layer of sand or ash over polythene sheet, sowing of pre-germinated seeds at a thin seed rate of one grain thickness and the application of urea in liquid form to the seedlings from the 8th day onward on alternate days have been described by the authors as follows :

(*i*) The height of the seedlings can be increased to facilitate easy handling at time of transplantation.

(*ii*) The seedling can be planted in the field with advantage even when there is water upto a depth of 4 cm or more.

(*iii*) The seedlings will be more healthy and robust and can withstand rough handling at the time of pulling and transplanting, just as in ordinary method.

(*iv*) Watering twice a day, both morning and evening will be enough.

(*v*) When seedlings are raised in sand or ash medium there is no difficulty in pulling the seedling and washing out the sand and ash.

(*vi*) The seedlings will be ready for transplanting from 12th day onwards. If however, water is not readily available in the main field for transplantation, the life of the seedlings can be prolonged for more than three weeks by keeping them in the nursery.

Care of Seedlings in the Nursery : Healthy seedlings are the must for getting healthy crops, weeding should be done 2–3 times alternately after a week. Excess water of the seed bed is to be drained out after 14–15 days in Aman Nursery and 20–25 days in Boro nursery and top dressing is to be done. Spraying the seedlings with any one of the following insecticides should be done to protect the seedlings from the attack of insect pest.

Name of Insecticides	*Amount Per litre of Water*
Endosulphan 35 EC	2 ml.
Phosphamidoon 40 SL	1.5 ml.
Methyl Parathion 50 EC	1.0 ml.
Triozophos 40 EC	2.0 ml.
Monocrotophos 36 SL	1.5 ml.
Imida chloroprid 17.8 S.L.	0.2 ml.

In wet seed bed, application of any one of the following insecticides before seven days of uprooting the seedlings checks the spread of insectpests in the main field.

Name of Insecticides	*Amount In*	
	1000 sqm.	1500 sqm.
Phorate 10 G	4000 gm.	6000 gm.
Carbofuran 3 g	1500 gm.	2250 gm.
Quinalphos 5 g	2000 gm.	3000 gm.

Spraying the seedling with any one of the following fungicides is required to check the spread of the diseases in the main field.

Name of Fungicides	*Amount Per litre of Water*
Mancozeb 75 WP	2.5 gm.
Copper Oxychloride 50 WP	5.0 gm.
Captan 50 WP	2.0 gm.

2. Vegetables : The nursery bed should be raised 15–20 cms. from ground level, specially during rainy season.

Measurement :

Breadth : 4 ft. (1.20 m)

Length : 6–8 ft (1.80 – 2.4 m).

A space of 1 ft. may be left around each bed for ensuring drainage of the bed. Each bed must be worked throughly to obtain a good tilth. The clods should be broken into a fine places. Weeds, stones and other undesirable material should be removed. The soil of nursery bed should be sterilised by any of the following methods to prevent the infection of "damping off" diseases.

(*i*) Dry refuges are spread on the surface of the bed and burnt. The weeds, insectpests and diseased materials which are likely to be present are killed by heat.

(*ii*) Dr. Metha (1959) recommends the following method of sterilizing the soil, one part of formalin mixed with 100 parts of water should be applied @ 4.55 litres per 0.09 square metres of surface soil as to saturate it upto a depth of 152.40 mm. Then the soil mixture is covered with gunny bags or thick paper for a day or so that fumes may penetrate the soil and all the fungi, insect pests are killed. The soil is then uncovered and loosened and then left for few days till the smell has disappeared. Basic organic manures such as compost, F.Y.M. etc. should be mixed with the soil of the bed. The surface of the bed should be fairly smooth and friable and free from clods in order to plant at a uniform depth and to have a good coverage of the seed.

Vegetables seedlings may be raised in boxes and pans. Dr. Mehta (1959) advocates that a good soil mixture for filling boxes and pans etc should consist of 2 parts of garden soil, one part of sand and one part of leaf mould. These are mixed throughly and sterilized well. Individual seedlings are raised in earthen pots, polythene tubes and paper pots so that they can be transported and transplanted to any distant place causing without much injury to their root system.

Generally seeds are sown by broadcasting method in the nursery bed. But it is not satisfactory method of sowing seed in the nursery bed. It is, therefore, recommended that seeds should be sown in rows of lines or a tiny 'U' shaped furrows drawn widthwise in the nursery by hands or wooden boards, seeds are sown in the furrows at a depth of 0.6–1.2 cm and a space of 2.5 – 5.0 cm which should be covered with fine sieved soil and pressed properly. The depth of seed will vary according to the size of seed. And the kind of the soil. The sown seeds should be covered with dried leaves, straw or old gunny bags. Watering is done over this covering at regular internal. As soon as the young germinated plant has come up, the coverings are removed.

Seedlings of some plants such as cabbage and cauliflower suffer from scorching sunshine and rains in early stages of their growth. The over head cover should be so provided that seedling get sunlight in the forenoon and afternoon for a few days until they become handy enough tolerate full sunshine. The seedlings should be sprayed with fungicides and insecticides as and when necessary.

Pulling or Uprooting of Seedlings

(i) Paddy Seedling : The seedlings of paddy become ready for uprooting at the four to five leaves stage or when they are three weeks old and about 20–25 cm. high. The seedlings of local aman paddy are uprooted at the optimum age (4–5 weeks old). The seedlings of boro paddy at the age of 40-45 days having 4–5 leaves becomes ready for uprooting and transplanting. The seed bed should be irrigated just before uprooting to seedlings. The water is allowed to stand in the seed bed and uprooting of seedling is done in standing water in the seed bed. The condition facilitates the uprooting of seedling with minimum damage of roots. The seedlings are pulled and washed to remove soil from their roots. The individual seedling are uprooted with sufficient care so that they do not get damaged except for tearing of some roots.

(ii) Vegetables Seedlings : The seedlings of 4–6 weeks old having 4–5 leaves becomes ready for transplanting. The seed bed should be given light irrigation about 7–8 hours before uprooting the seedlings from the seedbed. Under this conditions, roots of the seedlings are not damaged during their uprooting. The seedlings are pulled up with a lump of soil in the roots, sometimes seedlings are uprooted and transplanted temporarily in a bed, known as *Nursery Bed,* to make them harder. The seedlings of desired height should be uprooted first leaving the smaller ones to grow further. The uprooted seedlings should not be allowed to state. If staling is unavoidable, the seedlings should be kept in moist and cool shady place with frequent watering under loose condition having a lump of moist soil in their roots. The seedlings should be handled with care. Any injury in vital parts of seedlings leads their death and creates condition for disease infestation.

Calculation of Seed Rate : The seed rate depends on test grain weight, germination percentage, spacing and number of seedlings to be planted. The seed rate for paddy can be calculated by the following formula.

Seed rate (kg/ha) =

$$\frac{\text{1000 grain weight} \times \text{number of seedlings per hill}}{\text{Spacing}\left(\text{cm}^2\right)} \times 100$$

Example : Calculate the seed rate for paddy for an hectare of land on the basis of the following data.

Spacing : 20 × 10 cm.

No. of seedlings per hill : 3

Germination percentage of seed : 80.

1000 grain weight : 20 gm.

Solution :

$$\text{Seed rate}\left(\text{kg/ha}\right) = \frac{20 \times 3}{20 \times 10} \times 100 = 30$$

Seed rate after correction for germination

$$= \frac{30 \times 100}{80} = 37.5 \text{ kg.}$$

Seed rate after allowance for risk $= 37.5 + \left(37.5 \times \frac{1}{5}\right)$ kg

(20% extra) $= 37.5 + 7.5$ kg.

$= 45.0$ kg.

5.6 TRANSPLANTING

Transplanting is the process of removal of actively growing plant from the nursery bed and planting them in the main field for further growth, development and production.

Seedling are transplanted in a well prepared field for raising crop plants. According to Bailey, "It is a general term to designate the removal of living plants whereby they become established in new quarters. It is a violent operation which may be performed when the plant is in a dormant condition or it is actively growing."

Plants which are able to endure shock of uprooting and transplanting and have the capacity to produce newer roots or root branches rapidly can be used for transplanting. Those plants which have no such behaviour, may also be used for transplanting by lifting them with a lump of soil in the root zone.

Transplanting should be done in the late hours of day so that the plants may established in cool weather at night and may recover from shock of transplanting before the sun comes up in next morning. But the transplanting can be done at any time of day if it is cloudy or raining or drizzing or about to rain. Transplanting should be completed as early as possible after removing the plants from the nursery. The soil around the roots may be compressed lightly to provide more root soil contact and to remove any air gap and consequently the accumulation of more water in the rhizosphere. Watering in the morning and afternoon for 4–5 days helps in quick establishment of seedlings, gap filling may also be done where the seedlings are dried.

Advantages of Transplanting

(*i*) Saving and proper utilisation of costly seeds is possible.

(*ii*) It is easy to nourish the young and tender seedling in nursery bed.

(*iii*) It stimulates the seedling to produce more roots.

(*iv*) Due to transplanting, shorter time from planting in the field to harvest is needed.

(*v*) It is possible to grow crop under unfavourable climatic condition.

(*vi*) It is possible to produce more plants of high quantity through transplanting.

(*vii*) Some vegetables can grow well when transplanted.

(*viii*) Through transplanting higher yields with more uniform maturity are obtained.

Disadvantages of Transplanting

(*i*) It increases the labour and power requirement in a peak period.

(*ii*) It increases the cost of cultivation.

(*iii*) Cap filling operation is necessary.

(*iv*) Directly sown crop do not grown well if transplanted.

(*v*) According to Loomis (1925), it is harmful but frequently necessary operation.

5.7 MANURING

Crop cultivation needs manuring at proper time with suitable quantity of manures and fertilisers. The recommended fertiliser schedules for field crops under West Bengal condition is given in the following table:

Name of the Crops	Plant Nutrient(kg./Ha)		
	Nitrogen	Phosphorus	Potassium
1	2	3	4
1. **Paddy :**			
Aus (Broadcasted)	50	25	25
Aus (Transplanted)	60	30	30
Amon (120–120 days)	50	25	25
Amon (120–130 days and 130–145 days)	60	30	30
Amon (more than 150 days)	60	30	30
Amon (Masuri, O.C. 1393 N.C. 1281, CR–1014 etc.)	40	20	20
Boro (Short duration variety)	100	50	50
Boro (Medium variety)	120	60	60
Boro (Desi improved variety)	80	40	40
2. Wheat (Irrigated)	100	50	50
Wheat (Non-irrigated)	60	30	30
3. Jute (Capsularis	80	30	30
Jute (Olitorius)	60	30	30
4. Mustard (Rai : Irrigated and non–irrigated)	50	25	25
Mustard (Tori : Irrigated and non-irrigated)	40	20	20
5. Sesamum	25	25	25
6. Groundnut	15	30	45
7. Sun-flower	80	60	40
8. Linseed :	50	25	25
9. Pulses : Gram, Pea, Lentil, Red gram etc.	20	40	20
Black gram, Green gram etc.	10	40	20
Soyabean	25	80	50
Cowpea	20	20	20
10. Sugarcane	160	60	60
11. Potato	200	150	150
12. Maize	85	60	60
13. Oat	50	30	-

	1	2	3	4
14.	Jowar (Irrigated)	70	30	15
	(Non–Irrigated)	25–30	12.5	10
15.	Barley (Irrigated)	40	20	20
	(Non–Irrigated)	20	20	20
16.	Bajra	80–100	40	20–30
17.	Safflower	60–90	30	30
18.	Castor	90	40	30
19.	Mesta	40	20	20
20.	Cotton	37.5	37.5	37.5
21.	Sugarbeet	50	50	50
22.	Fodder Crops :			
	Berseem	20	80	40
	Lucern	25	62.5	30
	Napier grass	45	-	-
	Rhode grass	50	20	-
	Anjan grass	20	20	-
23.	Tobacco :			
	Cigarette	25	75–100	50
	Cigar and bidi	125–150	75–100	50
	Hookah and chewing	50	75–100	50

5.8 INTERCULTURAL OPERATION

Different agricultural operations adopted in between the sowing or transplanting and prior to harvesting of crops are collectively known as 'Intercultural operation' or after cultivation practices. Generally the followings are the important activities of intercultural operation *viz*:

***(i)* Weed Control :** Weeds interfere with man's utilization of land and water in various way. Weed compete with crop plant for water, space, light and mineral matters. Weeds in this may reduces the yield of crops. The weeds should be controlled as and when comes up with the crops and it is included in the intercultural operation. Hand hoeing is done in the newly planted sugarcane field with the optimum soil moisture during which none of the sugarcane selts would have germinated. This operations is termed as 'blind hoeing.' Blind hoeing helps in loosening the hard soil, aeration and easy sprouting of sugarcane selts.

(*ii*) Moisture Conservation : The water losses from the soil by evaporation can be controlled by intercultural operation. Natural mulching is very important process of preserving soil moisture.

***(iii)* Good Aeration :** Good aeration of soil is beneficial for the germination of seed and respiration of plant roots and soil organism. Aeration is not good in hard and compact soil. Loosening of soil by different agricultural implements helps to incorporate the air into the soil.

***(iv)* Manure Incorporation :** Top dressed fertilisers and agri-chemicals are incorporated during intercultural operation.

(*v*) Control of Disease and Pests : Crops in many times suffers from insectpests and diseases. They should be controlled by adopting different methods.

Intercultivation is to be done with light tools and implements and it should be scheduled after rain or irrigation. In some crops during intercultivation, irrigation and drainage channels are prepared and earthing up is done.

The following activities are also include in the intercultural operation :

1. Earthing Up : Earthing up operation is usually practiced for some crops such as tuber crop, sugarcane, maize, cotton, sunflower, ground nut etc. This operation is done to provide anchorage to the crops and to facilitate drainage during rainy season. In potato, the first earthing up should be done when the plants are about 10-15 cm high. A second earthing up may also be required to cover up the tuber properly to prevent greening of tubers. In sugarcane, partial earthing up is given 45 days after planting and subsequent earthing up is given as per requirement. Spade, ridge plough etc. is used for earthing up operation. In ground nut, earthing up is done at the time of peg formation at about 45 days after sowing. This practice helps easy penetration of peg into the soil and better pod formation.

2. Topping and Desuckering : The practice of topping and desuckering is commonly followed in Tobacco. The operation of topping consist of removal of terminal bud with or without some of top leaves just before or after emergence of flower. The terminal growth is arrested due to this operation and the auxillary buds become active one out forth shoot known as 'suckers'. The suckers are removel 2-3 times as and when the suckers reach convenient size for handling. This operation is known as 'desuckering.'

3. Nipping : In case of luxuriant growth of gram, shoot tips are nipped to encourage branching and flower formation and thereby increases the yield. Nipping is done at about 30-40 days after sowing and it stops the apical growth and promotes the lateral branching.

4. Trashing : As the sugarcane grows, the lower leaves gradually dry and the dried leaves are called trash. The operation of removing loosely adhering dried and drying leaves from the sugarcane is known as 'trashing.' It is generally done to clean the sugarcane. The trashing is recommended to be done twice *i.e.* first at fifth month and second at seventh months after planting.

5. Propping : When the crop growth is lisely to be heavy, special attention is needed in earthing up after the monsoon follwed by tieing and propping at proper stage to prevent lodging particularly in coastal areas where cyclones are common. Because lodging of sugarcane reduces its quality very much. Tieing three to four plants from adjacent rows by means of their own trash and old leaves and propping up the stalks with bamboos is good to prevent the lodging of sugarcane.

5.9 HARVESTING

Removal of entire plant of economic parts after maturity from the field is called harvesting. Harvesting is an important arts of techniques of crops production. Delay in harvesting is sometimes deteriorates the quality of the product and the market price decreases accordingly. The crop is harvested early, contains high moisture and more immature grains. The yield will be low due to unfilled grains and this product are subjected to insectpest and disease infection during storage. Harvesting of crops at correct time to get good quality grains and higher yield depend on the practical experience of the farmer. Harvesting time and method will vary with the crops as follows:

(*a*) Cereals : Paddy, Wheat, Maize, Jowar, Bajra etc. are the cereals. The cereals crop are harvested when ears are nearly ripe and the plants are dried after yellowing and the seed contain 15 per cent moisture in case of paddy and 12 per cent moisture in case of Wheat, Maize, Jowar, Oat, Bajra etc. The wheat is harvested when the straw is golden yellow and brittle. The local aman variety, Jower, Bajra etc. are harvested when the plants are completely dried. The maize is harvested at fully mature stage and green cob stage depending on the use. The cereal crops are cut close to the ground with the sickle.

(*b*) Fibre Crops : Jute is one of the most important fibre crops of our country. Generally the crops get ready for harvesting at 120 days after sowing. Jute can be harvested at flowering stage, pod stage and pod ripe stage. Both the yield and quality of fibre becomes good if jute is harvested at pod stage. The plants are cut close to the ground with the sickle.

(*c*) Sugar Crops : Sugarcane is one of the most important sugar crop of our country. The crop matures within 10 – 12 months after planting. Crops maturity is judged by the following symptoms :

(*i*) The leaves become yellow and the lower leaves are withering up gradually, leaving progressively fewer leaves at the top.

(*ii*) Plant stops growing.

(*iii*) The internodes of mature cane become short in size and produce metallic sound.

(*iv*) The cane become brittle and break easily at nodes.

(*v*) The buds swell out at nodes and starts sprouting.

(*vi*) Sucrose content is 20 per cent or more. The crops are cut at ground level preferably after digging down the earthed up ridges.

(*d*) Tuber Crops : Potato, Ginger, Onion, Turmeric, Elephant foot etc. are the tuber crops. They are ready for harvest when haulms starts yellowing and falling on the ground. Potato may be harvested at immature stage in order to get higher prices in the market or to vacant the land for the next crop to be grown. But this potato is totally unsuitable for storage as the skins of immature potato are tender and subjected to rotting. The harvesting is done by spade or country plough.

(*e*) Oil Seed Crops

(*i*) ***Mustard :*** Mustard is normally ready for harvest when the pods turn yellow and the plant shed leaves. Delay in harvesting causes shattering of seeds. Harvesting of this crops early in the morning minimizes the shattering loss of seed. The crop is harvested by sickle or uprooting the plants.

(*ii*) ***Sesamum :*** The crop is harvested when the leaves, stems and capsule begin to turn yellow and lower leaves starts shedding and the flowering is also ceased. Harvesting is delayed for the sake of immature pods results in shattering of mature pods. Delay in harvesting causes the shedding of grains. The crop is harvested by sickle.

(*iii*) ***Linseed :*** The linseed harvested when the plants turn golden yellow and the crop is dead ripe stage for oil seeds. But the crop is harvested at flowering stage when grown for fibre crops. The harvested by uprooting the plants or cutting them close to the ground by sickle.

(*iv*) ***Ground Nut :*** The yellowing of leaves, shedding of older leaves and the development of the proper colour of testa and a dark tint inside the shell are the prominent symptom of maturity. The bunch type variety is harvested by pulling out the plants when there is adequate moisture in the soil. The spreading variety is harvested by digging out the plant with the help of khurpi, spade or by ploughing the field. The pods are then picked up and dried for a week in sun.

(*v*) ***Sunflower :*** The sunflower crop matures in 90–100 days. The crop is harvested when the lower side of the heads turns yellow and some of the bracts dry up. The mature heads are cut with sickle and the seeds are separated by beating the heads with the help of wooden or bamboo sticks.

(*vi*) ***Safflower :*** The crop matures in about 130–150 days after sowing depending upon the variety. The plants are harvested when they hard enough by uprooting or cutting the plants with sickles.

(*vii*) ***Castor :*** The crops takes six months to mature. The crops is harvested when one or two capsules in a bunch show sign of drying. The harvested capsules are dried in a sun for 4 – 5 days and thereafter the seed is beaten out with sticks or mallets and winnowed.

(*f*) Pulse Crops : Gram, Pea, Red gram, Lentil, Green gram, Black gram, Cowpea, Soyabean etc. are important pulse crop. Most of the pulse crops are harvested when most of the leaves are dried and shed, the plants are drying and the pods ripe. Harvesting is done by cutting the plants with sickle or uprooting the plants. The pods of some pulse crops such as pea, cowpea etc. are picked at periodical intervals, when the pods are well filled and the colour of pods are changing from dark to light green. The pods of some pulse crops such as green gram, black gram, red gram etc. are picked when the pods become mature. These crops after third picking are harvested by cutting the crops with sickle or uprooting the plants. Delay in harvesting causes the shattering of seeds.

(*g*) Vegetable Crops

(*i*) ***Cabbage :*** The harvesting time after transplanting depends on the variety. *viz* :

Early Variety : 70 – 80 days.

Late Variety : 100 – 120 days.

(*ii*) Cabbage is harvested when the heads are large and firm enough for use. Harvesting may be done by sickle or any other convenient implements. The head is grasped on one hand, bent slightly and cut off with a heavy knife or sickle or hatchet.

(*iii*) ***Cauliflower :*** The harvesting time after transplanting varies according to the variety *viz* :

Early variety : 60 – 70 days.

Mid season variety : 90 – 100 days.

Late Variety : 110 – 120 days.

Cauliflower should be harvested when the curds reaches a proper size, compact and bright in colour but has not broken into segments. The plant is cut off well below the head and leaves are timed with knife.

(*iv*) ***Knolkhol :*** Knolkhol should be harvested when the swollen stem reach diameter of 5 to 7 cm. before it becomes tough and woody. The plants are pulled out from the land, it is generally marketed after removing both leaves and roots.

(*v*) ***Brinjal :*** Brinjals are harvested when they attains a good size and colour and still they are immature. Mature fruits are not so desirable. Fruits are allowed to attain a good size and colour till they do not lose their bright and glossy appearance. Harvesting is done by detaching the fruits from the stem.

(*vi*) ***Tomato :*** Tomato are harvested in the following stages according to the uses of the fruits :

(*a*) *Green Stage* : Tomatoes in this stage are harvested for distant places.

(*b*) *Pink Stage* : The fruits in this stage becomes red. But the fruits are not fully ripe. The tomatoes in this stage are harvested for local market.

(*c*) *Ripe Stage* : The surface of most of fruits in this stage is red and the softening of fruits begins.

Tomatoes in this stage are harvested for home or table purpose.

(*d*) *Full Ripe Stage :* Maximum colour development of the fruit at this stage occurred and they feel soft to the touch. The fruits are ordinarily used within 24 hours of picking and are consumed or used for canning and pickling.

(*h*) Root Crops

Radish : Radish becomes ready for harvesting within one month of sowing and they should be harvested according to the variety when the roots are still tender and crisp but they big enough for the market or home consumption. Delay in harvesting makes the root pithy and quite unsuitable for marketing.

Turnip : Turnip takes about 70 to 100 days to mature and becomes ready for harvesting. The turnip is harvested when the roots of turnip become 5 to 7.5 cm in diameter. Delay in harvesting makes the root hard and fibrous and their quality is deterioated rapidly.

Carrot : The carrots are harvested when their roots are 2.5 to 3.7 cm in diameter at the upper end. The crops are harvested by pulling out the plants.

Beet : Beet becomes ready for harvesting within 2 – 3 months of sowing. The beets are harvested when their roots are 5.0 – 7.5 cm in diameter. The crops is harvested, by pulling out the plant digging with spade.

Sweet Potato : Sweet potato becomes ready for harvesting from 105 – 150 days after planting and they are harvested when the leaves turn pale and later slightly yellow and cut surface dries up quickly and does not turn black as that of immature tubers. The vines are removed by cutting and then harvesting of tuber is done with the help of a spade.

(*i*) Lady's Finger : Flowering begins from 35–40 days after sowing and fruits are ready for harvest four to five days after flowering. The pods may be harvested continuously at some intervals. Delay in harvesting may make the fruits fibrous and they lose their tenderness and taste.

(*j*) Cucurbitaceous Vegetables : Pumpkin, bottle gourd, bitter gourd, snake gourd, ridge gourd, sponge gourd, cucumber, pointed gourd, water melon, muskmelon etc. are the important cucurbitaceous vegetables. These vegetables are harvested when the fruits are still young and tender with the exception of pumkin, which is harvested either in green stage or mature stage according to the demand of the market. The water melon and muskmelon should be harvested at proper stage of maturity.

(*k*) Fruit Crops : Mango, Jackfruit, Banana, Guava, Papaya, Grape, Pineapple, Litchi, Apple, Pear etc. are well known fruit crops in our country. Most of the fruits are harvested at ripe and half ripe stage depending on their mode of uses. The full ripe fruits are harvested at full ripe stage if they are consumed at home. The banana and papaya are harvested at green stage when they are big and firm enough for using as vegetables.

5.10 THRESHING

The process of separating the seeds from the ears is known as threshing. In cereals, straws and grains are separated and in pulses the seeds are separated from pods. Threshing in generally done immediately after harvest.

Following are the common method of threshing :

(*i*) Threshing by Manual Labour : It is most common method of threshing and is done mainly by beating against stones or any other hand materials or beating by sticks. It is a slow and labour consuming device, but it is followed for quantity of harvests.

(*ii*) Threshing by Animals : It is also very common method of threshing used in village. In this method, the threshing is done by treating the crop by means of a team of animals. A strong pole is fixed in the centre of the threshing space in which bullocks are tied in line one after the other. The harvest is spread on the threshing space and bullocks move round the harvest and trample them continuously till the grains are completely separated from straw. One man drives the bullock from the back. Threshing by treading is also done by tractor. It is most expensive method.

(*iii*) Threshing by Machines : It is most improved method of threshing. The threshing machines are used for this purpose.

Threshing is followed by winnowing. Winnowing is the method of separating grain or seed from chaff.

The seeds or grains are dried in sun to reduce their moisture content. Moisture content for safe storage is 14 per cent for most of the crops. High moisture in grain as well as high humidity in the atmosphere caused sprouting and moulting of grain.

5.11 STORAGE

The agricultural produce must be stored properly for the continuous consumption as well as to get the higher price. Because the market value is generally low at harvesting time. Generally the foodgrains, oilseeds, seeds etc. are stored for future use.

Factors Affecting Storage : The following factors influences the storage of food grains :

(*i*) ***Moisture Content of the Grains :*** The seeds that contain high moisture are subjected to the attack of insects and microorganisms. Moist seeds are amenable for easy biting or chewing by insects. Sometimes moist grains may even germinate and become unfit for consumption.

(*ii*) ***Climatic Condition :*** Grains are hygroscopic and absorb moisture from the atmosphere. Under moist climates, the seeds are subjected to the attack of insects and microorganisms as this condition is favourable for the growth and multiplication of insects.

(*iii*) ***Quality of Produce :*** The crops that are harvested early contain more illfilled and shrivelled grains. The shrivelled and broken grains are the predisposing causes for insect attack.

(*iv*) ***Storage Conditions :*** The product is generally stored in gunny bag which is more prone to insect infestation of the grains. The storage condition should be such to provide the grain protection from insects. The ill storage condition results in loses of food grains.

Method of Storage : Food grains and oil seeds are stored either in bags or in bulk (*i.e.* without bagging). Pusa bins may also be used for storage of foodgrains. Godowns are the most common structures for above ground bag storage. The godowns should be made free of insects and microorganism by fumigation using EDB (Ethylene-di-bromide). The save grain campaign organisation recommends the uses of EDB ampules as follows:

Quantity of Foodgrains	*EDB Ampule*	
(Quintal)	*Quantity*	*Number*
1 or less	3 ml	1
2	6 ml	1 or 2 of 3 ml.
5	6 ml	3 or 6 of 3 ml.
10	10 ml	3 or 5 of 6 ml.
20	15 ml	4 or 6 of 10 ml.

Several pests attack the produce during storage which can be controlled by spraying malathion and dichlorovas. Fumigants such as Aluminium phosphide, Methyl bromide etc. are used in godowns. The pest is killed by the poisonous gas released from the fumigants.

The following points are important to remember for the storing of agricultural produce :

(*i*) The stores should be at least 0.5 km away from the place which are a source of infection like kilns, flour and bone crusting mills, garbage dumping ground, slaughter house and tanneries. It should be constructed, as far as possible, away from dwelling houses.

(*ii*) The stores should be, as far as possible, situated near a transport head or main road.

(*iii*) There should not be any tree near the stores. Otherwise birds will be a nuisance.

(*iv*) The walls of the stores should be made smooth and the cracks and cravices and rat borrows should be filled up completely. The plinth should be kept 0.75 m. above the ground level.

(*v*) The floor of the stores should be constructed of either cement concrete or stone with a slope of 4 cm. from the wall of its outer edge to prevent rain water from getting inside the godown through the doors.

(*vi*) The ventilators and windows need to be fitted in such a manner that the store is open to air made air light for desired period.

(*vii*) The floor and loose matters of Katcha stores should be scraped annually in April and replastering should be done the store should be rat proof.

(*viii*) The roof of the godown should be water proof.

(*ix*) The debris of the godown should be collected and burnt before storing the agricultural produce.

(*x*) The old grains from the store should be disposed of or removed and the store should be throughly cleaned before storing the new produce. White washing the walls and cellings of pucca godown should be done.

(*xi*) The store should be disinfected with Malathion 50 EC or Pyrethin (Dilution–1 : 100) @ 3 litres per 100 sq. metres after cleaning the store. The store may be fumigated with Aluminium Phosphite @ 25 tablets (3 gm. each) per 100 cubic metre space in place of malathion spraying.

(*xii*) The bags should be disinfected by dipping 0.5 per cent malathion (Dilution–1 : 50) –50 EC or in boiling water for a minute in case of bag storage.

(*xiii*) The completely dried grains (Moisture content ranging from 12–14 per cent) should be stored.

(*xiv*) The grains meant for seed should be treated with Malathion 5 per cent 250 gm. per quintals of grains. The existing practise of mixing D.D.T. or B.H.C. should be strictly avoided as this hamper the germination of seed.

(*xv*) The new and old stocks should not be kept in the same godown.

(*xvi*) The dunnage comprising either timber pallets, timber squares, palm matting or still better, a layer of polythene sheet sandwiched between two layers of malting should be used. The floor space between slacks as well as the dunnage should be sprayed with malathion 50 per cent EC to avoid insect infestation.

(*xvii*) The bags should be stacked 60 – 70 cms. away from walls with proper alleys around for inspection and other operations (*i.e.* fumigation, spraying etc.).

(*xviii*) It is essential to inspect the store fortnightly for pest infestation and rodent damage and also for taking necessary measurement to avoid the losses.

5.12 COMMON FARM IMPLEMENTS AND THEIR USES

Quite a variety of implements and machinery are required for carrying out different agricultural operations, such as ploughing, harrowing, levelling, sowing, intercultivation, application of manures and fertilizers, harvesting, thrashing and winnowing. The implements are generally operated by men and animals. Now-a-days, the farmers are using power operated machines and equipments such as tractors, power-tiller, oil engines, pumps etc.

Following are the different kinds of implements used in different agricultural operations :

1. Implements Used for Tillage

(*i*) ***Hand Operated Implements :*** Spade, shovel, wheel hoe etc.

(*ii*) ***Animal Drawn Implements :*** Country Plough, Mould board plough, Bose plough, Harrow, Ladder etc.

(*iii*) ***Tractor Drawn Implements :*** Disc plough, Cultivator, Leveller, Spike tooth harrow, Spring tooth harrow, Disc harrow, Star pulverizer etc.

2. Implements Used for Seed–Dressing : Seed-dresser etc.

3. Implements Used for Sowing of Seed : Seed-drill (hand and bullock drawn), Dibbler etc.

4. Implements Used for Interculture Operations

(*i*) ***For Weeding :*** Country bida, Rake, spade, Nirani, Khurpi, Wheel hoe, Japanese Paddy weeder, Improved harrow (Spike tooth harrow, Spring tooth harrow, Disc harrow etc.)

(*ii*) ***Spraying and Dusting Machines :*** Sprayer (compression sprayer, Akella Sprayer, Knapsac sprayer, Power sprayer, Bucket sprayer, Foot sprayer etc.), Duster (Rotary duster) etc.

5. Implements Used for Irrigation : Don, Lata, Mote, Swing Basket, Persian Wheel, Egyptian Wheel, Windmill, Centrifugal pump etc.

6. Implements Used for Harvesting : Sickle, Hasuwa, Reaper, Harvester etc.

7. Threshing and Winnowing Machine : Paddy thresher, paddy winnower, wheat thresher, cane crusher etc.

5.12.1. Different Kinds of Farm Implements and their Uses

5.12.1.1 Implements Used for Tillage Operations

1. Indigenous or Country Plough : An indigenous plough is one of the most common implements used by Indian farmers. Different types of indigenous ploughs are found in our country which are basically the same, but with large variation in their shape, size, weight etc. This is due to the variation of soil type, tillage requirement of various crops in different parts of our country and weight and height of draft animals in the locality.

This plough is made of wood except the share which is made of a mild steel rod or bar.

Structure and Function of Different Parts of Country Plough

Name of the parts and their description	*Function*
(*i*) **Wooden is Body :** The body and shoe are in one piece, made of wood in a country plough used in West Bengal. It is made from mature wood Babul or gum tree (*Acacia arabica willd*).	Body is the main part of a plough to which share handle, beam etc. are generally attached.
(*ii*) **Shoe :** Shoe of different shapes and sizes are used in the plough. The shoes are sturdy and long and provide a greater depth of ploughing in some states like Punjab, Bihar, U.P. etc. The ploughs have long narrow shoes fitted at an angle for shallow ploughing in hilly tracts of our country.	The shoe and the share enter the soil and separate the furrow slice from the main body of the soil, when the plough is pushed ahead. A portion of the soil rides over the shoe, but the larger portion is pushed aside to both sides.
(*iii*) **Share :** Share is the working part the plough and is attached to the shoe. It is made of a mild steel rod or bar. When the plough is in working condition and remains tight over the shoe automatically by the soil pressure.	Share penetrates into the soil and protect wooden wedge against wear and tear. It cuts the soil forming 'V' shaped furrow and opens the soil but there is no inversion. A portion of soil is always left unploughed in between two 'V' shaped furrows. This is reduced by cross ploughing, but even then small squares remain unploughed.
(*iv*) **Beam :** The beam is generally made with a long wooden piece. Beams generally very in length from 2.7 to 4.5 metres. One end of the beam is fixed to the body of the plough by means of moties and tenon joint. The other end of the beam is provided with notches or holes to admit wooden or mild steel to facilitate easy yoking of bullock at a suitable point according to the height of the bullocks.	Beam connects the main body of the plough to the yoke and acts as power transmitter.

Name of the parts and their description	*Function*
(*v*) **Handle :** A wooden piece which is attached vertically to the body to enable the operator to control the plough is called the handle. It is provided separately only in those ploughs where both shoe and body are prepared from one piece of wood. In certain types of ploughs, the handle is separately fixed to the body by means of nails and lap joints.	The operator controls the ploughing by means of handle.

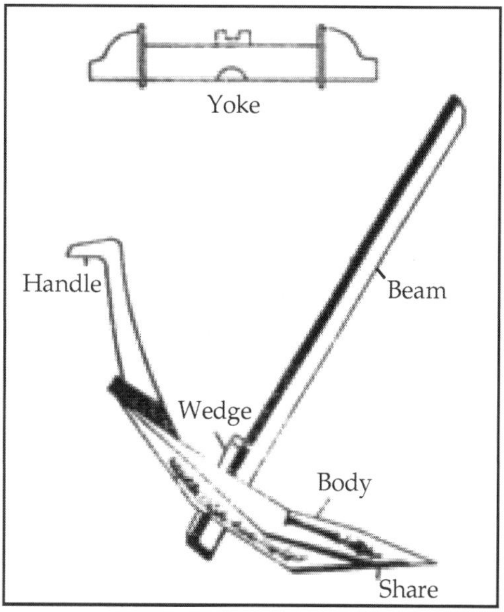

Fig. 5.2 : Country Plough

Uses of Country Plough : The country plough is one of the most important implements used mostly for tilling the land. The country ploughs are also used for making shallow furrow for depositing seeds, for intercultivation between rows of crops and to harvest some crops such as potatoes, sweet potatoes and groundnut and for puddling the land for transplanting the seedling of paddy, onion etc. Most for transplanting the seedling of paddy, onion etc. Most of the country ploughs are light in weight and can be carried on the shoulder by the farmers or on the yoke of the bullocks. Since the country plough is a multipurpose implement, it is difficult to be replaced.

The depth of ploughing is usually from 10 – 15 cm, and width from 12.5 – 17.5 cm. The depth of the plough can be adjusted by means of a wedge provided in the body of the plough or by fixing the yoke either forward or backward on the beam. If the yoke is fixed forward, the plough goes deep and if the yoke is taken backward the plough goes shallow. (Ramiah et. al 1967). The size of plough varies according to the type of land to be ploughed. The dry land plough is bigger in size and wet land plough is the smallest of wooden plough. A worn out dryland plough is used as a wetland plough. The country plough covers an area of 0.15 to 0.25 hectares in a working day of 8 hours. The wet land plough covers about 0.15 hectare for first pudding and about 0.24 to 0.28 hectares at the last puddling in a working day of 8 hours. The plough man have to walk 8.25 miles (*i.e.* 43560/1760 × 3) to plough an acre of land, when the plough makes a furrow of one feet breadth.

Yoke is an associate part of a plough. It is made of long (5'7" or 170 cm) and flat wood. The animals (*e.g.* cow, buffalo etc.) are hitched up in the yoke and the ploughing is done with the help of the animals. The country plough needs less draft power and is drawn by a pair of bullocks. The speed of a bullock drawn plough is 1.5 miles per hour.

2. Mould Board Plough : Mould board plough is one of the most important improved type of ploughs and is a very common implement used for primary tillage operations. Mould board plough is made of iron or steel except the beam which is made of a long wood. Mould board plough needs more draft power than that of a country plough and sturdier animals are required for drawing the plough. This plough has several advantages at a time as follows :

(*i*) Cutting the furrow slice.

(*ii*) Lifting the furrow slice.

(*iii*) Turning the furrow slice.

(*iv*) Pulverising the soil.

Structure and Functions of Different Parts of Mould Board Plough

Name of the parts and their description	*Function*
1. **Frog or Body :** It is the important parts of the plough bottom to which the mould board, the land-side and the share are attached rigidly. It may be made of cast iron for cast iron ploughs or welded steel for steel plough.	The main function of the frog is to hold the different parts of the plough.
2. **Mould Board or Beast :** The mould board is most important parts of this plough. It is made of steel or Iron and is attached to the share by bolts. Mould boards are designed in different types to suit different soil condition and crop requirements as described below :	Mould board receives the furrow slice from the share and lifts, turns and breaks the furrow slice.
(*i*) ***General Purpose Breast :*** It has a fairly long mould board with a graduals twist and the surface is slightly convex. It turns the well defined furrow slice and pulverised the soil throughly.	
(*ii*) ***Digger Type Breast :*** It is a short mould board with an abrupt curvature and its surface is concave. It is suitable for deep ploughing and pulverising the soil as soon as the furrow slice is turned.	
(*iii*) ***Semidigger Type Breast :*** It is a long mould board with gentle curvature and the surface is slightly concave. It is suitable for ploughing all types of soil. It is used largely in tractor drawn ploughs.	
3. **Share :** Share is an important part of the plough which is a sharp, well polished and pointed component. It consist of the following parts.	When the plough is pulled ahead, the share penetrates into the soil and cuts the soil horizontally below the soil surface.

Name of the parts and their description	*Function*
(*i*) ***Share Point :*** It is the forward and of share edge or the cutting edge, which actually penetrates into soil and support the plough bottom.	
(*ii*) ***Share Edge :*** It is the front edge of the share which makes horizontal cut in the soil.	
(*iii*) ***Wing of the Share :*** It is the outer end of a share edge which spreads the furrow slice on either side and support the plough bottom. Shares are of different types such as slip share, slipnose share. Shin share, Bar share and Bar point share.	
4. **Land Side :** It is a long and flat rod like structure attached to the left side bottom of the plough. It is made of steel or Iron. The rear-bottom end of the landside is known as heel which rubs against the furrow sole. The clearance that is provided in the mould board plough for proper penetration and efficient work by plough is called suction of the plough. It is of two types as follows :	Landside slides along the face of the furrow wall and helps to resist the side pressure exerted by the furrow slice on the mould-board. It also helps in stabilizing the plough while it is in operation.
(*i*) ***Vertical Suction or Vertical Clearance :*** It is the vertical distance from the ground, measured of the joining point of share and landside when the plough is resting on horizontal surface in the working position. It helps the plough to penetrate into the soil to a proper depth and decreases the frictional force and also energy requirement.	
(*ii*) ***Horizontal Suction of Horizontal Clearance :*** It is the horizontal distance measured at the junction	

Name of the parts and their description	*Function*
of share and land side when the plough is inverted and rests on the share edge and heel of the land side. It helps the plough to cut the proper width of furrow slice and it also decreases the draft of plough. Both the suction varies according to the size of the plough. The mould board plough is forced into the ground by suction of the plough.	
5. **Standard :** It is a part of the plough that is slightly bent, flat and long rod like structure. It has three consecutive holes.	It connects the plough with beam. The beam may be made upward and downward by adjusting with the beam to the hole of the standard and accordingly the depth of the plough is being adjusted.
6. **Handle :** It is a long, flat and narrow rod like structure, which is attached to the landside. It is made of iron having slight bent at the end.	The operator controls ploughing by means of the handle.
7. **Beam :** The beam is generally made with a long (2.7 to 4.5 metres) wooden piece, one end of the beam is fixed to the handle by bolts and the other end is provided with notches or holes to admit wooden or mild steel to facilitate easy yoking of bullocks.	Beam connects the main body of the plough to the yok and acts as power transmitter.
8. **Wheel :** Wheel is provided in heavy mould board plough such as Bijoy, Kilosdar–100 etc. It is usually placed in hanging position.	It helps to maintain uniformity in respect of depth of ploughing in different soil condition.
9. **Coulter :** Coulter is an accessory part of mould board plough and mainly used in tractor drawn plough. It is may be Rolling type and sliding type. The edge of coulter may be either smooth or notched and it can be adjusted up-down and side ways. The up-down and side ways adjustment is meant for taking care of depth and width of cut respectively.	It is device used to cut the furrow slice from the land and leaves a clear wall. It is also cut trashes which are covered under soil by plough.

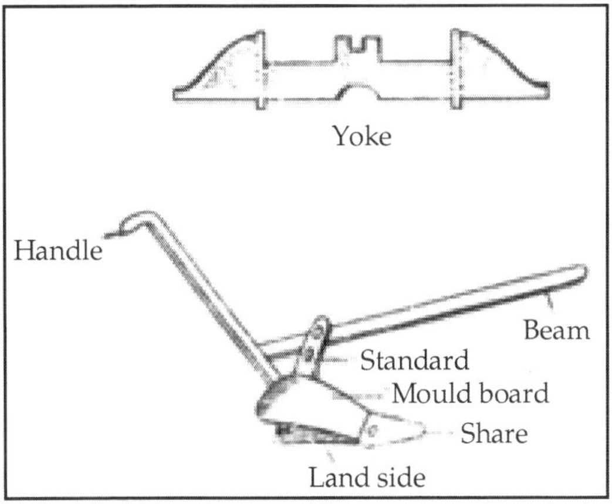

Fig. 5.3 : Mould board Plough

Uses

(*i*) Mould plough is most suitable for primary tillage of almost all types of soil.

(*ii*) It cuts square furrows and the soil is inverted completely depending on the curvature given to the mould board plough. Mould board ploughs are used where soil inversion is necessary and it leaves no unplough land. M.B. plough ploughs to a depth of 15 cm having width from 15 to 35 cm.

(*iii*) Light mould board ploughs are also used for puddling of light soil.

(*iv*) Mould board plough is being adopted where green manure crop or surface vegetation has to be burried into the soil to hasten decomposition.

(*v*) Mould board plough is also used for eradication of deep root crops like kans (*Saccharum spontaneum*) and Kunda (*Cyperus rotundus*).

(*vi*) Mould board plough cuts the furrow slice clean which is inverted to one side resulting in better pulverisation.

(*vii*) It covers 0.2 – 0.4 hacters of land in a working day of eight hours. The speed of bullock and tractor drawn plough is 1.5 and 2.5 miles per hour respectively.

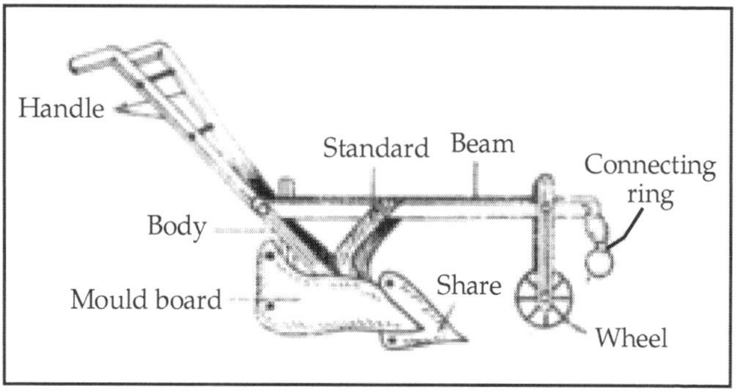

Fig. 5.4 : Heavy mould board plough

3. Bose Plough : Bose plough is a modified form of mould board plough. Bose plough is made of wood except the share which is made of steel or iron. The structure and function of this plough is almost similar to mould board plough. This plough is gaining popularity among the farmers and most of the farmers of our country are using this plough.

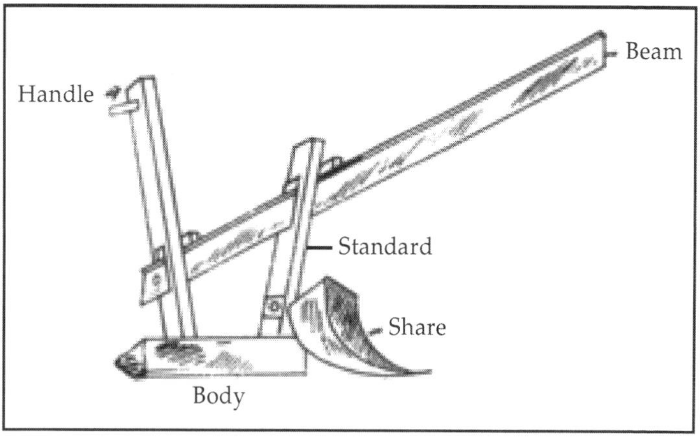

Fig. 5.5 : Bose plough

Table : 5.3 : Difference between a Country Plough and Mould Board Plough

Country Plough	*Mould Board Plough*
(*i*) Country ploughs have no mould board. So this plough cannot turn or pulverise the soil.	(*i*) Mould board plough can turn or pulverise the soil due to having mould board.
(*ii*) Country plough cuts 'V' shaped furrow leaving unploughed area in between two adjacent furrows.	(*ii*) Mould board plough cut square shaped furrows leaving no unploughed area in between two adjacent furrows.
(*iii*) Country plough ploughs to a depth of 10–15 cm having a width of 12.5–19.5 cms.	(*iii*) Mould board plough, ploughs to a depth of 15 cm. having a width of 15.35 cm.
(*iv*) Ploughing may be done in any type of soil with this plough.	(*iv*) Light or heavy mould board plough is needed for different kinds of soil.
(*v*) Country plough is not very much suitable for burring green manure crop or surface vegetation.	(*v*) Mould board plough is very much suitable for burring green manure crops and surface vegetation.
(*vi*) The depth of the furrow can only be controlled by a country plough.	(*vi*) Both the depth and width of the furrow can be controlled by Mould board plough.
(*vii*) Country plough is light in weight and needs less draft than that of a Mould board plough.	(*vii*) Mould board plough is heavy in weight and needs more draft than that of a country plough.
(*viii*) The village carpenters can make and repair the country plough. Its price is less.	(*viii*) The village carpenters can not make and repair the mould board plough. Its price is high.
(*ix*) Country plough covers *0.15–0.25* hectares in a working day of eight hours.	(*ix*) Mould board plough covers 0.2–0.4 hectares in a working day of eight hours.

4. Disc Plough : Disc plough is also used for ploughing which have little resemblance to common mould board ploughs. A large revolving, concave steel disc replaces the disc and mould board. Disc ploughs may be of two types as follows :

(*a*) *Standard Disc Plough :* These may be animal drawn and tractor drawn ploughs.

(*i*) *Animals Drawn Disc Plough :* This plough is attached to the universal frame which is mounted on wheel. 2–3 discs are present on this plough. The diameter of disc varies from 40–50 cm having a weight of 30 kg. A rear furrow wheel is provided with the plough to take care of the side thrust of the plough. The frame is pulled by a pair of bullocks.

(*ii*) *Tractor Drawn Disc Plough :* This plough may have 5–7 disc blades, the diameter of which varies from 60–80 cm having a weight of 180–540 kg. These plough are provided with three wheels such as front wheel, rear furrow wheel and land wheel, drag links, deep adjustment levers and scrappers. The scrapper is a device meant for removing the soil that tends to adhere on working surface of the disc. The discs are sharpened on the edge and are bolted to the cast iron supports which are individually suspended from the main frame. The individual disc rotates on its own axle towards the direction of pull.

(b) Vertical Disc Plough : This plough have more discs, nearly 30. The diameter and curvature of the discs are comparatively less. The disc blades mounted on a common axle and they rotate as one unit which throws the furrow slice in only one direction. This plough is generally used for shallow ploughing. The disc plough is forced into the ground by its own weight.

The penetration of the plough is improved by increasing the disc angle, and also by decreasing the till angle. Besides these, addition of additional weight on the ploughs help them to penetrate into soil.

The bullock drawn disc plough covers 2 acres of land in a working day of 8 hours.

Advantages of Disc Plough

(*i*) Disc plough is suitable for ploughing the land which are hard, dry and stony.

(*ii*) It is more useful for deep ploughing and plough pans or plough soles are not formed on the bottom of the furrow.

(*iii*) It requires less energy as compared to mould board plough.

5. Ridging and Harvesting Plough : This plough has double mould board, one for turning the soil to the left and another to the right. Double winged share is attached to the mould board. This ploughs are available in three sizes, such as heavy, medium and light. The heavy ones is used for cultivation of sugarcane and the lighter ones for cultivation of vegetables and other crops. This plough is used for making furrow and ridges for planting of sugarcane, potato etc. and also for earthing up of crops. This plough have arrangement for adjusting the width of furrow by changing the distance between two mould boards. This plough is also used for harvesting of some crops such as potato, ground nut etc.

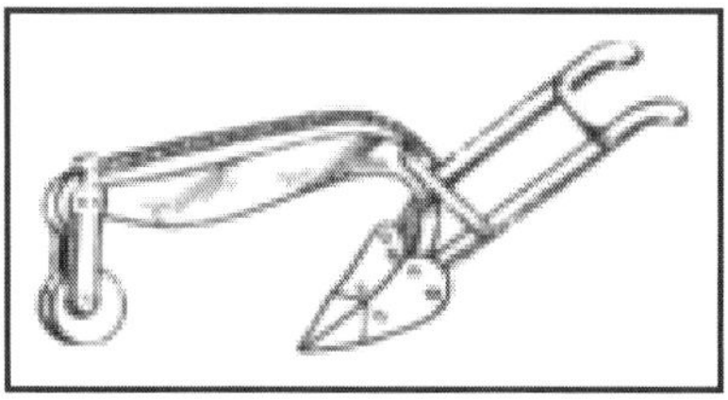

Fig. 5.6 : Ridging and harvesting plough

6. Ladder : Ladder is very simple implement. It is generally made with two pieces of bamboo or a heavy wooden beam of two metres in length. The ladder is used after ploughing to crush the clods and to make the soil level. It is also used for micro levelling and slight compaction necessary after sowing of seed.

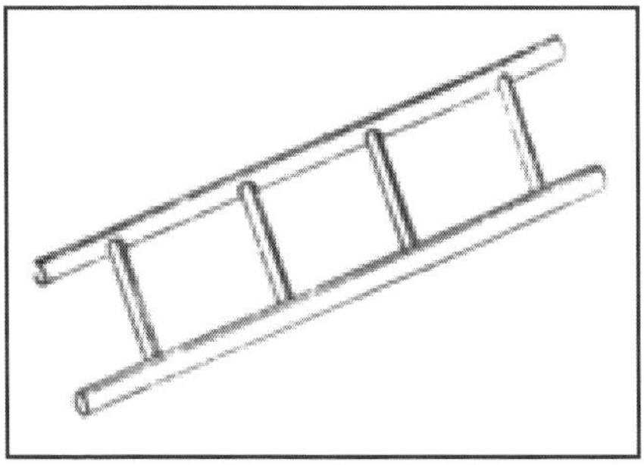

Fig. 5.7 : Liadder

7. Tractors : Tractors are important tools for machanisation of various farm operations and it has become a vital part of modern agriculture. Four wheeled tractor having average horse power of 30 to 35 can be used for ploughing, harrowing, sowing, harvesting, transport and belt work (*i.e.*, cutting of chaff, lifting water etc.) The tractor is a wonderful piece of invention for the agriculture. The main units of the tractor are as follows :

(*i*) ***General Construction :*** The transmission case, clutch housing, engine, and front axle support are bolted together to form of rigid unit.

(*ii*) ***Engine :*** This is a four cylinder, four stroke water cooled engine equipped with electrical glow plug for cold starting.

(*iii*) ***Clutch :*** The basic clutch is single plate dry clutch.

(*iv*) ***Transmission :*** This has eight forward speeds and two reverse speeds.

(*v*) ***Differential :*** This is mounted on taper roller bearings and incorporates a differential lock.

(*vi*) ***Rear Axle :*** These are mounted on ball bearings and enclosed in removable housing which are bolted to transmission case.

(*vii*) ***Front Axle :*** This is of hollow square section design mounted on a central pivot pin.

(*viii*) ***Wheels :*** The front wheels are mounted on taper roller bearing on the stearing knuckles.

(*ix*) ***Brakes :*** Independent disc brakes operated by two pedals which can be locked together for road travel. A hand brake is fitted for parking.

(*x*) ***Steering :*** Recirculating ball or peg and worm type steering gear has a single drop arm with a fixed length drag link.

(*xi*) ***Cooling System :*** The engine is cooled by pressure cooling system with forced circulation by belt driven centrifugal pump. The radiator is a tube type with multiple rows of tubes.

(***Source :*** Mahindra and Mahindra Ltd. Mumbai–400101)

Fig. 5.8 : Tractor

Fig. 5.9 : Power tiller

8. Power Tiller : A power tiller is a machine having two wheels on one axle. The power tiller is generally used for ploughing, harrowing, puddling, and pumping of water. There are different types of power tiller produced in the country. The power tiller consist of the following parts :

(*i*) An engine,

(*ii*) Transmission Belts,

(*iii*) Clutch,

(*iv*) Gear Box,

(*v*) Wheels,

(*vi*) Chain Drive,

(*vii*) Rotavator (Tilling blades).

The engine mainly consist of :

(*i*) Fuel Tank

(*ii*) Fuel pump

(*iii*) Air cleaner

(*iv*) Cylinder

(*v*) Piston

(*vi*) Connecting Rod

(*vii*) Crank shaft and

(*viii*) Fly wheel.

The power tiller works on diesel oil, when the starting handle of the engine is turned, some air gets sucked into the cylinder through the Air cleaner. When the piston which is contained inside the cylinder moves forward, the air in the cylinder get compressed. Because of this compression, the air becomes very hot. At that time, the fuel pump injects some diesel oil into this hot air inside the cylinder. The diesel thus gets ignited due to the high temperature of the air and burn with great force and pushes back the piston. The piston is connected to the crankshaft by means of the connecting rod. Therefore when the piston moves in the cylinder, the crank shaft starts turning. On the crank shaft Fly-Wheel is mounted which is quite heavy. The weight of Fly-wheel helps it to continue to rotate some more and thus the process of sucking air, compressing it, injecting diesel and burning is repeated. This process takes place once when the Fly-Wheel rotates twice. On the Fly-Wheel is fitted the Engine pulley which is connected to the main clutch by means of three V-belts.

The power and revolutions of the Engine are transmitted through the belts and clutch to the Gear-box. The wheels are connected to the Gear-box through the Gears and Axles. The Rotavator is also connected to the Gear box through the sprocket and chain-drive. The clutch is capable of being "engaged" and "disengaged". By engaging the clutch, the Engine becomes connected to the Gear box and the wheels will starts turning and the Rotavator will start rotating. When the clutch is disengaged, the Engine gets disconnected from the Gear Box and the wheels will not turn and the Rotavator will not rotate.

Power tiller may be used for tilling on dry land and wet land for puddling. When the power filler is used for puddling, the pneumatic wheels and rear wheel should be replaced by steel wheels and tail skid respectively.

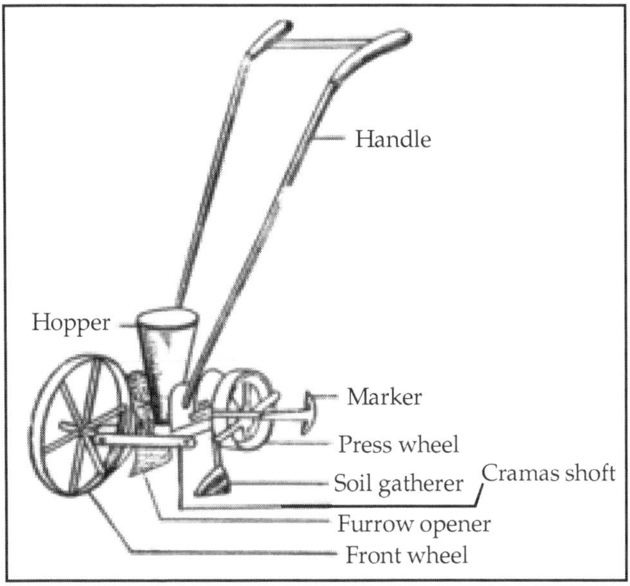

Fig. 5.10 : Seed drill

Some important information of Kampco Power Tiller.

(*i*)	Model	: ER 90 ENGINE
(*ii*)	Horse Power	: 9 – Continuous
(*iii*)	Engine Revolution	: 2000 R.P.M.
(*iv*)	Weight of Engine	: 145 kgs.
(*v*)	Length × Width × Height of Engine	: 82.0 × 51.2 × 64 cm.
(*vi*)	Total Weight of Tiller	: 485 kgs.
(*vii*)	Overall Lenght × Width × Height of Tiller	: 225 × 82 × 103 cm.
(*viii*)	Diesel consumption	: 1.25 to 1.5 Lit/hr.

(*ix*) Type of Engine : Single cylinder, 4 Stroke-horizontal (95 × 105).

(*x*) No. of Tilling Blades : 20.

(*xi*) Tilling capacity : 0.25 Acre/hr.

(*xii*) Average tilling depth : 19 cm.

(*xiii*) Tilling Width : 60 cm.

(*xiv*) Travelling speed.

	Gear position	*Speed (km/hr)*
Low	1	1.5
	2	2.4
	3	3.3
	Reverse	1.3
High	4	5.7
	5	8.9
	6	14.0
	Reverse	4.9

(*xiv*) Revolution Blade shaft

H–Position : 310 R.P.M.

L–Position : 215 R.P.M.

(*Source* : Operation Manual, Kamco)

The Power packet Tiller. Kerala Agro Machinery Corporation Limited).

5.12.1.2 Implement Used for Sowing Seed

(*a*) **Seed-Drill :** The machine that is used for sowing seed in rows at uniform rate and at controlled depth is called seed-drill. Seed drills are classified as follows :

Seed drills are classified depending upon the seed metering device.

(*i*) Manually metered drills

(*ii*) Mechanically metered drills.

(*b*) Seed drills are classified according to their function

(*i*) Plain seed drill.

(*ii*) Seed cum fertiliser drills.

In this machine, a seed drill is fitted with fertiliser dropping attachment which distributes fertiliser uniformly on the ground. There is a seed box in this–drill which is divided lengthwise into two compartment. Out of these, one is meant for seeds and the other for fertilisers.

(*c*) Seed drills are classified depending upon the nature of power employed:

(*i*) Tractor drawn seed drills.

(*ii*) Animal drawn seed drills.

Functions of Seeds-Drills : Seed drills perform the following functions :

(*i*) Carrying of seed in hopper.

(*ii*) Opening of furrow to an uniform depth by the furrow opener.

(*iii*) Metering the seeds.

(*iv*) Placing of seed in furrows in an acceptable pattern.

(*v*) Covering of the seed by the soil by soil gatherer.

(*vi*) Compacting of soil around the seed by press wheel.

Hand Seed Drill : The hand seed drill is employed for sowing different types of seed such as Paddy, Wheat, Jute, Mustard etc. in a row at uniform depth. A seed drill is suitbale for sowing different types of seed by changing the disc only according to the size of seeds. Different parts and their functions are given below :

Different parts of a seed drill	Functions
(i) **Frame or Body :** It is usually made of angle iron with suitable braccs and bracks which are strong enough to withstand all types of loads in working condition.	It connects all parts of a seed drill.
(ii) **Front Wheel :** Seed drill is provided with a wheel (Diameter : 36 cm) which is fitted in the front side of a seed–drill.	The seed drill is brought forwarded on prossing the front wheel and power is transmitted to the seed droppings mechanism from front wheel. It acts as the transport wheel.
(iii) **Hopper :** It is funnel like part of a seed drill and fitted in the frame by screws.	Hopper carries seed for sowing.
(iv) **Felt :** It is usually made by felt or rubber. Provided with rectangular aperture, which remains in between the base of hopper and disc, the seed dropping aperture.	It reduces the friction of hopper and disc, seed rates can be controlled by adjusting the felt.
(v) **Disc :** It consists of two *semi-circular* or only one circular Iron Disc. Disc is provided with some pit or groove on regular intervals on its periferix. Discs are of different types according to the size of seed. Disc is attached to a rod by screw.	The disc collects the seed from the hopper and delivers seed to the furrow when the seed drill is brought forward.
(vi) **Furrow Openers :** Furrow openers is provided in a seed drill for opening the furrow before dropping the seed. It remains attached on the base of the disc and it is meant for opening a furrow. There are four types of furrow openers such as ordinary type or hoe type, shoe type, single disc type and double disc type.	It opens the furrow, when the seed drill moves forward and the seed falls into the furrow from the disc.
(vii) **Soil Gatherer :** It is a device to refill a furrow after the seed has been placed in it and this is attached on both sides of the seed drill.	It is meant for refilling the open furrow and cover the seed with soil.

Different parts of a seed drill	*Functions*
(*viii*) **Marker :** It is a long iron rod fitted with the seed drill having an arrangement of adjusting the row distance.	The marker helps in maintaining uniform distance between a row.
(*ix*) **Handle :** Handle is provided in a seed drill and two handles are attached on two sides of the seed drill.	The operator pushes the seed drilling by holding the handle with hand. The seed drill moves forward causing the seed to drop behind the furrow opener.
(*x*) **Scraper :** Scraper is a part of a seed drill that is attached behind the front wheel and press wheel.	It removes the soil from the wheel.
(*xi*) **Connecting Rod :** It is made of iron.	The different parts of the whole seed drill is connected properly by it.
(*xii*) **Press Wheel :** It is attached at the back side of a seed drill. It is smaller in size than that of the front wheel.	The press wheel presses the soil of furrow after refilling by a soil gather. It performs the function of laddering.
(*xiii*) **Crank Shaft :** It is made of iron and connects the front wheel with a rod on which a disc is attached. When the front wheel rotates, the disc is also rotated and the seeds are automatically dropped in the furrow.	Crank shaft acts as a medium of power transmission.

Mechanism : Seed drill is an automatic seed sowing machine. When the operator pushes the seed drill by hand and moves forward, the front wheel rotates and the rotation power is transmitted to the disc through the crank shaft. The disc then rotates and the seed feeds the pits of the disc coming through the felt. Seed dropping aperture may be of gravity feed or forced feed. Forced feed may be of three types. Such as *(i)* Brash *(ii)* Flutted roller and *(iii)* Disc. The furrow opener opens the furrow in which seeds fall from the pits of disc due to its rotation. The soil gatherer refils the furrow which is pressed by a press wheel. The next row is demarcated by the marker at the time of sowing seed. The seed rate can be controlled by the arrangement of felt.

Calibration of Seed Drill : The procedure of testing the seed drill for correct seed state is called calibration of seed drill, which is necessary before putting it in actual use to find the desired seed rate. Calibration is done to get a predetermined seed rate of the machine.

The method of calibrating a seed drill before using it in the field is as follows :

(*i*) At first, the diameter of driving wheel *i.e.* front wheel is to be measured.

(*ii*) The distance covered by front wheel for one revolution is to be calculated. Put a mark on the wheel so that the revolution may be counted easily.

(*iii*) The opening of seed drill should be fixed.

(*iv*) The seeds are kept in the hopper and the front wheel is rotated for one revolution. Then the seed are collected under each seed tube on a paper or a cloth and weighed.

(*v*) The seed metering mechanism (*i.e.* the mechanism of a seed drill which delivers seeds from the hopper at selected rates is called seed metering machanism) is changed and the same process is repeated until the desired seed rate is obtained.

(*vi*) After calibration, the seed drill is taken to the field for actual sowing.

The speed of bullock drawn and tractor drawn seed drill is 1.5 and 2.5 miles per hour respectively. The seed drill covers 0.15 hectares for sowing Aus paddy and 0.12 hectares for sowing wheat seed per hour.

Planter : Planter is normally used for those seed which are larger in size and can not be down by the usual seed drill. The function of planter is more or less similar to the seed drill. The planter consist of the following parts:

(*i*) Hopper.

(*ii*) Feed metering device.

(*iii*) Knock out arrangement.

(*iv*) Cut off mechanism.

(*v*) Furrow opener.

(*vi*) Other accessories.

Planter is usually used for those seeds, which are required to be sown at equal intervals between plant and a plant.

Potato Planters : Potato, an important crop in India, requires a lot of labour for sowing the seed tuber in the field. The slow laborious hand dropping method of planting potatoes has been largely supplemented by the mechanical planter. The potato planter opens the furrow, drops and spaces the seed pieces at various distance, place the fertilisers to sides and below the level of the seed and covers both the seed and fertilisers at desired depth.

Potato planters have the following types of potato planting mechanism :

(*i*) Automatic.

(*ii*) Semi-automatic.

(*iii*) High speed automatic.

Paddy Transplanter : Manual transplanting is a labour–intensive operation comprising nursery raising, uprooting of the seedlings, transporting and transplanting the uprooted seedlings in the main field, with a total labour requirement of about 280–350 man hrs per hectare. High labour demand during the peak periods adversal effects the timeliness of operation, thereby reducing the crop yield. The steady drift of agricultural labour to industrial sector is adding more to the woes of the rice farmer. Because of drudgery and notion that the farm operations are below the dignity, labour availability, in general, has decreased considerably in farm operation.

To offer these problems, mechanical transplanting is the solution. Many transplanters were developed in the past involving the use of traditionally grown paddy nursery for mechanical transplanting. However, several on-farm attempts made with different models of transplanting using the traditional root washed nursery were not successful because of the mechine related problems to use the nursery. This called for the necessity of raising the nursery in a special way, called Mat nursery technique. Among the existing commercial available designs suiting Mat nursery. Chinese model of paddy transplanter is found to have great promise.

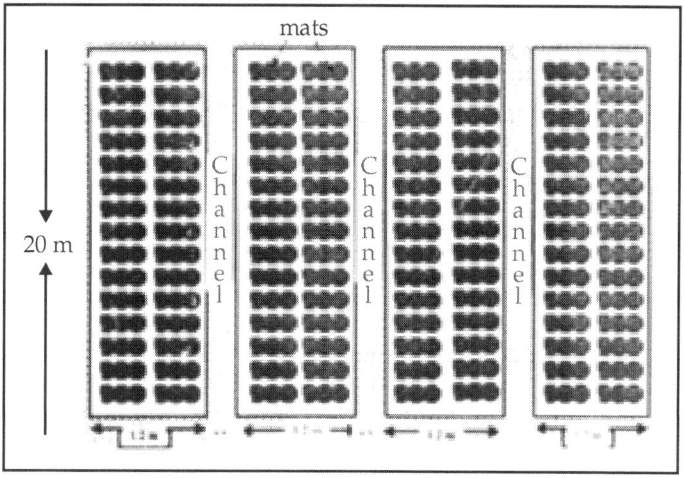

Fig. 5.11 : Layout of Mat Nursery for (for 1 ba)
(*Photo Courtesy* : Directorate of Rice Research, Hyderabad)

Features of the Machine

(*i*) Available with its own source of power of 3–HP diesel engine.

(*ii*) Covers eight row at a time with a row to row spacing of 23 cm.

(*iii*) Provision to vary the plant to plant spacing of 10 and 12 cm.

(*iv*) Provision for cage wheel for field use and rubber wheel for transport.

Feature of the Mat Nursery

(a) ***Seed preparation***

(*i*) Soak the seeds for one day.

(*ii*) Incubate the seeds for one day.

(*iii*) Apply a seed rate of 50 kg/ha.

Fig. 5.12 : Matnursery
(*Photo Courtesy :* Directorate of Rice Research, Hyderabad)

(*b*) ***Mat Preparation***

(*i*) Choose the suitable site for mat nursery raising (preferably within or nearer to the main field) and puddle the selected site.

(*ii*) Raise the puddled soil bed to a level of about 5 cm and level it properly. To lay more than two rows of mats, select each bed width as 1.2 m. To cover one hectare of land, selected plot should be of size of about 20 × 7.5 m with provision of channels of width.

(*iii*) Spread the plastic sheet evenly on the levelled and raised beds.

(*iv*) Mix the soil with FYM and water to proper slurry.

(*v*) Keep the DRR designed mat frame (having dimensions of 19 × 50 × 2 cm of each block) on the plastic sheet and spread the slurry mixture in each block of the farmer.

(*vi*) After proper settling, lift the frame to leave the mats behind. Each time, five mats can be prepared this way.

(*vii*) Take 100–110 g of sprouted seeds and spread it evenly on the prepared mats and press them gently. One hectare of main field needs 500–600 mats.

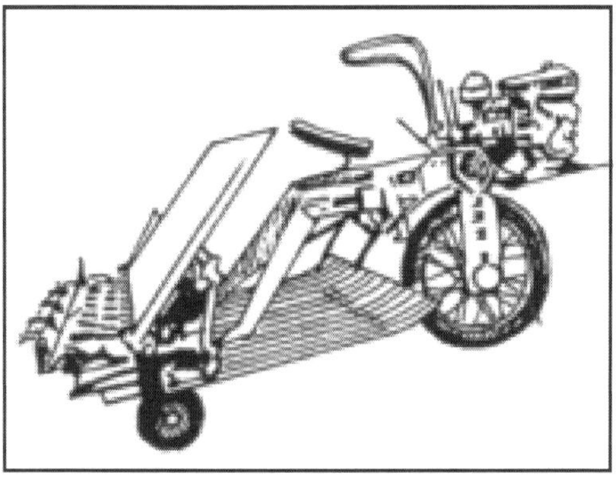

Fig. 5.13 : Transplanter

(*Photo Courtesy :* Directorate of Rice Research, Hyderabad)

(*viii*) The procedure is repeated to prepare all the mats. Spread either grass or plastic sheet on the mats overnight and take it out in the normal. Do not uncover if there is rain during the first week of mat raising.

(*ix*) Sprinkle with rose can on all the mats everyday to keep the mats moist.

(*x*) After a week when the nursery looks green, provide water through channels to the level just below the mats.

(*xi*) Apply fertilizer @ 30 g of DAP/litre of water and sprinkle on the mats with hand sprayer after 7–10 days of nursery.

(*xii*) The nursery will be suitable for transplanting after 15–20 days when the plant reaches a height of 12.5cm.

Paddy Transplanter Operation : For efficient operation of transplanter, the land preparation knowledge and mechanical provisions incorporated with the transplanter are essential to be known.

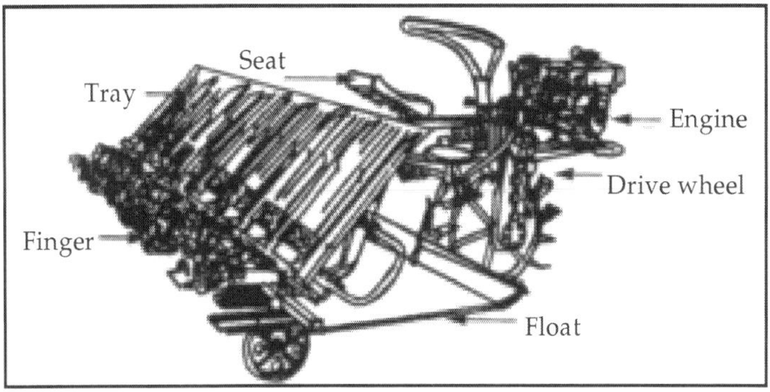

Fig. 5.14 : Line diagram of trnasplanter

(*Photo Courtesy :* Directorate of Rice Research, Hyderabad)

Land Preparation

(*i*) Puddle the land, level it and remove the water after 24 hours (for clayey soil) or 12 hours (for paddy/loamy soil) before the transplanting. In black or clay soils, the settlement is critical as the loosened soil can bury the seedlings planted.

(*ii*) Just before the operation of transplanter, a thin film of water is necessary to ensure the free movement of transplanter and avoid adhering of soil to the moving parts of transplanter.

Operation of the Transplanter

(*i*) Check oil level in gear box, fuel tank and grease the specified parts.

(*ii*) While operating observe for any missing hills in any particular row and diagnose the problem immediately. The missing hills can be due to presence of stones/ foreign mat that would have obstructed the fingers that pick the seedlings from the tray.

(*iii*) Choose the right speed of operation for a plant hill to hill distance of 10 or 12 cm.

(*iv*) Observe the number of plants per hill and adjust to the requirement using the lever operating the finger movement.

(*v*) Observe the depth of planting with the lever and adjust accordingly.

Performance : There will be a net saving of 40 per cent over the manual transplanting. It can transplant about one hectare in a day when 8 hours run. The transplanter performs with a missing hills of 2–3 per cent. The machine cost is about Rs.1,05,000 and is marketed by VST Agro Inputs, Mahadevpura, Whitefield Road, Bangalore–560 048.

Fig. 5.15 : Plant stand with transplanter
(*Photo Courtesy :* Directorate of Rice Research, Hyderabad)

5.12.1.3 Implements Used for Intercultural Operations

1. Spade : Spade is made of iron. The blade of the spade is broad. A wooden handle is attached to the body. The standard size spade is 25 cm. long and 16.5 cm. broad. The size of the spade varies according to their mode of uses. Spade is used for performing different agricultural operations as follows :

(*i*) To control weeds and to make the soil loose in open space of row crop.

(*ii*) To earthing up of some crops such as Potato, Sugarcane, Arum, Brinjal, Tomato, Chilli, Cabbage, Cauliflower, Beet, Turnip etc.

(*iii*) To make furrow for planting and harvesting of some crops such as Potato, Sweet potato, Groundnut, Sugarcane, Onion etc.

12–15 labour can interculture an acre of land in a working day of 8 hours.

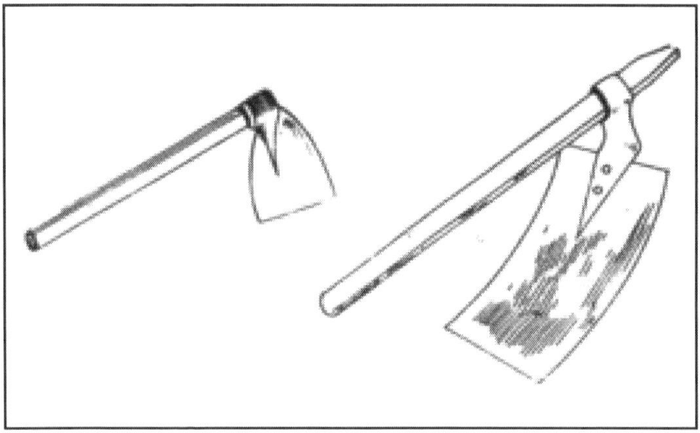

Fig. 5.16 : Spade

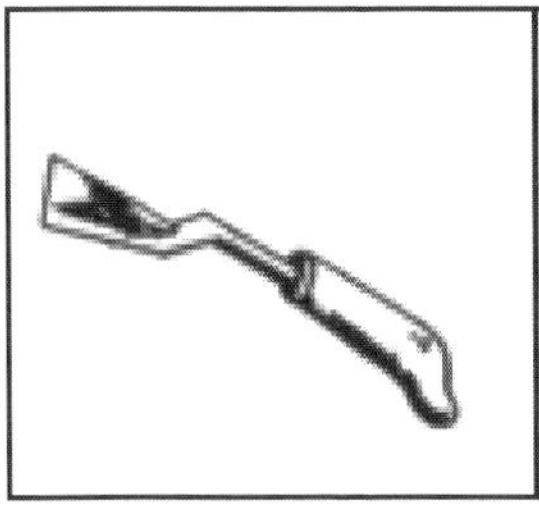

Fig. 5.17 : Khurpi

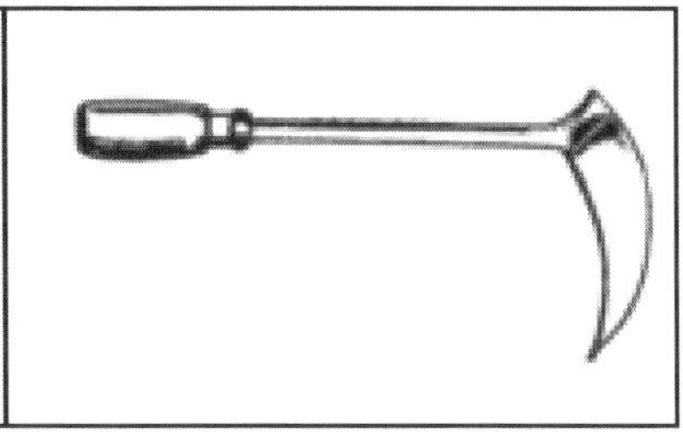

Fig. 5.18 : Nirani

2. Khurpi and Nirani : These two implements are made of iron and their handle is made up of wood. Khurpi consist of a saw-edged triangular blade where as nirani consist of sickle shaped blade. They are used for weeding, thinning, loosening the surface soil, making pit for planting the seedlings of some crops such as Cabbage, Cauliflower, Brinjal, Tomato, Chilli, Beet, Carrot and Sowing of seeds of some crops such as Curcubits, Mango, Litchi, Papaya, Guava etc. The users have to squat while working with these implements.

3. Country Bida : Country bida is more common implement. The body of this implement is made up of babul wood which is cut into rectangular shape. The length of the body is 3 to 6 feet (90 – 120cm.) and the width is 10–12 inches (25 – 30 cm.). This is simple in design, easy to make and cheap. The blades that are made up of bamboo are attached to the log on its underside by means of two slanting wooden or bamboo pegs. This implement is used for weeding, thinning and loosing of surface soil of broadcasted paddy and mesta field. It is also used for preparing seed bed and covering of seed already sown in the field.

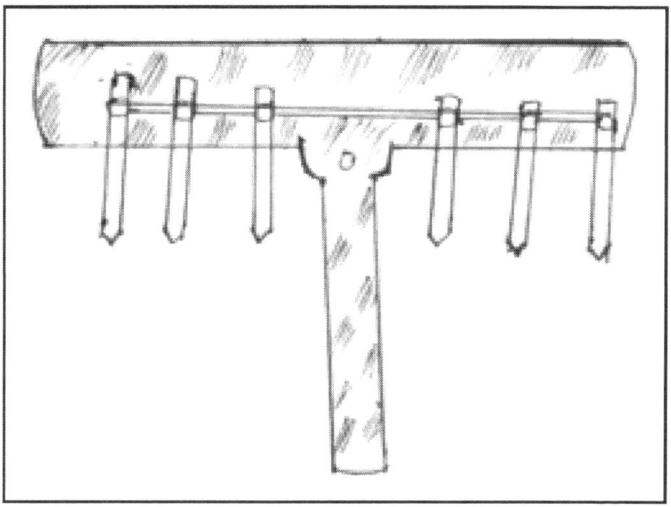

Fig. 5.19 : Country bida

4. Harrows : Harrow is an implement used for secondary tillage which pulverizes and smoothen the soil, levelling the soil, preventing and eradication of weeds and covering the seeds. It is particularly used for breaking the clods in hard soil. Following are the different types of harrows.

(i) Spike Tooth Harrow : Spike tooth harrows are so called because the teeth resembles long spikes. The spike tooth harrows are made in sections. Each section may be 4-5 feet (120 – 150 cm.) wide having 25, 30 or 35 teeth. The teeth are attached to bars which may be made of steel or wood. The teeth may be made in different shapes such as round, oval, square, triangular or diamond. The teeth are held firmly against the edges by means of clamp. The teeth should have wide range of adjustment varying from horizontal to vertical position. There is a tooth with head that is long and curved and fitted at each corner or section in such a manner that when the teeth are placed flat, these teeth serves as runners. These narrows consist of teeth, tooth bar, clamps, guard, liver, hooks and brakes.

Uses : This harrow is used to break clods, stir the soil, uproot the weed, break the soil crust, cover the seed and level the soil. Levelling the soil directly after ploughing is its principal use.

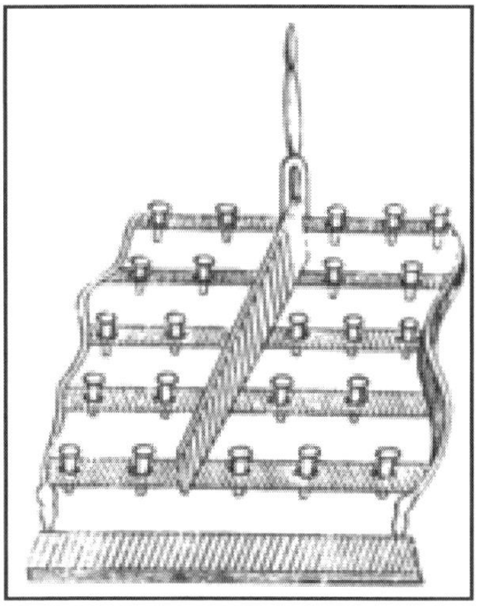

Fig. 5.20 : Spike tooth horrow

(*ii*) ***Spring Tooth Harrow :*** The Spring Tooth Harrow is designed in independent sections having tough flexible teeth. Since it gives springing action in working condition. It is suitable to work in hard and stony soil. The tyres pulverises the soil and helps in killing the weeds. The spring tooth harrow consists of frame, tooth bar, tooth, clamps, clevis, lever, links and guard rail. The levers are provided for setting the teeth for varying depth of harrowing.

Uses : It is used for ploughing the land full of weed and seed bed preparation.

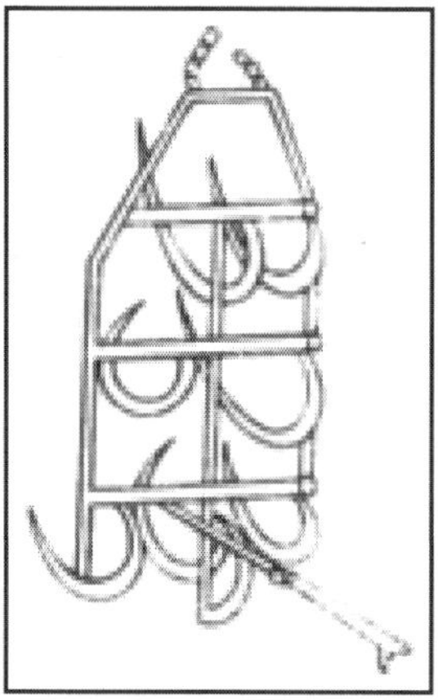

Fig. 5.21 : Spring tooth horrow

(*iii*) ***Disc Harrow :*** In this harrow, a set of rotating steel discs are mounted on a common shaft, which is called gang. Disc harrows are of two types according to the nature of power employed, such as *(i)* Tractor drawn disc harrow and *(ii)* Animal drawn disc harrow. The animal drawn disc harrow mainly consist of a disc, gang frame, gang angle mechanism, beam, scraper, saucer, clevis, axle, middle tyre, and bearing. In disc harrow, discs are smaller in size than disc plough.

Uses : Disc harrow is found suitable for hard soil, full of stalks and grasses. The discs cut through the soil and pulverise the soil very effectively.

Disc harrows are of two types depending upon the arrangement of disc.

(*a*) *Single Action Disc Harrow* : This harrow consists of two gangs placed opposite to each other which throws the soil in opposite direction *i.e.* the left side gang throws the soil towards the right and the left gang throw the soil towards the left.

(*b*) *Double Action Disc Harrow* : This harrow consists of two single disc harrows which are attached together, one behind the other. The front and back gang throws the soil in opposite directions.

(iv) Acme Harrow : Since the cutting blades of acme harrow consists of large knives, it is sometimes called knife harrow. This harrow consists of a series of knives which are attached securely to a frame. The knives used in this harrow may be straight or curved.

Uses : This harrow is found suitable for levelling the soil.

4. Cultivators : Cultivator is an implement used for tilling the soil between standing rows of crops, cultivator is also known as tiller or tooth harrow. A cultivator has two rows of tyres attached to its frames. Cultivators are generally used for intercultural operation, mainly for destruction of weeds, aeration of soil, conservation of soil moisture by natural mulching etc. Cultivators are of two types according to the power employed for the implements, *viz.*

(*i*) Tractor drawn cultivator and

(*ii*) Bullock drawn cultivator.

Followings are the different types of cultivators available in our country :

(a) Horse Hoe Cultivator : The hoes are attached at an interval of 45 – 90 cm. It covers 1.2 hectares for performing interculture operation and 0.08 hectares for making furrow for sowing of seed in a working day of eight hours. The five tined lever expansion cultivator is used for weeding in sugarcane and vegetables. In this implement, 8 – 10 cm. long hoes or tines are attached at an interval of 30 cm. This implement is provided with a lever for controlling depth of tillage. This is pulled by a pair of bullock.

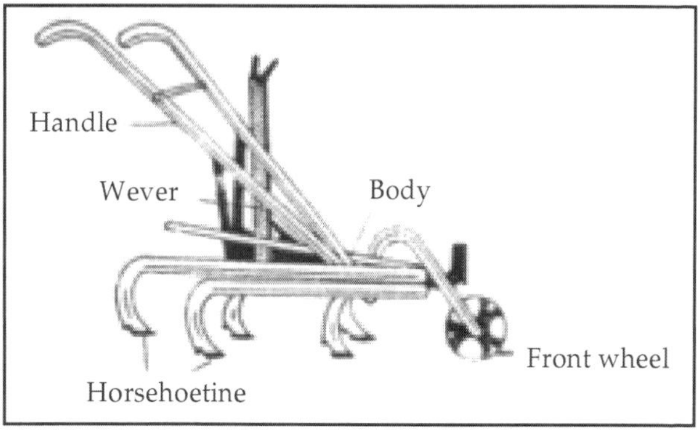

Fig. 5.22 : Horsehoe cultivator

Uses : This implement is used for shallow tillage and control of weeds in between the standing rows of crops.

(b) Planet Junior Cultivator or Wheel Hoe : Wheel hoe is a small implement pushed by manual labour. Wheel hoe consist of a body handle and a wheel. The duck foot type tynes of 15 cm. long are attached to its body at an interval of 10 – 15 cm. They may be adjusted according to spacing of crops. The prong fork reversible hoe, combination etc. can also be used in place of duck foot tynes. The implement is provided with a wheel and a handle.

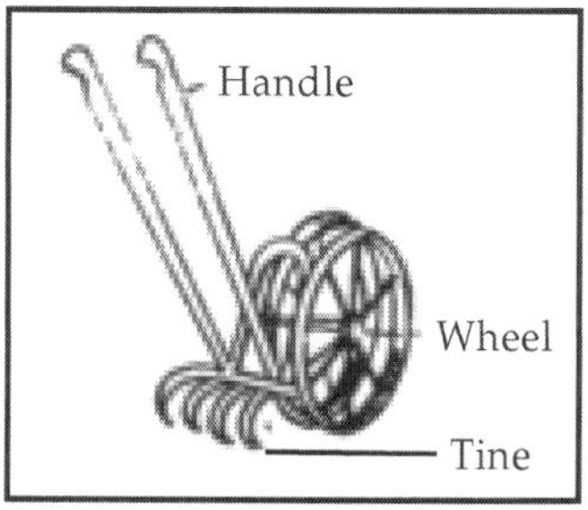

Fig. 5.23 : Wheel hoe

Uses : The implement is found suitable for shallow ploughing and control of weed in between the standing rows of crop. It is also used for making furrow for sowing of seed, and manures incorporation in the soil. The working with this implement is easier due to having a wheel.

(c) *Paddy Weeder* : Paddy weeder is also called Japanese rotary weeder. It is operated by manual labour. The implement consists of the following parts.

(*i*) *Body* : It is made of iron to which other parts are attached.

(*ii*) *Float* : A float is provided behind the spike tooth drum to avoid sinking of the implement in the puddle field.

Fig. 5.24 : Japanese Paddy Weeder

An adjustment is provided with it to control the depth of ploughing.

(*iii*) *Puddler of Rotating Drum* : Rotating drum consists of small spikes which stir the soil when this implement is pushed forward.

(*iv*) *Handle* : It is made of wood and is used for pushing the implement.

***Uses* :** Paddy weeder is useful for weeding and incorporation of fertilisers in line-sown low land Paddy. It covers 0.24 to 0.26 hectares of land for intercultivation of paddy.

Besides these, Akola, Trifali and Baroda hoe are also used for intercultural operation.

5.12.1.4 Implements Used for Harvesting and Threshing

Harvesting is the operation of cutting the standing crops in the field. Different implements are used for performing this function.

1. Sickle : Sickle is the most common implement used for harvesting of some crops such as paddy, wheat, jute, maize etc. and cutting other vegetables in our country. It is a simple harvesting tool. It is essentially consisted of a blade and wooden handle. Blade is the main metallic part of the sickle which may be plain and serrated. The blade is made curved shape. The inner side of the curved blade is made sharp for efficient working in the field which is called the cutting edge. The serrated sickles are generally used for cutting paddy.

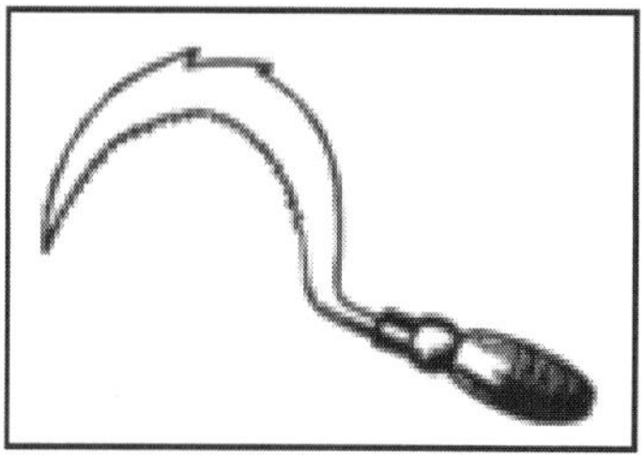

Fig. 5.25 : Sickle

2. Mower : Mower is a machine mainly used for harvesting grasses and forage crops. A mower consists of the following parts:

(*i*) ***Frame :*** The frame provides space for gear, clutch and bearings. A liver that is used for lifting the cutting bar is attached to the frame. It is provided with two wheels connected by a bar with a big gear.

(*ii*) ***Power Transmission Unit :*** In bullock drawn mowers, the power transmission unit consist of axle, gear, crank wheel, crank shaft and pitman. The transmission shaft (*i.e.* a long axis that connects the bevel gear) transmits the power to the crank wheel from the bevel gear. The power is transmitted through gear box. As the crank wheel revolves, it gives reciprocating motion to the Pitman, which operates the knife in the cutter bar.

(*iii*) ***Cutter Bar :*** The cutter bar made of high grade steel is used for cutting grasses and forage. The cutting mechanism consists of cutter bar, shoes, guards, ledger plate, wearing plate, knife, grass board etc.

(*iv*) ***Pitman :*** Pitman is the connecting rod between the crank wheel and the knife. Power is transmitted from the crankwheel to knife by Pitman.

Fig. 5.26 : Combine harrester

3. Reaper : Reaper is an improved implement for cutting of crops in the field. Reaper is generally an animal drawn implement and it is pulled by a pair of bullocks. Reaper consists of the following parts :

(*i*) Frame,

(*ii*) Cutter bar knife,

(*iii*) Wheels,

(*iv*) Bearings and other attachments.

4. Paddy Thresher : Paddy thresher is gaining popularity in an area where high yielding varieties of paddy is being cultivated extensively.

Fig. 5.27: Paddy Hiresher

The pedal operated paddy thresher consist of the following parts :

(*i*) *Threshing Drum* : The main part of machine is a wooden drum with peg-teeth all around its circumference. Threshing drum is a hollow drum having cylindrical shape. The diameter of the drum used on common thresher is about 45 cm and its width may vary from 40 – 76 cm.

(*ii*) *Body Frame* : It consist of the base, the side frame, the front grain shield and rear grain shield. They may be of mild steel angle section except front grain shield which is made of wooden plank of about 12 cm thick and is fitted suitably to the side frames. The base may be made of wood also.

(*iii*) *Cylinder* : The cylinder is made of mild steel and a series of slats supported on two sides. The cylinder may be 450 mm in length when the thresher is operated by one man. The threshing drum is suitably reinforced at its centre for proper rigidity.

(*iv*) *Threshing Teeth or Spikes* : Threshing teeth are fixed to the wooden battens of the threshing drum, Drum has been mounted on the drum shaft by cotter pin.

(*v*) *Driving Mechanism* : The thresher is operated by one or two man with the help of a pedal. Drive consist of a crank, one end of which is connected to a spur gear and the end is connected suitably to the pedal from fulcrum which is welded to the pedal frame. Operation of a thresher is just like a sewing machine *i.e.* by operating the pedal, the drum will revolve. Drum can move four time with one stroke of the pedal.

The threshing is done by holding the paddy sheaves with both hands a little away from the ears and spreading it as much as possible against the teeth of revolving drum. The ears are slowly pressed on the cylinder and the bundles are twisted side ways once. This finish up the bundle without lifting it up. The stack of bundles should be kept on the left of the operator to enable him to work quickly. The empty bundle of straws should be thrown away from the thresher. The normal operating speed is 300 – 400 RPM. The capacity of single man operated thresher is about 2.0 – 2.40 quintals per day.

Chaff Cutter : The implement that is used for chopping the straw and fodder is called chaff cutter. The fodders that are chopped properly become more palatable. The chaff cutter machine may be of different types depending on the power used for operating the implement as follows :

(*i*) Hand chaff cutter.

(*ii*) Animal operated chaff cutter.

(*iii*) Power chaff cutter.

Parts of Chaff Cutter

(*i*) ***Aproan :*** It is a tin plate and attached to the frame to serve as the support of the straws or fodder, which are stay here before chopping.

Fig. 5.28 : Chaff cutter

(*ii*) ***Fly Wheel :*** It is a big iron wheel in which two blades (*i.e.* sharpened knife) are fitted oppositely. The fly wheel is also provided with a handle.

(*iii*) ***Worm :*** It is a iron spindle provided with particular grove. It transmit the power in right angles.

(*iv*) ***Worm Wheel :*** There are two worm wheels or spur gears which are fitted to the frame.

(*v*) ***Fed Roller :*** There are two grooved roller, one is upper and other is lower. They are connected with worm wheel.

Machanism : The chaff cutter machine must be properly installed on a rigid foundation in a shed or otherwise protected from rain. Before the machine put into operation, its gears and bearings should be lubricated and the knives should be sharpened if it has become too blunt and the clearance between the knives and the share plate should be adjusted for most effective cutting of fodder. The nuts, bolts and screws should be tightened just before the using this implement to avoid accident.

The straws or fodders are fed first in between two rollers having the support or aproan and then the fly wheel is rotated. As a result, worm is rotated and so the worm wheel. The power is therefore transmitted to the roller via worm or worm wheel or spur gear from the fly wheel. The fed roller utilises the power to move the straws or fodder meant for chapping *forward* uniformly. The knives rotate due to rotation of the fly wheel and cut the straw or fodder uniformly which are collected below:

Sugarcane Crusher : Sugarcane cursher is a machine used for extraction of juice from the sugarcane. This machine is operated by bullock.

Parts of Sugarcane Crusher

(*i*) ***Roller :*** There are three roller in this machine which are fitted to the frame. The first and second roller are of about 10" diametre, grooved perpendicularly and connected with each other. The third roller is of

8" diameter, grooved horizontally and connected *with first roller.*

(*ii*) ***Trashturner :*** It is a steel plate and placed in between second and third roller. It diverses the sugarcane toward first and second roller.

(*iii*) ***Red Plate :*** It is a tin plate on which the three rollers are placed. It is made cone shaped at the front side from which the extracted juice come out and collect in the container.

(*iv*) ***Spur Gear :*** There are three gears placed on each of three rollers and they are called spur gear.

(*v*) ***Crank :*** Crank, a part of the frame, receives the power from the transmission shaft.

(*vi*) ***Transmission Shaft :*** It is a wooden part of the frame and serve as medium through which power is trans- mitted finally from spur gear to the roller.

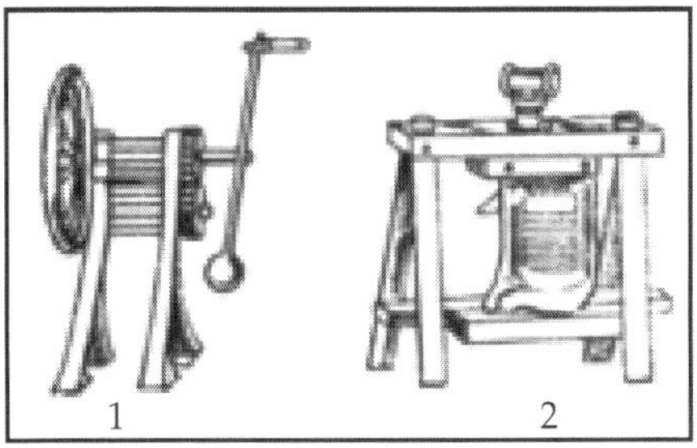

Fig. 5.29 : Sugarcane crusher 1. Hand driven 2. Animal driven

Mechanism : When the wooden transmission shaft rotates around the frame, the power is transmitted from it to the roller through crank and spur gears and as a result, roller began to rotate clockwise. The sugarcane that is fed into the first and third roller is crushed and about 65 per cent juice is extracted. The crushed cane then passes into first and second roller. They are again crushed and juice is extracted to its maximum. The juices come out through cone shaped bed plate and this is then collected in the container. The crushed cane passes out to the back of this machine through first and second roller.

Mechanical Extraction of Jute Fibre : The conventional method of jute retting is dependent on the availability of large volume of slow flowing soft water and the skill of the persons engaged for the purpose. Jute retting requires placement of the bundles in specified manner under water (Steeping). To obtain better quality of fibre, the recommended practice is to maintain a minimum volumetric ratio of plant into water as 1:20. Besides the requirement of slow flowing good water, there are other factors such as temperature, age and quality of plant, depth of steeping, care taken during retting etc. which as a rule are not adequate or optimum to render optimal quality of fibre. In practice, one or the other factor of retting becomes limiting, thereby producing inferior quality fibre.

Apart from these extraneous factors, non–uniformity of fibre quality lies in the jute stem itself which causes differential retting in a single reed. The basal diameter of a jute stem is much bigger and so bark at the base is thicker and tougher than that of the top; as a result tender top rets much quicker than the thicker and older bottom bark. This differential retting of a stem in a given time affects the quality and often leads to the production of rocky fibre (incomplete retting) at the base. This is the inherent and cultural defects of the plant retting. To circumvent such a situation, mechanical decortication of the green jute plant and water retting of the post decorticated material play a very significant role towards improvement fibre quality. Decortication leads to the sizeable reduction in volume of the plant material. Steeping of the decorticated material not only requires much lesser volume of water, but also consumes less time for retting as compared to the retting of whole plant. Moreover, decortication followed by water retting yields better quality of jute fibre.

Finally, with a view to reducing or eliminating the major constraints associated with the whole plant retting, decortication followed by retting may be restored too.

Decortication : Decorticator is a machine designed in such a way that it can extract and scrape out barky material from stem of any bast fibre crop. Stems of the crops which can be decorticated are jute, mesta, sunnhemp, urena etc. the machine breaks down the inner woody cores of such stems into pieces and scrapes the bark mostly at the bottom portion. Such machine extracted, material needs only 5 to 6 days to ret in water to produce fibre of commerce.

Capacity of the Machine : Capacity of the machine is about a ton of green jute plant per hour with a 5 Hp prime–mover. Five men are needed to operate the machine, two for feeding into the machine and three for arranging, handling and steeping of the extracted material. The capacity of this machine can further be increased, if required.

Fig. 5.30 : Jute Decorticator

Age of the Plant : The best stage of decortication of plant is at the age of 115 ± 10 days. Decortication should be made just after harvest of the crop and can be continued for 2 to 3 days after harvest if the material is kept under a shed. Defoliation is not necessary before feeding the stems into the machine. Feeding is done by the butt ends of the plants. Two to eight plants are fed into the machine in a lot, depending upon the diameter of the plant. It has been found that plants having around 12 mm diameter at base are better decorticated. For better result, plants are sorted out into two groups depending upon the diameter, one group of 12 mm and below and the other of 12 mm. above.

Table 5.4 : Cost Comparison between Mechanical cum Microbiological and Conventional Methods of Fibre Production per Hectare

Method	*Number of Labour*	*Energy Cost (Rs.)*	*Depreciation Cost (Rs.)*	*Other Cost (Rs.)*	*Total Cost (Rs.)*	*Total Fibre Price (Rs.)*
Whole Plant Retting	85	–	–	–	1190	5520
Mechanical cum micro-biological retting	63	28 (electrical)	160	10	1038	6000
		132 (Diesel)			1142	

Assumption

1. Carrying cost has not been included.
2. A hectare of plot yields 40 tonnes of green jute which yields 24 quintals of fibre.
3. Labour rate is @ 14 per day for both the method.
4. Qualities and price of fibres obtained by traditional and mechanical methods are W5 with Rs. 230/q and W4 with Rs. 250q respectively.

Retting : Steeping of the extracted material is done soon after decortication. Retting of such material may be carried out either in a cemented tank or in a pool of water. It is a pond, a bamboo structure like a platform is erected below water surface. The material is then placed flatly over the platform and covered the 'Jak' with water hyacinth or any equatic weeds. No weighting down material is, however, necessary. The material is kept under such condition till it is ready for wash. The 'Jak' should be examined after 2 to 4 days of steeping. As soon as the fibre is free from gummy matter, it needs to be washed immediately in the same water, if it is a concrete tank it is better to add some fresh water at the time of washing. One man can easily wash material out of a tonne of green jute in a day.

Fig. 5.31 : Feeding of Jute plants into decorticator

Fibre Quality : The unique feature of the machine extracted fibre is that it always produces fibre, free of 'roots'. Strength of the fibre is usually improved if retting is done properly. Fineness is generally maintained at per with that of the whole plant retted fibre. Mechanically extracted and retted fibres are generally classified between 1 to of the ISI eight point grading system. However, machine decorticated and retted fibre tends to retain some amount of gum which when normally treated with softener and spinned in the mills produced normal yarn. This fibre is likely to retain some broken and loose sticks.

Cost of Fibre Production : There are some inherent advantages in the mechanical method such as it requires less quantity of water and time, fibre becomes free from 'root'. Fibre does not possess any other body defect as a result this fibre remains at least one grade above than that of conventional one. If this gain due to grade advantage is taken into consideration, then there may be a total profit of Rs. 632.00 and Rs. 528.00 per hectare due to mechanical method with electric motor and diesel engine respectively. For initial trial and adoption it is advised to use this method only in places where scarcity of retting water exists.

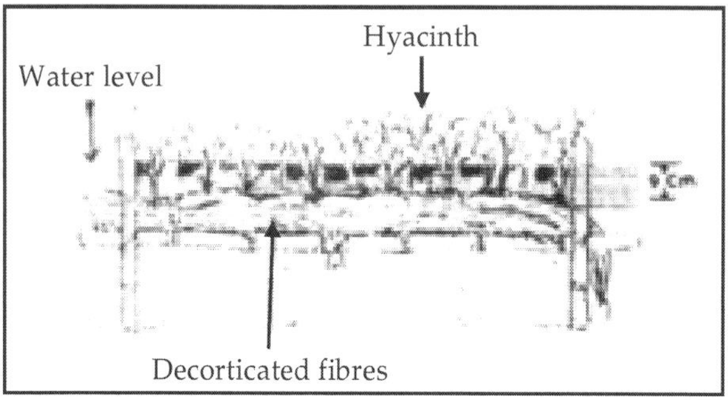

Fig. 5.32 : Steeping of decorticated material

Advantages

(*i*) Mechanical cum microbiological method produces 'root' free fibre not below grade 4.

(*ii*) Water requirement is reduced by more than 50 per cent.

(*iii*) Retting time is also reduced by more than 50 per cent.

(*iv*) Drudgery of working into foul water is reduced.

(*v*) About 5 per cent more fibre recovery is expected in this method.

(*Source:* Mechanical Extraction of Jute, J.A.R.I.).

5.12.1.5 Pesticide Application Equipment

1. Sprayer : The equipment which is used for dispersal of insecticide in liquid form is called sprayer. Sprayers are of two types depending on the source of energy for its operation as follows :

(*i*) Manually operated or Hand sprayers.

(*ii*) Power operated sprayers which are commonly termed as power sprayer.

Sprayers consist of the following essential parts :

(*i*) ***Tank :*** Tank is an important part of a sprayer in which spray materials are kept. It is generally made of galvanised iron or steel coated with anticorrosion material particularly on the inside. The tank is provided with a large filler hole fitted with strainer for pouring of liquid into the tank as well as for cleaning the spray materials. The tank is also provided with lid.

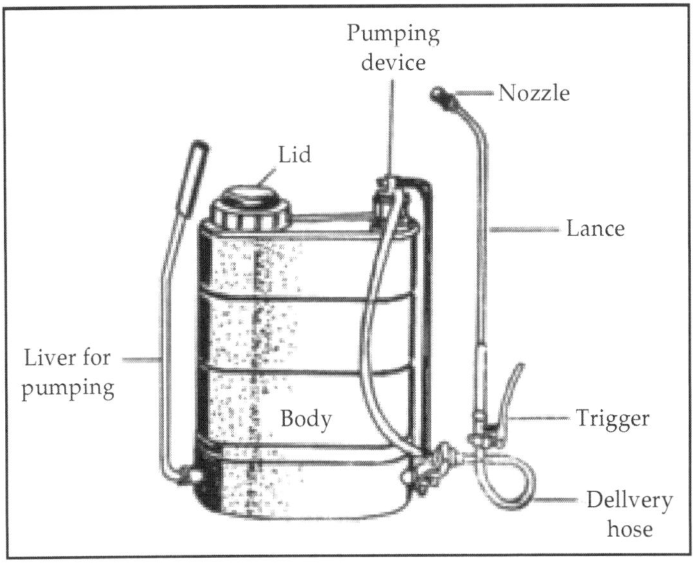

Fig. 5.33 : Different Parts of a Knapsack Sprayer

(*ii*) ***Pump :*** It is a most essential component of the sprayer as it is needed for atomisation of the liquid through air. Depending on the mechanism of action, the source of power etc. pumps is use in the sprayer may be broadly put into three categories. *(i)* Pneumatic, *(ii)* Displacement pumps, plunger, piston, rotary and diaphram and *(iii)* centrifugal or impeller (Chattopadhyay, 1980). Pump exerts pressure on liquid which is kept in the tank and forces it through the nozzle in knapsac sprayer, a lever is provided for pumping.

(*iii*) ***Agitator :*** Agitator is usually provided in the power sprayer for maintenance of an uniform dispersion of suspensions and emulsions.

(*iv*) ***Delivery Hose :*** It is fitted with sprayer at one end and spray lance or boom on the other and made of rubber or synthetic rubber or plastic tube or nylon. The hose is capable of withstanding high pressure.

(*v*) ***Spray Lance :*** It is brass tube of variable length. One end of the spray lance is fitted with a hose and other end with a nozzle. The nozzle end is usually bent.

(*vi*) ***Nozzle :*** Nozzle is most important part of a sprayer to break up the liquid coming out of the spray tank into fine droplets. Different types of nozzles are used in the sprayer.

(*vii*) ***Trigger :*** Trigger is provided in the sprayer to control the flow of spray materials.

DIFFERENT TYPES OF SPRAYERS

(*A*) Manually Operated Sprayers

(*i*) *Foot Sprayer :* The pump in the foot sprayer consist of a pump barrel and a pressure chamber. The pump is worked by the upward and downward movement of the foot used for the purpose. The plunger with a suction cup or piston, drives into the pump barrel, thus sucking the liquid in the pressure chamber and expelling it through the discharge line. The return stroke of the plunger pulls the liquid in through the suction hose for the next discharge. The sprayer develop a pressure of 60–80 psi and has a provision for attaching two discharge lines. It is an ideal all purpose sprayer suitable for both small and large scale spraying on field crops, *e.g.* in orchard, tea and coffee plantations and rubber estates.

(*ii*) *Rocking Type Sprayer :* Rocking type sprayer consist of a lever operated pump assembly which rest on a wooden plant form. The plunger is operated by means of a lever that is pused forward and backward by hand. Suction hose with a strainer is immersed in a separate container containing spray mixture and the delivery system consist of a separate pressure chamber, a flexible hose, spray lance and a spray nozzle. The rocker sprayer develops 60 – 80 Psi pressure.

The rocking type sprayers are ideal for spraying plantations, in orchards and farms.

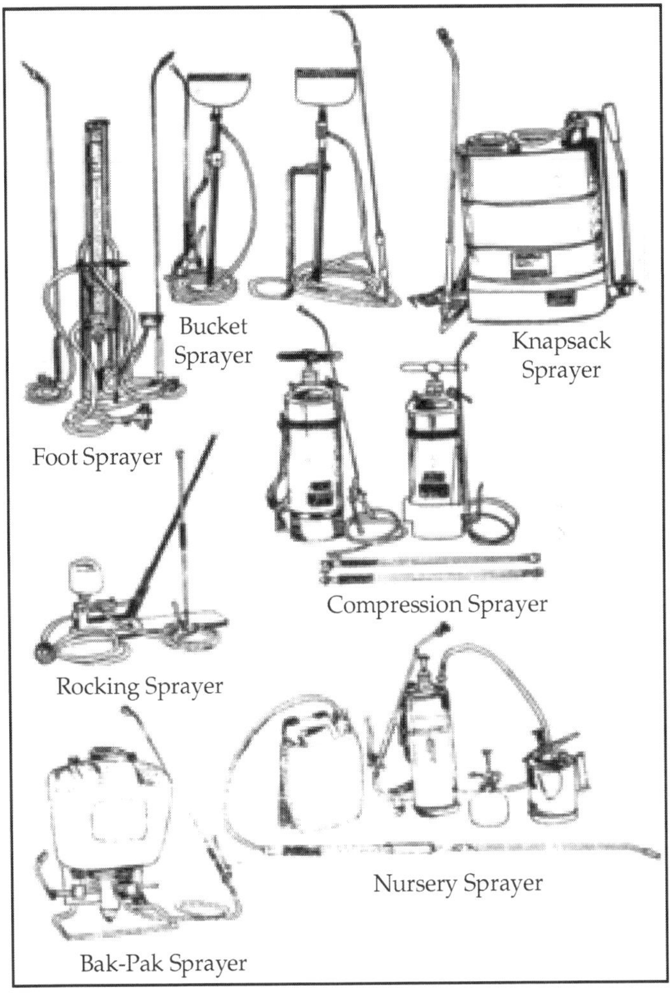

Fig. 5.34 : Different types of Sprayers

(iii) Compression Sprayer : In this sprayer, the tank is pressure vessel in which the liquid is filled to two-third capacity, leaving one forth volume for the compressed air. It is then pressurised to 60 psi by means of the air charge pump which is fitted vertically inside the tank (*i.e.* container). A pressure gauge or safety valve may optionally be fitted to the tank. A pressure regulator may also be used when the discharge pressure needs to be strictly controlled. The compressed air forces the liquid through the nozzle and desired type of spray is achieved. This sprayer is ideal for spraying all field crops and vegetables.

(iv) Knap-Sack Sprayers : In this sprayer, the tank is non–pressurised and made of brass or galvanised steel. The pump may be fitted inside or outside the tank and sucks the liquid from the tank and expels it through the delivery system consisting of lance and nozzle which is connected with the pump by a flexible hose. The knap-sack sprayer develops 30 – 40 psi pressure. The operator operates the lever with his left hand and spraying is done by his right hand. A spray boom or rig may be attached when wider areas have to covered. This sprayer is ideal for spraying all field crops at high pressure with low volume nozzle.

(v) Bucket Sprayer : This sprayer consist of a double action pump with a lance and nozzle outlet. Suction hose is placed in the bucket. In the single barrel type, the plunger is hollow and acts as a pressure chamber. In the double barrel type, one barrel is of smaller diameter than the other and acts as a pump while the bigger barrel serves as a pressure chamber to produce more continuous spraying. Liquid is discharged in both suction and delivery strokes. A pressure of 30 – 40 Psi is generated.

(vi) Hand Sprayers or Nursery Sprayers : Nursery sprayers are small and simple sprayers. These sprayer produce a fine mist spray and can be effortlessly operated. These sprayers are generally recommended for use in nursery or private garden. Varieties with a small compressor type or plunger type sprayer are also available.

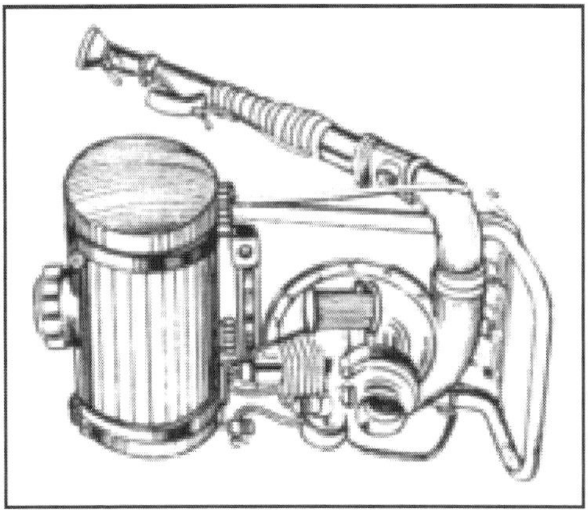

Fig. 5.35 : Monsoon winds over India

(*B*) Power Operated Sprayer

(*i*) *Hydraulic Sprayer :* Hydraulic sprayer may be engine or electric motor driven and are available with single, double and triple piston pumps, which is driven by an air cooled or water cooled engine. The single piston pump and double and triple pump develops a maximum pressure of 150 Psi and 300 – 400 Psi respectively. The single piston pump have only two discharge lines and the double and triple piston pumps can accomodate 4 – 6 discharge lines. This sprayer is operated by means of 1 – 2 H.P. electric motor or 2–3 H.P. Petrol, Petrol–kerosene or diesel engine, and this sprayer can also be driven by power tiller or tractor.

(*ii*) *Motorised Knapsack Mistblower Cum Duster :* This sprayer cum duster is fitted with a two stroke air cooled engine of 35 to 70 c.c. capacity connected to a centrifugal fan by a direct drive. The spray liquid is first pressurised by air generated by the blower. This air current achieves a velocity of over 275 Kmp at the nozzle and sprays the chemical in fine particles that can be measured in microns. The nozzle design enables even spraying at efficiency.

When dusting, the air blast enters the tank from air inlet which is connected to a tube with several hole on its surface. This agitates the power which is then thrust out by the velocity of the air coming out of the blower, through the pleated hose and out through the nozzle. The blower is mounted with Rubber elements on frame with Back cushion and Padded carrying straps and is fitted with villiers L–34–2 stroke 1.2. H.P. petrol or petrol/kerosine engine. This implement is ideal for quick application of pesticides in dust and liquid form in orchards, tea, coffee estates and other crops.

(iii) Tractamount Sprayers : As the name indicates, this sprayer is attached to tractor for use. The pump is driven by PTO shaft of the tractor and the sprayer unit sucks the chemical and discharges it through the spray boom or through the discharge line consisting of a delivery hose and spray guns. The boom has a swivel arm to direct the spray correctly. The main frame allows the spray boom to be adjusted according to the height of the crops being sprayed.

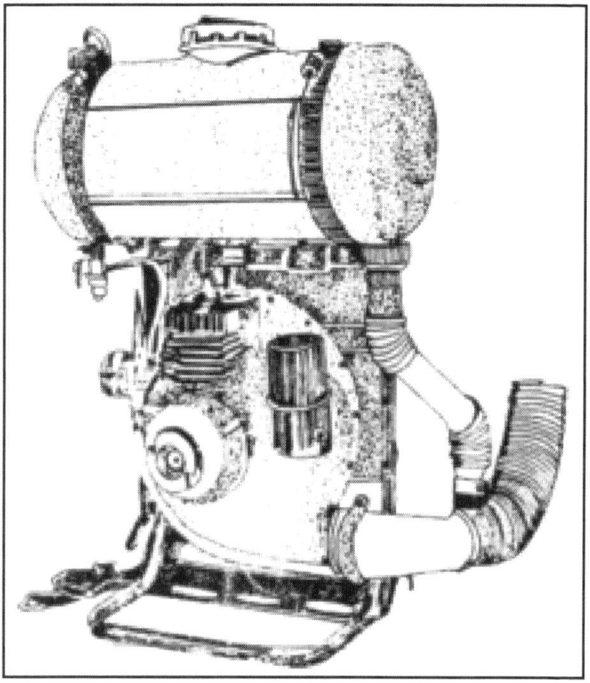

Fig. 5.36 : Misiblower cum duster

(iv) Spinning Disc Sprayers : This sprayer consist of a battery operated motor with a spinning disc, a liquid tank, a handle and a set of batteries. Liquid is fed from the tank on the spinning disc by the force of gravity. The spinning disc, which has 180 channels on the wall and 180 teeth on its periphery, operates at 4000–5000 rpm to stir the liquid and create very fine, even particles for low volume spraying. A smooth flow of liquid with highly controlled droplet (100–165 microns) application is thus achieved. The stainless steel disc is inter changeable.

DUSTERS

(A) Hand Rotary Dusters

(i) Rotary Type Dusters : Rotary dusters are provided with an agitator which stirs the powder and releases it evenly through the discharge vent. The blower sucks the dust or powder from the hopper through the connecting pipe and pushes it out forcefully to achieve efficient dispersal. The operator carries the duster by means of one or two shoulder straps and holds the lance in his left hand, cranking the handle with his right hand.

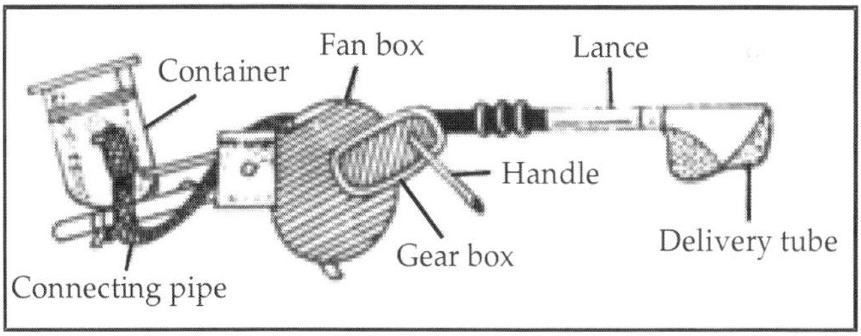

Fig. 5.37 : Hand Rotary duster

(*Photo Courtesy :* American Spring and Pressing Works Ltd., Bombay-400064)

(ii) Plunger Type Duster : This operates on the piston principle by generating an air blast which passes through the dust chamber and expels the dust through the discharge out let.

(iii) Bellows Type Duster : This type of duster also creates a blast of air through the dust chamber to discharge the powder, the significant difference being that the force is generated by means of bellow operation. This variety of duster is not popular in India.

(B) Power Operated Dusters

(*a*) ***Rotary Type :*** Here the air is sucked in by a motorised blower and discharge through the blower outlet.

(*b*) ***Air Jet Type :*** A jet of air from the blower enters the hopper, which agitates the dust and blows it out simultaneously with a jet of air.

(*Source :* American Spring and Pressing Works Ltd. Bombay–400 064).

REFERENCES

Aiyer A.K.Y.M. Field Crops of India (1985), The Bangalore Printing and Publishing Co Ltd. Bangalore.

Balasubramaniyan P and S.P. Palanippan, Principles and Practices of Agronomy (2004), Agribios (India), Jodhpur.

Das P.C. Ucchamadhyamik Krishi Bijna, Oriental Book Co. Pvt. Ltd., Kolkata.

Dey Gopal Chandra, Fundamental of Agronomy (1980), Oxford and IBH Puslishing Co. Pvt Ltd., New Delhi.

Ghosh R.K. and Swami S., Practical Agricultural Engineering (1990), Naya Prakash, Kolkata.

Khedekar, D.N. Agricultural Engineering (1990) in Hand Book of Agriculture. ICAR. New Delhi.

Reddy T. Yellamanda and G.H. Sankara Reddi, Principles and Practices of Agronomy (1992), Kalyani Publisher, Ludhiana.

Singh. S.S. Principles and Practices of Agronomy (1990) Kalyani Publishers, Ludhiana.

6

MANURES AND FERTILIZERS

6.1 INTRODUCTION

The Manures is originated from the French Word "Manoeurver" which refers to "Work with Soil", that is why the word tillage and manures whose synonyms as it clear by the statement of Jethro Tull (1700 B.C.)—"Tillage is manure". The word of manure is also originated from Latin word Manu (hand) and operare (works). Manures are substances which are organic in nature, capable of supplying plant nutrient in available form, bulky in nature, having the low analytical value and having no definite composition and most of them are obtained from animal and plant waste products.

Fertilizers are inorganic material which can supply plant nutrient in available form, having the high analytical value, and having a definite composition and mostly are industrial products. A fertilizers refers to a material added to the soil in order to supply chemical elements needed for plant nutrition and to improve soil fertility.

6.2 CLASSIFICATION OF MANURES AND FERTILIZERS

Manures

A. Bulcky Organic Manures

1. *Compost*

   (*a*) Village compost.

(*b*) Town compost, (made from farm waste products).

2. *F.Y.M.*

(*a*) Cattle manure.

(*b*) Ship Penning.

(*c*) Poultry manure.

3. *Sewage and Sludge :*

B. Green Manures

(*a*) Leguminous plant.

(*b*) Non-leguminous plant.

C. Oil Cake : Richest source of plant nutrient of all organic manures.

1. Edible cake i.e., used for cattle feeding.

(*a*) Mustard cake.

(*b*) Groundnut cake.

(*c*) Linseed cake.

(*d*) Sesame cake.

2. Non edible cake i.e. used as manures.

(*a*) Castor Cake.

(*b*) Neem Cake.

(*c*) Sunflower Cake.

D. Waste Product of Slaughter House

(*a*) Blood meal.

(*b*) Bone meal.

E. Fish Product

(*a*) Fish meal.

F. Guano : Material obtained from the excreta and dead bodies of sea bird.

Fertilizers

1. Nitrogenous Fertilisers : Sources of nitrogen.

(*a*) Fertilizer containing nitrogen in ammonium (NH_4) form: Such as Ammonium sulphate, Ammonium chloride etc.

(*b*) Fertilizer containing nitrogen in nitrate (NO_3) form : Such as Calcium nitrate, Sodium nitrate, Potassium nitrate etc.

(*c*) Organic fertilizer or fertilizer containing nitrogen in non–ionic form : Such as Urea, etc.

2. Phosphatic Fertilizers : Sources of Phosphorus :

(*a*) Fertilizers containing water soluble phosphorus : Such as Superphosphate, concentrated Superphosphate, Mono–ammonium Phosphate, Diammonium Phosphate, etc.

(*b*) Fertilizers containing citrate soluble Phosphate : Such as Dicalcium Phosphate, Basic Slag etc.

(*c*) Fertilizers containing insoluble phosphate. Such as Rock Phosphate, etc.

3. Potasic Fertilizer : Sources of Potassium : Such as Potassium Chloride, Potassium Sulphate etc.

4. Compound Fertilizers : Sources of more than one nutrient : Such as Gromor (Urea Ammonium Phosphate), Mono–ammonium Phosphate, Diammonium Phosphate, Ammonium Phosphate, Suphala 20:20:0 (Ammonium Phosphate Sulphate), Suphala 15:15:15 (Nitro–Phosphate–Potash), IFFCO 10–26–26 etc.

5. Mixed Fertilisers : Source of three nutrients : Such as Grade–1 (8:8:8:) Fertiliser.

6.3 MANURES

1. F.Y.M. : One of the most commonly used bulky organic manure is farm yard manure (F.Y.M.). F.Y.M. is produced from liquid and liquid excreta of cattle mixed with litter used for breeding purposes of cattle.

The following method has been recommended by C.N. Acharyya for preparing good quality of F.Y.M. and to avoid high nutrient loss. F.Y.M. is prepared in a trenches of 20 – 25 ft. (6 – 7.5 m) long, 5 – 6 ft. (150 – 180 cm) breadth and 3 – 3.5 ft (90 – 105 cm) deep. Farm waste mixed with earth should be under each animal in the evening for absorption of Urine. Each morning urine socked litter and dung should be mixed and taken to manure trench. A section of 3 ft. (90 cm) length is taken up from one end for filling with daily collection of refuses, when the trench is filled to a height of 1.5 – 2ft. (45 – 60cm) above ground level, the surface of the refuges is plastered with cow–dung mixed with soil. The manures becomes ready in about three months. By this time the second trench is being filled up. Generally two such trenches would be needed for 3–4 heads of cattle. It is possible to prepare by this process 250–300 cubic feet of manures (10–12 cartloads) per year per head of cattle.

2. Compost : Compost is an another organic manure artificially prepared from vegetables and animal waste products. Two method of composting waste organic materials, such as aerobic and anaerobic are usually recommended. In the aerobic process, the mixed farm residues are collected in a heap and in the an–aerobic process, the mixed farm residues are collected in a pits of a convenient size, say 15' x 5' x 3' (450 x 150 x 90 cm). Each day a collection is spread in a thin layer and sprinkled with a mixture of fresh cow–dung (4.50 kg.). Compost manure is reinforced the pits is filled to a height of 1.5 – 2 ft. (45 – 60 cm) above ground level, the surface is plastered with one inch layer of a mixture of mud and cow-dung. The compost becomes ready in about three to four months without any further attention.

Method of Compost Making : The main method of making compost are as follows :

(i) Adco Process : This process was introduced by Hutchinson and Richard in England in 1921. Adco–Powder, originated by Agricultural Development Company, England (Private concern operating at Herpenden, England) is used as a starter @ 7 kg. per 100 kg. of dry waste product.

(ii) Activated Compost Process : This process was introduced by Fowler and Redge in 1922 at Indian Institute of Science, Bangalore. They used sewage sludge as a starter *i.e.* the material used for decomposing plant residues.

(iii) Indore Process : This process was devised by Howard and Ward at the Institute of Plant Industry, Indore. The cow-dung in small amounts are used as 'Starter'. The indore process is aerobic and decomposition takes place in aerobic condition.

(iv) Bangalore Process : This process is devised by C.N. Acharya as a result of his experiment at the Institute of Science, Bangalore. Night Soil is used as a starter.

Note : 1 cft. refuge weighs roughly 20 lbs. 1 cft. of compost weighs roughly 40 lbs. One cart loads of refuse (30 cft.) weighs roughly 600 lb.

3. Compost Tea or Liquid Manure : Compost tea is prepared by decomposing the green leaves and leaves having pungant odour which are mixed with cow–dung or droppings of goat, sheep or poultry and taken in gunny bags which is kept in the drum having 200 litres of water. The compost tea is applied to the paddy in the morning or in afternoon. It can also be used as foliar spray.

4. Vermi Compost : Vermi compost is now gaining popularity in Agriculture. It is made and used to recover their existence in the crop field. There is no existence of earthworm where chemical fertilizers and pesticides are used extensively. Vermi compost is prepared in a pit specially by using earth worm. The available different kinds of farm waste are used for making vermi compost. It is also available in the market.

5. Green Manures : The practice of turning into the soil undecomposed green plant tissue is referred to as green manuring (Buckman and Bredy, 1980) and the manures obtained by this method is known as green manure.

7. Characteristics of Green Manure Crops : The green manure crop should possess some desirable characteristics :

(*i*) The green manure crops should have profuse leaves and rapid growth.

(*ii*) The green manure crops should have abundance and succulent tops.

(*iii*) The green manure crop should be able to grow well on poor soils.

(*iv*) The green manure crop should have deep root system.

Other factors remaining the same, use of leguminous green manure crop is more useful in comparison to non–legumes as more nitrogen is added by legumes which will be advantageous for the soils and crops grown after green manuring.

Crops Suitable for Green Manuring : Crops suitable for green manuring are divided into two groups :

1. *Non–Legumes :* The non–legumes used as green manuring crops provides only organic matter to the soil.

Examples : Mustard, Wheat, Radish, Carrot etc.

2. *Legumes :* The legumes used as green manuring crops provides nitrogen as well as organic matter to the soil.

Examples : Sannhemp (*Crotolaria juncea*), Dhaincha (*Sesbania aculeata*), Senji (*Melilotus parviflora*) cow–pea (*Vigna catjang*) etc.

Recently the leaves of *Glyricidia maculata* and *Sesbania speciosa* (Perennial and shrubs) are used as green manuring materials. These plants are planted in a bank of tank or fallow land and their leaves and succulent tops are added to the soil for green manuring.

Benefits of Green Manuring : There are numerous advantages of green manuring such as :

(*i*) ***Supplies of Organic Matter :*** Green manures supply organic matter to the soil. Humus formed from green manures increases the adsorptive capacity of soil, promotes aeration, drainage, and granulation of soil particles which is helpful for plant growth.

(*ii*) ***Addition of Nitrogen :*** The green manures crops supplies additional nitrogen in addition to organic matter, if it is a legume crop, which has the ability to fix nitrogen from the air with the help of nodule bacteria. So all the legumes crop leave the soil in better physical condition and richer in nitrogen content.

(*iii*) ***Nutrient and Soil Conservation :*** Green manure crops act as cover crop. They prevent the soil erosion and nutrient loss by taking up soluble nutrients which might otherwise been lost in drainage water or due to erosion. Green manuring crop makes the availability of Phosphorus and other nutrients to the succeeding crops. Green manures have a marked residual effect also.

(*iv*) ***Increases the Biochemical Activity :*** The organic matter added to soil by way of green manures acts as a food for microorganism and they stimulate bio-chemical changes.

Table 6.1 : Nutrient Content of Manures and Oil Cakes

Manures	*Nutrient percentage*		
	Nitrogen (N)	*Phosphorus (P_2O_5)*	*Potassium (K_2O)*
1	*2*	*3*	*4*
A. Bulky Organic Manure			
Cow dung (Fresh)	0.3–0.4	0.1–0.2	0.1–0.3
Horse dung (")	0.4–0.5	0.3–0.4	0.3–0.4
Sheep dung (")	0.5–0.7	0.4–0.6	0.3–1.0
Night soil, Fresh	1.0–1.6	0.8–1.2	0.2–0.6
Poultry manure, fresh	1.0–1.8	1.4–1.8	0.8–0.9
Sewage, Sludge, dry	2.0–3.5	1.0–5.0	0.2–0.5
Sewage, sludge, activated dry	4.0–7.0	2.1–4.2	.5–0.7
Urine, Cattle	0.9–1.2	Trace	0.5–1.0
Urine, Horse	1.2–1.5	Trace	1.3–1.5
Urine, Human	0.6–1.0	0.1–0.2	0.2–0.3
Urine Sheep	1.5–1.7	Trace	1.8–2.0
F.Y.M.	0.5–1.5	0.4–00.8	0.5–1.9
Compost (Town) dry	1.2–2.0	1.0	1.5
Compost (Rural) dry	0.4–0.8	0.3–0.6	0.7–1.0

1	2	3	4
Water hyacinth compost	2.0–3.0	1.0–2.0	3.0–4.0
Green manure (Various, average)	0.5–0.7	0.1–0.2	0.6–0.8
(*a*) Dhaincha	0.62	0.15	0.58
(*b*) Cowpea	0.71	0.15	0.58
(c) Sannhemp	0.75	0.12	0.51
B. Manures of Animal Origin			
Dried blood	10.0–12.0	1.0–1.5	1.0
Fish manure	4.0–10.0	3.0–9.0	0.3–1.5
Bone meal (raw)	3.0–4.0	20.0–25	–
Bone meal (steamed)	1.0–2.0	25.0–30.0	–
C. Wood Ash			
Ash, Coal	0.73	0.45	0.53
Ash, Household	0.5–1.9	1.6–4.2	2.3–12.0
Ash, Babul	0.1	0.8–1.3	1.5–3.1
D. Crop Residues			
Paddy husk	0.3–0.5	0.2–0.3	0.3–0.5
Straw	0.36	0.08	0.71
Groundnut shell and stem	1.6–1.8	0.3–0.5	1.1–1.7
E. Oil Cakes			
Edible oil cakes			
Mustard cake	5.1–5.2	1.8–1.9	1.1–1.2
Linseed cake	5.5–5.6	1.4–1.5	1.2–1.3
Sesame cake	6.2–6.3	2.0–2.1	1.2–1.3
Safflower cake (Decorticated)	7.9	2.2	1.9
Groundnut cake	7.0–7.3	1.5–1.6	1.3–1.4
Coconut cake	3.0–3.2	1.9–2.0	1.7–1.8
Cotton seed cake (decorticated)	6.4	2.9	2.2
Non-edible cake			
Cotton seed cake (Undecorticated)	3.9–4.0	1.8–1.9	1.6–1.7
Castor cake	4.3	1.8	1.3
Karanj or honge cake	3.9	0.9	1.2
Mahua cake	2.5–2.6	0.8–0.9	1.8
Neem cake	5.2–5.3	1.0–1.1	1.4–1.5

7. Oil Cakes : Oil cakes are the by-products of oil seed crops. Oil cakes are the important and quick acting organic nitrogenous manures. It also contains small amount of phosphorus and potassium. A large variety of oil cakes are produced in the country and they can be grouped into two classes as follows :

(*i*) ***Edible Oil Cakes :*** This types of oil cakes are mainly used as cattle feed. But it can also be used as organic manure in the field.

Examples : Mustard cakes, groundnut cakes, Sesame or til cakes, Linseed cakes, coconut cake etc.

(*ii*) ***Non-Edible Oil Cakes :*** This types of oil cakes are not suitable for feeding cattle and mainly used for manuring crops.

Examples : Castor cake, Neem cake, Karanj cake, Mahua cake, Safflower cake etc.

Uses : Oil cakes are usually used as organic nitrogenous manures. They can be applied as basal dressing (*i.e.* a few days prior to rowing, at sowing time of the crop) and as a top dressing after the crop has made a certain amount of growth. Oil cakes should be well powdered before application so that the manure is spread uniformly (Daji, 1955).

8. Blood Meal : Blood meal is obtained from slaughter house. It is a very quick acting manures and is effective on all crops and on all types of soil.

Uses : Blood meal is a organic manure of animal origin. It is used as manure and also used as ingredient of poultry feed. It should be applied to crops like oil cakes.

9. Fish Manures : Non-edible fish carcasses of fish and offal are used to prepare fish meal. It is available either as dried fish or fish meal or powder. It contains 4–10 per cent organic nitrogen, 3–9 per cent phosphorus and 0.3–1.5 per cent potassium.

Uses : Fish manure is a quick acting organic manure. It is suitable for application to all crops and on all soils. It is also used as an ingredient of poultry feed.

10. Wood Ash : Wood ash, Cattle dung ash etc. are the indigenous source of potassium. Wood ash is a good source of potassium. But it also contains Nitrogen and Phosphorus.

Uses : Wood ashes are applied in all crops. It is a good manure for onion, garlic etc. Cowdung ashes alone or mixing with kerosine oil are dusted on vegetable crops as repellent for the insects.

11. Sewage and Sludge : Sewage and sludge is the product of sewage system of sanitation, a modern systems of sanitation adopted in cities. In general, sewage has two components as follows :

(*i*) Solid portion, technically known as sludge.

(*ii*) Liquid portion commonly known as sewage water. It is known as treated effluents.

The sludge is collected separately, dried and used as manure.

Uses : Sludge can profitably used as organic manure for producing crop. Sewage water can be used for irrigation. Both sewage and sludge are used in the compost making by activated compost process.

12. Poultry Manure : The excreta of bird and manures obtained from deep litter system of poultry farming are the good source of organic manures.

Uses : Poultry manures are a good organic manure for all crops. It should be applied as basal dressing.

13. Night Soil : Night soil is the human excrement. Dehydration of night soil can be done as such or admixture of absorbing materials such as soil, ash, charcoal and saw dust. Saw dust has high dehydration capacity as well as property of absorbing fuel smell.

Uses : Night soil is a good manure of producing crop. But there is some prejudice of using night soil for crop production. Night soil is used as starter in compost making by Bangalore process.

14. Bio-fertilizers : The term "Bio-fertilizers" includes selective micro organism like algae, fungi and bacteria. This microorganisms are capable of atmospheric Nitrogen or converting insoluble phosphorus into soluble form. Bio-fertilizers are ecofriendly and renewable sources of plant nutrients to supplement chemical fertilizers.

Classification of Bio-fertilizers : The bio-fertilizers can be classified depending on the types of microorganisms as follows:

1. **Algal Bio-fertilizers**

   Example : Azolla, Blue Green Algae (BGA) etc.

2. **Bacterial Bio-fertilizers**

   Example : Azospirillum, Azotobacteria, Phosphobacteria (*i.e.* Phosphorus solubilizing bacteria), Rhizobium etc.

3. **Fungal Bio-fertilizers**

   Example : Mycorrhiza etc.

4. **Actinomycetes Bio-fertilizers**

   Example : Frankia etc.

Uses : Biofertilizers play a vital role in maintaining long term soil fertility and sustainability. Bio-fertilizers are used in different ways in the production of field crops as follows :

(*i*) ***Seed Treatment or Seed Innoculation :*** Bio-fertilizers (*i.e.* Azotobacter, Azospirillum, Rhizobium, Phosphobacteria etc) are used for seed treatment.

(*ii*) ***Dipping of Seedlings :*** The seedlings are dipped in the solution of bio-fertilizers before transplanting. The treated seedlings are then transplanted in the field. Azospirillum is used for dipping of paddy seedlings.

(*iii*) ***Soil Application :*** The bio-fertilizers are mixed with well dried, powdered F.Y.M. (Farm yard manure) or compost and then broadcasted in the field just before sowing of seed or transplanting of seedlings.

6.4 FERTILIZERS

(*a*) Nitrogenous Fertilizers

1. Ammonium Sulphate $(NH_4)_2SO_4$: It is most widely used fertilizers in our country. It is white crystalline salt and contain 20.5–21 per cent ammonical nitrogen. It is quick acting fertilizer with good storage quality. Ammonium sulphate can be applied either at sowing time or as a top dressing *i.e.* during growing period of the crop. It should not be applied along with the seed. It is a suitable fertilizer for all crops and for a wide variety of soils. Ammonium sulphate physiologically is acidic in nature. Acid equivalent is 107 kg of Calcium Carbonate per unit of nitrogen *i.e.* 107 kg Calcium Carbonate is required to neutralize the acidity produce by application of 20 kg. nitrogen as Ammonium sulphate.

2. Ureas $Co(NH_2)_2$: Urea is a white, crystalline product, soluble in water, hygroscopic and it has a tendency of caking. It is highly concentrated nitrogenous fertilizers containing 44–46 per cent organic nitrogen. It is very soluble in water and therefore, subject to rapid leaching. It may be applied at sowing time or as a top dressing. When applied at sowing time, it should not be allowed to come in contact with seed. Now–a–days Urea is being applied as foliar spray. It is suitable for application to all crops and soil when the soil is not wet. Physiologically Urea is a acid forming fertilizer and its acid equivalent is 36 kg $CaCO_3$ per unit of nitrogen.

3. Ammonium Chloride NH_4Cl : It is a white crystalline substances and contain 26 per cent nitrogen. It is very much soluble than ammonium sulphate. In India, it is largely used for industrial purposes. Its behaviour is same as Ammonium Sulphate.

4. Sodium Nitrate $(NaNO_3)$: It is a white crystalline substances and very hygroscopic in nature. It contain 16 per cent nitrogen. It is not suitable fertilizer for waterlogged condition, because large portion of fertilizer is lost by leaching. It is a good for wheat, maize, barley, cotton, sugarcane etc. It is physiologically basic in nature and its basic equivalent is 27 B lb. or calcium carbonate per unit of nitrogen.

5. Ammonium Nitrate (NH_4NO_3) : It is a white crystalline salt, very soluble in water, and hygroscopic in nature. It contains 33–35 per cent nitrogen, half as nitrate nitrogen and half in the ammonium form. It is also a acid forming fertilizers.

6. Ammonium Sulphate Nitrate [$(NH_4)_2SO_4 + NH_4NO_3$)]: It is a mixture of ammonium nitrate and ammonium sulphate. It is available in white Crystalline form or as granular form of a dirty white colour. It contains 26 per cent nitrogen, three forth being in ammonical form and the rest (6.5 per cent) in nitrate form. It is soluble in water and is readily available to crops. It can be applied before sowing, at sowing time or as top dressing. It is useful for all crops and is suitable for application to all types of Soil. It also produces acidity in soil but to a much less extent (about half) than ammonium sulphate.

7. Calcium Cyanamide ($CaCN_2$) : It is a dark grey powdered substances. It is not hygroscopic and therefore easy to store. It is a poisonous substances. So care should be taken during handling. It contain 20–21 per cent nitrogen. The basicity is 57 lb of calcium carbonate per unit of nitrogen.

(*b*) Phosphatic Fertilizers

1. Superphosphate : It is the most widely used phosphatic fertilizer in India. It is a grey ash like powder with good keeping quality. It is manufactured largely by treating ground phosphate rock with almost an equal quantity by weight of sulphuric acid. The fertilizer is manufactured in three grades : *viz.* Single, double and triple superphosphate. At present only single superphosphate is available in the market. Double and triple superphosphate is prepared by reaching rock phosphate with a mixture of phosphoric acid and sulphuric acid respectively. The phosphoric acid in the superphosphate is in the water soluble form which is readily available for nutrition of plants. Although superphosphate readily dissolved in water, it does not wash out from the soil. It is suitable for all crops and can be applied to all soil except acid soil. In acid soil, it is immediately converted into insoluble phosphate due to precipitation as iron and aluminium phosphate. It should be applied before or at the time of sowing.

2. Bone Meal : It is both phosphatic manures and fertilizer. It contains a small quantity of nitrogen also. It is available in two forms. *(i)* raw bone meal and *(ii)* steamed bone meal. Bone meal having particles not large than 3/32 inch in size is considered suitable for the use as fertilizers. Bone meal may be applied to the soil either at sowing time or just before it. It is a most suitable fertilizers for acid soil and is considered a safe manure for all crops.

(*c*) Potassic Fertilisers

1. Muriate of Potash or Potassium Chloride (KCl) : It is a white or reddish to white crystalline or powdered materials. It contains 50–60 per cent potash, the whole of which is readily available. It is soluble in water but it is not lost from the soil as it is absorbed by the soil particles. It can be applied at sowing time or prior to sowing. Now a days, it is being applied as top dressing.

Table 6.2 : Nutrient Contents of Different Fertilizers (In Percentage)

Name of Fertilizers	*Nitrogen*	*Phosphorus*	*Potassium*	*Lime (Cao)*	*MgO*	*Equivalent acidity*	*Equivalent basidity*
1	*2*	*3*	*4*	*5*	*6*	*7*	*8*
1. **Nitrogenous Fertilizers**							
Ammonium Sulphate	20.6	–	–	–	–	110	–
Ammonium Chloride	25.0	–	–	–	–	148	–
Ammonium Sulphate nitrate	26.0	–	–	–	–	93	–
Calcium Ammonium nitrate	25.0	–	–	10-20	7-7.5	–	–
Ammonia anhydrous	82.0	–	–	–	–	148	–
Sodium Nitrate	16.0	–	–	–	–	80	–

1	*2*	*3*	*4*	*5*	*6*	*7*	*8*
Urea	44-46	–	–	–	–	80	–
2. **Phosphatic Fertilizers**							
Super Phosphate (Single)	–	16-20	–	25-30	0.5	–	–
Super Phosphate (Double)	–	35-38	–	–	–	–	–
Super Phosphate (Triple)	–	46-50	–	17-20	0.5	–	–
Basic Slag	3.0-8.0	–	–	–	–	–	–
Bone meal (raw)	3.0-4.0	20-25	–	–	–	–	–
Bone meal (steamed)	1.0-2.0	25-30	–	–	–	–	–
Rock Phosphate							
(*i*) Mussouri Phos	–	23.0-24.0	–	–	–	–	–
(*ii*) Purulia Phos		23.0					
3. **Compound and Complex Fertilizer**							
Monoammonium Phosphate	11.0	48.0	–	–	–	–	–
Diammonium Phosphate (D.A.P.)	18.0	46.0	–	–	–	–	–
Ammonium Phosphate (Paramphosh)	16.0	20.0	–	–	–	–	–
Suphala 20:10:0	20.0	20.0	–	–	–	–	–
Suphala 15:15:15 (Nitro-phosphate potash)	15.00	15.0	15.0	–	–	–	–
Gromor 28:28:0	28.0	28.0	–	–	–	–	–
(Urea ammonium Phosphate)							
IFFCO 10:26:26	10.0	26.0	26.0	–	–	–	–

2. Potassium Sulphate or Sulphate of Potash K_2SO_4 : It is a another potassic fertilizer made by treating potassium chloride with magnesium sulphate. It contain 48 to 52 per cent potash. It dissolves readily in water and becomes available to the crop almost immediately after application. It can be applied at any time upto sowing. It is considered better than Muriate of potash for crops the tobacco, fruit trees and potato etc.

Mixed Fertilizer : Mixed fertilizer is one which contain two or three fertilizers and available in the market in a particular grade (*e.g.* Grade – 1–8: 8: 8). The fertilizer grade refers to the minimum guarantees of plant nutrient content in terms of total nitrogen, available phosphoric acid and water soluble potassium.

We can make the fertilizer mixture by adopting the following formula :

Calculation :
$$\frac{\text{Amount of nutrient in mixed fertilizer} \times \text{Amount of mixture to be prepared.}}{\text{Amount of nutrient in the ingredient.}}$$

Example : Preparation of fertilizer mixture of 1000 kg.

Grade – 5: 5: 10

Ammonium Sulphate : $\frac{5 \times 1000}{20} = 250$ kg.

Super Phosphate : $\frac{5 \times 1000}{16} = 312.5$ kg.

Muriate of Potash : $\frac{10 \times 1000}{50} = \frac{200.0}{762.5}$

Filler : 237.5

If the fertilizers is used immediately after preparation, no filler is added.

(N.B. Filler is "make weight materials" added to the mixed fertilizers or fertilizers material. It is added to make up the difference between weight of the added ingredient required to supply the nutrient in a ton of given analysis).

One fertilizer can not be mixed with other fertilizer. Because uneven mixing of incompatible fertilizers leads to a loss of some of the fertilizer nutrients in the form of gas, convert soluble nutrients into insoluble ones or produces caking. Certain fundamental principles are to be followed in mixing fertilizer such as :

1. Ammonium Sulphate, Ammonium Chloride and other ammonical fertilizers and nitrogenous organic manures should not be mixed with lime.
2. Calcium cyanamide, Basic slag, quick lime, slaked lime should not be mixed with fertilizers containing nitrogen in ammonical form.

3. Urea should not be mixed with superphosphate and superphosphate should not be mixed with lime or Calcium Carbonates or wood ashes.
4. Sodium nitrate or potassium nitrate should not be mixed with superphosphate.
5. Ammonium sulphate nitrate should not mixed with lime.

6.5 METHODS OF FERTILIZER APPLICATION

The fertilizers are generally applied to crop by different methods as follows :

A. Application of Fertilizers in Solid Form

1. Broad Casting : The fertilizers are spread by hand in the soil with last preparatory tillage just before sowing of seed or planting of seedlings. This method is known as "broadcast application" there are two types of broadcasting method of fertilizer application depending on the time of application as follows :

(*i*) ***Broad Casting at Planting :*** The fertilizers are broad casted to soil just before planting. The applied fertilizers are incorporated to soil by ploughing followed by planting.

(*ii*) ***Top-Dressing :*** The method of application of fertilizer in standing crop is known as 'Top–dressing' generally nitrogenous and potassic fertilizers are applied to crop as top dressing.

2. Placement : It is a method of placing fertilizer in the soil before sowing or after sowing the crops. The followings are the common methods of this category.

(*i*) ***Plough Sole Placement :*** The fertilizers are placed in the plough sole after opening the furrow by plough and these furrows are covered immediately as the next furrow it is turned.

(*ii*) ***Deep Placement :*** The nitrogenous fertilizers is placed deep in reduced layer to check denitrification.

(*iii*) ***Sub-Soil Placement :*** The phosphatic and potassic fertilizers are placed in the sub-soil with the help of heavy machinery to avoid their fixation in strongly acidic soil.

3. Localised Placement : It is a method of placing fertilizer into the soil close to seed or plant. The roots of young plant can get nutrition as per their requirement from the fertilizer applied by this method. This method is usually employed when relatively small quantity of fertilizer is to be applied. Localised placement reduces fixation of phosphorus and potassium (Yawalker *et. al.*). It is two types as follows :

(*i*) *Contact Placement or Combined Drilling or Drill Placement :* Contact placement refers to the drilling of seed and fertilizer together during sowing. The phosphatic and potassic fertilizer in small quantities are applied to cereal crops and cotton. But this method is not suitable for pulse crop. Seed-cum-fertilizer drill is used for such placement. In the absence of seed-cum-fertilizer drill, small doses of granular fertilizer may be mixed with seed and drilled together. However where using higher dose of fertilizer, adoption of this practice, even under irrigated condition, will not advisable (Sharma *et. al.*, 1980).

(*ii*) *Band Placement :* In this method the fertilizers are placed in bands on one side or both sides of the row at about 5cm away from the seed or plant in any direction. Band placement are of two types as follows :

(*a*) *Hill Placement :* The fertilizers are placed in bands on one or both side of wide spaced plant such as cotton, castor, cucurbitaceous crops and fruit crops. Nitrogenous and phosphatic fertilizers are applied by this method. This method is suitable for applying small quantities of fertilizer.

(*iii*) *Row Placement :* The fertilizers are placed on one or both sides of the row in continuous band by hand or seed drill. This method is practised in row crops (*e.g.* Sugarcane, Potato

Tobacco, Cotton and Vegetables) with wide space between rows. This method is suitable for applying large quantities of fertilizers.

(iv) Side Dressing : In this method, fertilizers are applied along the side of row or around the plant and mixed them with the soil by Spade or Khurpi or Nirani. It is of two types as follows :

(*a*) *Spot Placement :* The fertilizers are applied in spaces in between the plants and mixed them with soils by Khurpi or Nirani. This method is employed for applying fertilizer in vegetable crops.

(*b*) *Basin Placement :* The fertilizers are applied in the shallow basin already made around the base of the fruit trees. The applied fertilizers are mixed with the soil by Spade, Khurpi or Nirani.

4. Pellet Application : Small pellets of convenient size are made after mixing the nitrogenous fertilizers specially urea with the soil and they are applied one to two inches deep between the rows of paddy crop. The pellets are deposited in soft mud of paddy field. This method of fertilizer application decreases the nitrogen loss through leaching or by run off of water. Paddy response well to pellet application of Nitrogenous fertilizers compared to top dressing which is the most common method followed by cultivators. This is the findings of field experiment at C.R.R.I. cuttack.

B. Application of Fertilizers in Liquid Form

Liquid form of fertilizers are applied by the following methods :

(*a*) Starter Solution : Starter Solution is prepared by mixing N.P. and K. fertilizer in the ratio of 1:2:1 or 1:1:2. This is applied to vegetable seedlings at the time of transplanting and it helps rapid establishment and quicks early growth of seedlings.

(*b*) Foliar Application : This is a method of spraying on leaves of growing plant with suitable fertilizer solution having low concentration. Application of one fourth to one third of

Nitrogen through foliage has been found to increase wheat and rapeseed yield to the extent of 10–15 per cent under unirrigated condition and over drilling the entire dose of nitrogenous fertilizer at sowing of seed. Deep placement of super granule urea (S.G.W.) resulted in one third saving in Nitrogen dose in Paddy over broadcasting urea of ordinary grade (Sharma *et. al.*, 1980).

There are certain difficulties associated with foliar application of nutrients as enumerated by Yawalker *et. al.*, 1977. These are as follows :

(*i*) Marginal leaf burn or scorching may occur if strong solution are used.

(*ii*) As solution of low concentration (usually 3–6 per cent) are to be used, only small quantities of nutrients can be applied in one single spray.

(*iii*) Several applications are needed for moderate to high fertilizer rates.

(*iv*) Foliar application of fertilizer is costly compared to soil application, unless combined with other spraying operation taken up weed, insect or disease control.

(*c*) Direct Application to the Soils : Liquid fertilizers such as Anhydrous Ammonia and Nitrogen solution are directly applied to the soil with the special injection equipment. Liquid manures such as urine, sewage water and cowshed washing are let into the field (De, 1990).

(*d*) Fertigation : Fertilizers, both straight and mixed fertilizers are easily soluble in water. They are allowed to dissolve in irrigation water and applied to the soil through irrigation water. Nitrogenous fertilizers are generally applied through irrigation water.

6.6 TIPS TO GET BEST EFFICIENCY OF APPLIED FERTILIZERS

1. The fertilizer scheduling must be based on soil test.
2. Selection of fertilizers should be done according to the soil reaction *viz.* acidic fertilizer for alkaline soils and basic fertilizers for acidic soils.

3. Surface application through broadcasting should not be adopted, but fertilizer should be placed about 3–4 cm by the side or below the seed. This discourages weed growth also.

4. The phosphorus and potassic fertilizers should be basal dressed, because P_2O_5 and K_2O are not lost from the soil but they are absorbed by soil particles. Their poor mobility restricts them to the place of application, therefore, they must be placed in the root zone.

5. Home mixing of fertilizers should be in accordance with fertilizer mixing guide and such fertilizer mixture must be applied as soon as possible.

6. In case of heavy soil type, half of nitrogenous fertilizers should be basal placed and rest half should be top dressed in one split only, but in case of light soil, Nitrogen should be applied in three equal splits *i.e.* 1/3 as basal dressing, 1/3 after 30 days of sowing and rest 1/3 about 50–60 days after sowing.

7. Flooding with too deep water or poor drainage should be avoided after application of fertilizers at least for a week time.

8. Top dressing should be done after draining out the water and weeding so that loss of nutrient's is minimum.

9. The Paddy field, used for transplanting should be puddled, and fertilizers should be applied at the time of puddling, because this will help the fertilizers to reach and get store in reduced zone of the soil.

10. Light sandy, calcarious and soils under very high cropping intensity become deficient in micronutrients like zinc and sulphur. The deficient plants become sicky, and cannot absorb nutrients, thus the fertilizer is not absorbed, therefore such soils must be supplied with Zinc sulphate ($ZnSO_4$) at the rate of 10–15 kg/ha 2–3 years.

11. The acidic soils should treated with liming materials as and when required.

12. Deep placement of fertilizer along with foliar feeding of Nitrogen (*i.e.* urea) through spraying of Nitrogenous fertilizer in place of top dressing should be done at least once in 3–5 years of time.

13. Weed growth should not be permitted in cropped area during any part of the year.

14. In case of flooded soil or calcarous soils, used of slow release Nitrogenous fertilizers like U.F.-30, Sulphur coated urea, Super granules, Neem coated or neem blended urea should be done, so that loss of Nitrogen can be minimised.

16. Mud bolls contains urea, should be used in case of deep water crops, because it helps in proper placement and also reduces the loss of Nitrogen from the field.

17. An appropriate plant protection measures and proper tillage practices should be adopted, so that plants remain healthy and absorb the applied nutrients from the field. (*Source* : Principles and Practices of Agronomy (1988), by S.S. Singh (p. 81–82).

REFERENCES

Balasubramaniyan P. and S.P. Palaniappan, Principles and Practices of Agronomy (2004). Agrobios (India) Jodhpur.

Das P.C. Manures and Fertilizers (1999), Kalyani Publishers, Ludhiana.

D.C. Gopal Chandra, Fundamentals of Agronomy (1990), Oxford and IBH Publishing Co. Pvt. Ltd., New Delhi.

Gupta, P.K., A Hand Book of Soil, Fertilizers and Manures (2004), Agrobios (India), Jodhpur.

Gustafson A.F. Hand Book of Fertilizers (2003), Agrobios (India), Jodhpur.

Rai M.M. Soil Science, plant chemistry, manures and fertilizers (1985), Ramprasad & Sons, Agra–3.

Reddy S.R. Principles of Agronomy (1999), Kalyani Publishers, Ludhiana.

7

WATER MANAGEMENT OF CROPS

7.1 INTRODUCTION

Water is one of the most important inputs essential for production of crops. Water plays the major role for successful crop production. Water is essential for the maintenance of turgidity, absorption of nutrients and the metabolic process of plants. Plants need it in huge quantities, continuously throughout their life. Water management is the planned use of water for better utilisation in agriculture. It includes irrigation and drainage (De, 1990). It is essential to maintain ideal soil moisture level for better growth, yield and quality of the crops.

7.2 ROLE OF WATER FOR GROWTH AND DEVELOPMENT OF PLANTS

(*i*) Water is a constituent of protoplasm.

(*ii*) Water acts as a Solvent. The plant can absorb nutrients when these nutrients are dissolved in water. Water is essential to make the nutrients soluble for absorption by plants.

(*iii*) Water is used for transportation of nutrients from the soil to green plant tissue where they are used for Photosynthesis and the carbohydrate, the end products of Photosynthesis, are conveyed in water solution to the different parts of the plants.

(*iv*) Water forms over 90 per cent of plant body weight when they are green.

(*v*) Plant can synthesize food through Photosynthesis only in presence of water in their system.

(*vi*) Water helps in maintaining turgidity of cell walls of plants. Water helps in cell enlargement due to turgor pressure and cell division which ultimately increase the growth of plant.

(*vii*) Water is essential for the germination of seeds, growth of plants roots, and nutrition and multiplication of soil organism.

(*viii*) Water is essential in the hydrolytic process in the plant. It helps in the conversion of starch to sugar.

(*ix*) Water helps in the transpiration process which is very essential for maintaining the absorption of nutrients from the soil.

(*x*) Water helps in chemical, physical and biological reaction in soil.

(*x*) Water cools the plant and prevents the injury due to high temperature.

7.3 SOIL WATER

(A) Physical Classification of Soil Water

This classification is based on moisture coefficient. Such as:

1. Gravitational Water : It is a free water in excess of capillary water. Gravitational water may be defined as the quantity of water removed from the soil by the forces of gravity. This water saturates the soil and usually percolates downward to the water table under the influence of gravity. Gravitational water results in the loss of plant nutrients from the soil. This type of water is not available to plants. Gravitational water remains in macropores and is held at a tension of 1/3 atmosphere or less.

2. Hygroscopic Water : Hygroscopic water may be defined as the percentage of water occurs as a thin film (4–5 μ thick; 1μ or micron = 0.0001 millimeter) surrounding the particles of soil or organic matter. It is absorbed by vacuum dried soil from an atmosphere of water is held with a tension of 31 atmosphere or greater and it can be driven out from a soil by heating the soil at a temperature of 100°C to 110°C for 8–10 hours. This water is not available.

3. Capillary Water : Capillary water is held by the soil due to capillary forces. It is available form of water and is held between field capacity (0.33 atmosphere) and hygroscopic co–efficient (31 atmosphere) in micropore spaces. The availability of capillary water depends on the following factors–Soil texture, soil structure, and humus content of the soil.

(B) Biological Classification of Water

1. Available Water : It is the quantity of water present in the soil from the field capacity to permanent wilting point. This water is available to plant for its growth and development. The amount of available water that can be stored in the root zone of the soil depends on its depth, soil type and soil structure.

2. Unavailable Water : The soil water below the wilting co-efficient is not available to plants and is termed as unavailable water. It is held by the soil particles so tenaciously that plants can not extract it from the soil.

3. Superfluous Water : It may be defined as the quantity of water in between maximum retentive capacity and field capacity. Gravitational water is known as superfluous water. This water is subjected to gravitational forces and is removed easily from the soil. It has no use for the plants.

7.4 SOIL MOISTURE CONTENTS

1. Maximum Retentive Capacity or Maximum Water Holding Capacity : It may be defined as the percentage of moisture in the soil when all the porespaces are completely filled with water. The cropped soil remaining at this stage for a long period suffers from good aeration and this condition is consequently harmful to most crop plants.

2. Field Capacity : The maximum amount of capillary water remaining in a soil after the removal of gravitational water is called its field capacity. Thus the field capacity may be defined as the percentage of water that soil can retain against gravity when adequate drainage is provided.

3. Hygroscopic Co-efficient : It may be defined as the percentage of moisture in the soil when the soil is kept in an atmosphere having 100 per cent relative humidity. Dry soil kept in an open place absorb water vapour from the atmosphere. The water absorbed in this way is called hygroscopic water.

4. Wilting Co-efficient : It may be defined as the percentage of moisture in the soil at which the soil can not supply water to the growing plant at a sufficient rate to maintain their turgidity and the plants wilt permanently.

The difference between the field capacity and the wilting co-efficient is the maximum available water wilting co-efficient may be calculated as under.

$$\text{Wilting Co-efficient} = \frac{\text{Hygroscopic Co-efficient}}{0.68} \text{ or}$$

$$\frac{\text{Moisture equivalent}}{1.84}$$

5. Transpiration Ratio : The ratio of the quantity of water used to the quantity of dry matter produced is called transpiration ratio. This ratio varies from 200 to 500 in humid region to twice that in arid region. Abundance of soil moisture, dry atmosphere, high temperature and strong winds increase transpiration ratio (Prasad, 1967). Crops differ in transpiration ratio as mentioned below :

High Transpiration Ratio : Rice, Sugarcane, Potatoes and Banana.

Medium Transpiration Ratio : Wheat, Barley, Linseed and Cotton.

Low Transpiration Ratio : Sorghum and millets.

It can be expressed in the following manner :

Transpiration Ratio :

$$\frac{\text{Quantity of water transpired by the crop}}{\text{Quantity of the produce}}.$$

Crops having low transpiration ratio are suitable for cultivation in dry soil and low rainfall tracts.

Soil Moisture Deficit and Plant Growth

(*i*) Plant fails to absorb nutrient if soil moisture become deficit.

(*ii*) Decrease in soil moisture, decreases the absorption of water and the opening of stomata decreases and the stomata begins to close much earlier than the permanent wilting point of the soil.

(*iii*) The entry of carbondioxide is decreased if the stomata closes due to lack of absorption of water and the photosynthesis is reduced due to decrease in entry of carbondioxide.

(*iv*) The rate of photosynthesis will also decrease due to dehydration of protoplasm as a result of decrease in soil moisture. Dehydration of protoplasm reduces the photosynthetic capacity of a plant.

(*v*) The respiration will increased with the decrease of the soil moisture. But lastly the respiration will decrease because of the limitation of respiratory substrate.

(*vi*) Net growth of the plant will also decrease due to the deficit of soil moisture.

Net growth = Photosynthesis – Respiration.

(*vii*) Under deficit soil moisture condition, the maturity of plant is hastened and the succulency is decreased. But the protein content in grain and other plant parts is increased.

Table 7.1 : Practical Guide for Soil Moisture

Available Moisture in Soil	*Coarse Textured Soil*	*Moderately Coarse Textured Soil*	*Medium Textured Soil*	*Fine and Very Fine Textured Soil*
75 per cent of field capacity.	Stick together slightly, may form a week ball under pressure.	Form a week ball, that break easily, does not sticks.	Form a ball, very pliable, sticks readily if relatively high in clay.	Ribbon's out between finger easily, has a sticky feeling.
At field capacity (100 per cent)	On squeezing no free water appear on soil but wet line of a ball is left on hand.	Same as for coarse textured soil at field capacity.	Same as for coarse textured soil at field capacity.	Same as for coarse textured soil at field capacity.
Above field capacity.	Free water appears when boil bounced at hand.	Free water is released with kneading.	Free water is squeezed out.	Puddle, free water forms on surface.
0 per cent	Dry, 100 sc and single grained, flows through fingers.	Dry, and loose, flows through fingers.	Powdery dry, in some places slightly crushed but breaks down easily into powder.	Hard, baked, and cracked, has loose crumbs on surface in some places.
50 per cent or less.	Appears to be dry, does not form ball under pressure.	Appears to be dry, does not form ball under pressure.	Some what crumbly but hold together under pressure.	Some what pliable balls under pressure.
50–75 per cent.	Appears to be dry : does not form a ball under pressure.	Forms a ball under pressure but seldoms hold together.	Form a ball under pressure, some what plastics, sticks slightly, under pressure.	Forms a ball, ribbons out between thumb and fore–finger.

Source : Hand Book on Irrigation and Water Management (1971), Water Management Division, Min. of Agri., New Delhi.

(*viii*) The sugar content of a sugarcane crop is increased when the irrigation is withheld prior to harvesting of the crop.

Forces of Retention of Moisture in Soil

(*i*) ***Adhesive Force :*** It is the attraction of force between soil particles and water molecules.

(*ii*) ***Cohesive Force :*** It is the attraction between the molecules of water.

(*iii*) ***Imbibitional Force :*** The soil colloids absorb water and they swell due to the absorption of water. This is the process of imbibition.

Expression of the Forces of Retention

Soil Moisture Tension : It may be defined as the total force with which water is held in the soil. It is expressed in atmosphere.

1. Atmosphere = 14.7 lbs/sq. inch

   = 1036 cm height of water column.

2. *PF* : PF is the logarithmic values of soil moisture tension expressed in centimeters height of a water column. Soil moisture is commonly expressed in terms of P.F. values, whose numerical values are the same as pH values denoting hydrogen ion concentration. The PF value varies according to soil condition.

Soil condition :	PF value
Oven dry soil :	7.5
At field capacity :	2.5
At hygroscopic point :	4.5
At wilting point :	4.2

1 atmosphere = 1036 cm

= 1000 cm (approx)

= 10^3

1. Atmosphere = 3 $\log^{10}$

= 3 PF.

3. Atmospheric Tension

(*i*) Maximum retentive capacity : 0

(*ii*) Field capacity : 0.1 – 0.3 atmosphere

(*iii*) Permanent wilting point : 15 atmosphere

(*iv*) Hygroscopic co-efficient : 31 atmosphere

7.5 IRRIGATION

Plants need water for their growth and development. Plants must be supplied adequate amount of water for successful crop production.

Hansen *et. al.*, (1980) defined irrigation as artificial application of water for the purpose of supplying moisture essential to plant growth. They further stated that irrigation is the application of water to the soil for other purposes as follows:

(*i*) To add water to the soil for supplying the moisture essential for plant growth.

(*ii*) To provide crop insurance against short duration droughts.

(*iii*) To cool the soil and atmosphere, thereby making more favourable environment for plant growth.

(*iv*) To wash out or dilute salt in the soil.

(*v*) To reduce the hazard of soil piping.

(*vi*) To soften tillage pan.

Simply irrigation is the artificial application of water in the soil for the normal growth and development of the plants and their production.

Table 7.2 : Sourcewise Calculation of Total irrigated Land of India

Medium of irrigation	*Irrigated area (in percentage)*
Well and tubewell	38
Water reservoir	18
Canal	40
Others	4
Total	**100**

Table 7.3 : Water Wealth of India

Item	*Million Ha. metre*
Total annual precipitation	400
Annual evaporation loss	70
Annual surface runoff	115
Annual ground water recharge	67
Total annual soil moisture	170
Ultimate irrigated area by 2025 AD	110 m.ha.

Source : Agri water Management (1991) by R.A. Singh and S.R. Singh, Tara Book Agency, Varanasi.

Table 7.4 : Water Holding Capacity of Different Soil

Soil type	*Field Capacity*	*Permanent wilting point % (d.p.)*	*Depth of available water per unit of soil (Cm/m depth of soil*
Fine sand	3–5	1–3	2–4
Sandy loam	5–15	3–8	4–11
Silty loam	12–18	6–10	6–13
Clay loam	15–30	7–16	10–18

Table 7.5 : Approximate Amount of Irrigation to be Added to the Soil During each Irrigation (cm)

Depth of irrigation	*Sandy soil*	*Loamy soil*	*Clay soil*
Shallow upto 60 cm	2.5–5.0	5.0–7.5	7.5–10.0
Medium between 60–90 cm.	5.0–7.5	10.0–15.0	15.0–20.0
Depth more than 90 cm.	10.0–15.0	20.0–25.0	25.0–35.0

Source : Irrigation – Theory and Practice (1981), by A.M. Michal, Vikas Publishing House, New Delhi.

7.5.1 Objective of Irrigation

Following are the objectives of irrigation :

(*i*) To supply the water for the nutrition and growth of plant. Irrigation is essential for crop production.

(*ii*) To hasten the decomposition of organic matter in the soil and to keep the activity of soil microbes at the optimum level.

(*iii*) To remove the stress condition of plants.

(*iv*) To release nutrients in soil solution for absorption by plants.

(*v*) To leach or wash out injurious salt from the soil.

(*vi*) To maintain the temperature and humidity of the soil.

(*vii*) To control the harmful soil insects.

(*viii*) Irrigation is necessary to reclaim fallow land and to bring soil condition for tillage.

7.5.2 Necessity of Irrigation

Plants must be supplied with water as required by them for successful crop production. Irrigation is essential for crop production due to the following reasons.

(*i*) **Uncertainty of Monsoon Rainfall :** Monsoon winds causes 80% rainfall in India. Monsoon rainfall is very uncertain. So irrigation is very important to supply water to plants as and when needed.

(*ii*) **Uneven Distribution of Rainfall :** Rainfall is uneven in all the places of India. Irrigation is essential for crop production in an area having low and uncertain rainfall.

(*iii*) **Effect of Winter Rainfall :** Rainfall is a rare in winter season in our country. Rabi crops need much water for their growth and development. Irrigation is very important for the cultivation of rabi crops.

(*iv*) **Cultivation of High Yielding Crops :** The high yielding crops are being cultivated in our country which requires sufficient water for their growth and development. Irrigation has the key role for the successful cultivation of high yielding crops.

(*v*) **Difference in Water Holding Capacity of Soil :** There are different types of soil in our country having difference in water holding capacity. Irrigation must be provided depending on the type of soil.

7.5.3 Sources of Irrigation

Rainfall are the primary sources of water for irrigation. But it is not available throughout the year in all arable areas and in appropriate quantities required by each crop variety at the different stages of its growth. Rainfall reserved in other water bodies is used as resources of irrigation water for an indefinite time. Followings are the sources of irrigation water :

1. Surface Water : Surface water is provided by flowing water of river or from still–waters of tanks, ponds, lakes, ponds or artificial reservoirs such as dams, barrage and diversionary bunds with different storage capacity. Irrigation through canals under different river valley project in our country is done profitability as the water from the river is carried to the field by flow due to gravity and is therefore cheap. Water from other sources is carried to the field by lifting them with the help of water lifting devices used for irrigation. River lift irrigation project is also functioning in our country. The water of rivers or streams is lifted to irrigate a limited command area adjacent to such water courses.

2. Ground Water : Water is reserved underground enriched by percolation of the surface water through the porous strata of earth's crust and this type of water is known as ground water. The subterranean water is tapped by digging or drilling wells. It may be shallow (30–40 feet) or deep (85–100 feet or more) The water from wells is lifted before it can be used for irrigation. The command area for irrigation of shallow tubewells and deeptube wells is 10–12 acres and 100 acres respectively.

7.5.4 Methods of Scheduling Irrigation to Crops

Methods of scheduling irrigation to crops may be decided on the basis of soil, crop and its growth stage and atmospheric condition (*i.e.* climate).

1. Soil : Soil moisture have the influence on physiological process, growth and development of plants and these can be affected as the soil moisture content decreases from field capacity towards permanent wilting point. Irrigation should be scheduled when the soil temperature is moderate. So it is better to irrigate the field in hotter season during late afternoon or at night. The plant and soil get sufficient time to absorb water before dry and desiccating weather prevails the next day. Irrigation given in hot soil causes the greater loss of water through evaporation. On the other hand, water should be applied during morning hours in cold season when the soil starts warming up.

Table 7.6 : Optimum Soil Moisture Tension for Irrigating Different Crops

Crops	*Optimum Soil Moisture Tension*
Potato, Berseem, Cabbage and Cauliflower	0.30 atmosphere
Jute, Maize and Mustard	0.60 atmosphere
Sugarcane	0.70 atmosphere
Cotton	1.00 atmosphere
Wheat	0.50 atmosphere
Paddy	0.5 cm standing water.

2. Crop : Crops have different stages of growth and they require sufficient moisture in the soil at different growth stages for their proper growth and development. This stage is known as the critical stages of water requirement. Critical stages are those stages of growth of a crop when deficit of soil moisture decreases its growth, development and ultimately the yield is reduced appreciably. Water requirement of a crop varies with the stages of its growth.

Table 7.7 : Critical Growth Stages of Crops

Crop	*Critical stages*
Rice	(*i*) *Seedling stage.*
	(*ii*) Maximum tillering stage.
	(*iii*) Panicle initiation stage.
	(*iv*) Flowering stage.
	(*v*) Soft dough or milk stage.
Wheat	(*i*) Crown root initiation stage (18–22 days after sowing).
	(*ii*) Tillering stage (40–45 days after sowing).
	(*iii*) Flowering stage (70–75 days after sowing).
	(*iv*) Milk and dough stage (90–95 days after sowing).
	Besides these jointing and boot stage is critical.
Maize	(*i*) Seedling stage (Two weeks after sowing).
	(*ii*) Tasseling stage (8 weeks after sowing).
	(*iii*) Silking stage (10 weeks after sowing).
	(*iv*) Cob development stage (12 weeks after sowing).
Sorghum	(*i*) Seedling stage (2–4 weeks after sowing).
	(*ii*) Pre–flowering stage (12–14 weeks after sowing).
	(*iii*) Flowering stage (15-16 weeks after sowing).
	(*iv*) Grain formation stage (17 weeks after sowing.)
Barley	(*i*) Early tillering stage.
	(*ii*) Flowering stage.
	(*iii*) Grain filling stage.
Oat	(*i*) Ear emergence stage.
	(*ii*) Grain filling stage.
Mustard and Rape seed	(*i*) Pre-flowering stage.
	(*ii*) Pod filling stage.
Sunflower	(*i*) Batoning stage.
	(*ii*) Grain filling stage.
Ground nut	(*i*) Emergence stage.
	(*ii*) Pegging stage.

Crop	*Critical stages*
	(*iii*) Pod development stage.
Linseed	(*i*) Pre-flowering stage.
	(*ii*) Capsule development stage.
Potato	(*i*) Pre-emergence stage.
	(*ii*) Stolonization stage.
	(*iii*) Tuberization stage.
	(*iv*) Bulking stage.
Pulses	(*i*) Pre-flowering stage.
	(*ii*) Post flowering or Pod flowering stage. (Excess soil moisture at flowering stage of gram and lentil is detrimentral as it induces shedding of flowers. So irrigation at flowering stage should be avoided)
Sugarcane	(*i*) Emergence stage.
	(*ii*) Tillering stage.
	(*iii*) Growth stage.
Cotton	(*i*) Branching stage.
	(*ii*) Flowering stage.
	(*iii*) Boll formation stage.

3. Soil and Climate : Irrigation is scheduled on the basis of I.W. (*i.e.* depth of irrigation water) and C.P.E. (*i.e.* cumulative Pan evaporation value) ratio. A ratio of 0.9 with 6 cm. IW and 0.4 with 6 cm. IW is optimum for wheat and gram respectively.

7.5.5 Water Requirement of Crops

Water requirement may be defined as the ratio of water transpired during its growth to the weight of dry matter produced in the plant. The ratio varies for different crops. Besides water is also lost from the soil surface by evaporation. So the absolute water requirement may be termed as consumptive use of water. Consumptive use of water may be defined as the quantity of water in hectare centimeter utilised by the crop in crop season together with evaporation, from the field where the crops is grown, transpiration and metabolic activities of plant. For all practical purposes, the total water requirement of the crop is equal to consumptive use of water plus water lost at various stage such as water lost during distribution, surface run-off, percolation etc.

Water requirement and irrigation requirement of crop are not same. Irrigation requirement may be defined as the quantity of irrigation water to be applied to the filed for raising a successful crops whereas water requirement may be defined as the total quantity of water required to raise a crop successfully.

Irrigation requirement = WR – ER – GWC – SMS

(where ER = Water requirement.

ER = Effective rainfall.

GWC = Ground water distribution.

SMS = Soil moisture storage)

Sometimes, water requirement is also called as transpiration ratio.

Factor Affecting Water Requirement of Crops

1. **Season :** Water requirement varies with season. Paddy grown in different seasons have different water requirement.

2. **Crop :** Water requirement varies with crops as different crops have different water requirement. Water requirement is also varied with the variety of crops. Long duration variety of crop has more water requirement than short duration variety of the same crop. Water requirement also varies with the stage and duration of the crop.

3. **Soil :** Water requirement varies with the soil type. Water requirement for sandy soil is more than the clay soil. Level lands require less water than unlevel lands. Soil having impervious layer in root zone prevents the percolation loss and this helps to reduce water requirement.

4. **Depth of Irrigation Water :** When more water is applied in the field than necessary, the excess water will be lost

as deep percolation and this will increase the water requirement of a crop so it is essential to maintain the proper depth of irrigation water.

5. **Method of Irrigation :** Wastage of water depends on the method of irrigation. It is more in case of surface method of irrigation as compared to sprinkler method of irrigation. So water requirement for surface method of irrigation is more than sprinkler irrigation.

Table 7.8 : Water Requirement of Different Crops

Crop	*Growth period (days)*	*Water requirement (Hectare centrimetre*	*Daily water requirement (Hectare centimetre)*
1. Paddy			
(*a*) Aus	90–100	30–35	0.43
(*b*) Aman	110–120	40–50	0.50
(*c*) Boro	120–150	60–70	0.60
2. Wheat	90–100	14–15	0.17
3. Sugarcane	365	95	0.26
4. Jute	90–100	20–24	0.18
5. Potato	88–100	25–36	0.30
6. Mustard	88–90	10–12	0.12
7. Groundnut	120–125	26	0.21
8. Maize	100	18	0.18
9. Pea	88	12	0.14
10. Chilli	200	38–39	0.19
11. Onion	90	15–20	0.15
12. Cotton	190–200	42	0.20
13. Berseem	175	45	0.20

7.5.6 Proper Utilization of Irrigation Water

(*i*) Major quantity should be applied on the basis of soil moisture deficit and irrigation efficiency. Application of more quantity of water than this calculated value will lead to wastage of irrigation water.

(*ii*) Utilization of irrigation water will be better, if the water is applied when the crops have high demand of water.

(*iii*) The irrigation water should be provided into a suitable channel which is made according to the slope of the land.

(*iv*) Proper banding around the plots is essential to reduce run off loss of water.

(*v*) The land should be levelled properly to reduce water requirement.

(*vi*) Crop rotation ensure better utilization of irrigation water.

(*vii*) Wastage of irrigation water is minimum when it is applied in a field near to the source of irrigation water.

(*viii*) Pucca or concrete irrigation channel is ideal for preventing wastage of water.

(*ix*) Plant can do better use of irrigation water when it is applied at proper time with suitable amount. If excess water is applied, it is lost from the root zone by downward percolation. The loss represents the loss of irrigation water.

(*x*) Proper utilisation of irrigation water is possible if the improved method of cultivation (*i.e.* application of organic manures, control of weeds etc.) is followed.

(*xi*) Irrigation efficiency depends on the texture of soil. Irrigation efficiency of different soil is as follows :

Sandy soil – 60 per cent

Loamy soil – 70 per cent

Clay-loam soil – 75 per cent

Clay soil – 40 per cent.

7.5.7 Quantity of Irrigation Water

Irrigation aids in the growth and development of crops. The application of required quantity of irrigation will ensure optimum growth of the plant. In addition leaching losses of materials will be reduced. There are some factors that effects the quantity of irrigation to be applied as follows :

1. **Physical Property of Soil :** Moisture retention capacity of soil depends on the type of soil. The moisture retention capacity of the clay soil is more than loam and sandy soil. For this reason, less water is required for clayey soil.
2. **Available Moisture in the Soil :** The quantity of irrigation water to be applied depends on the moisture context of the soil. Dry soil requires more irrigation.
3. **Crops :** Water requirement vary according to the type of the crop. Rice needs more water than other crops.
4. **Doses of Fertilizers :** The quantity of irrigation water depends on the quantity of fertilizers that are applied in the field. The quantity of irrigation water should be more when the doses of fertilizer is high in comparised to field having less fertilizer dose.
5. **Water Taste :** The land having high water taste needs less water. Therefore, the application of irrigation water will be less.
6. **Season :** The water requirements vary according to the season. The crops grown in summer season required more irrigation water than other season.
7. **Method of Irrigation :** The water requirement vary accord to the method of irrigation. Flooding method requires more water than other methods.
8. **Quantity of Irrigation Water :** The irrigation water containing salt should be applied in larger quantity in comparison to irrigation water free from salt.

Frequency of Irrigation : The frequency of irrigation depends on some factors as follows :

1. **Character of the Soil :** The frequency of irrigation depends on the character of the soil. The water retention

capacity of clay soil is more than other soils. For this reason irrigation may be given at a larger interval in clay soil.

2. **Crops or Varieties :** Water requirements vary according to the crops or varieties. Paddy needs more frequent water than other crops. The dwarf high yielding varieties of the crops require more frequent water than tall traditional varieties.
3. **Season :** In summer season, irrigation is needed more frequently than other season. The evaporation lossess of water is more when the weather is hot, dry and windy.
4. **Growth Stages of Crop :** Water requirements vary according to the growth stages of the crop. Paddy requires more water in the growth phase and sugarcane requires heavier irrigation from sixth or seventh month onward.
5. **Topography of the Land :** The sloping land requires more frequent irrigation than level land. Because, topography influences the frequency of irrigation.
6. **Types of Root of the Crops :** The deep rooted crops absorb water from the lower depth of the soil even when upper layers of the soil are not moist. So, a longer interval of irrigation can be allowed in deep rooted crops.

7.5.8 Depth of Irrigation

The depth of irrigation depends on some factors as follows:

1. Type of the Soil : The depth of soil to be wetted depend on the types of soil especially the soil texture as follows:

Textural classes	*Depth of irrigation (mm.)*
Sandy soil	25–50
Sandy loam soil	45–80
Loam	70–110
Clay loam	80–120
Clay soil	100–140

2. Plant Characteristics : The water requirements vary according to the plant. The depth of irrigation be decided by the extent of the root system of different crops. The depth of soil to be wetted by irrigation water should be more in case of deep rooted crops. The root system of important crops on a normal well drained soil are as follows :

Shallow Rooted Crops : Onion, Garlic, Leafy vegetables etc.

Medium Rooted Crops : Rice, Wheat, Maize, Tobacco, Groundnut, etc.

Deep Rooted Crops : Sugarcane, Cotton, Potato, Gram, Pea, Castor, Safflower etc.

3. Doses of Fertilizers : The nutrients in fertilizers becomes available to plant when they dissolve in water. The depth of irrigation depends on the amount of fertilizer applied in the soil. Light and frequent irrigation is found better than heavy irrigation and longer irrigation.

7.5.9. Losses of Irrigation Water

Losses of irrigation water may occur due to some factors as follows :

(*i*) **Percolation :** It is the downward movement of water. Percolation losses of water depends on some factors such as soil character, amount of water, present of crop, soil compaction and cracking.

(*ii*) **Seepage :** It is the lateral sub-surface movement of water. It is to some extent and it may be due to crack or hole in the bund, drying and cracking of the soil and soil texture.

(*iii*) **Transpiration Process of the Plant :** It is the process of releasing excess water from the plant through the stomatal aperture. The transpiration losses depends on the atmosphere condition, soil moisture content, and plant characters. It is physiological process of plant and in this process the plants give off excess water from their body in a vapour form.

(*iv*) **Evaporations :** It is the loss of moisture in the form of vapour from the soil. It is the physical process of losses of water. It is subjected due to high atmospheric temperature associated with lower relative humidity.

(*v*) **Unsuitable or Uncontrolled Material of Irrigation :** The unsuitable or uncontrolled method of irrigation leads to a loss of water through deep perception and seepage.

The factors affecting the quantity of irrigation water to be applied depends on some factors as follows :

(*i*) Physical properties of soil.

(*ii*) Available moisture in the soil.

(*iii*) Amount of fertilisers applied.

(*iv*) Water taste of the land.

(*v*) Seasons of the growing crops.

(*vi*) Crops and their varieties.

(*vii*) Method of irrigation.

7.5.10 Irrigation Water Efficiency

Irrigation water has to be used efficiently as it is an expensive input in crop cultivation. The losses of irrigation water may occur due to run off, seepage deep percolation and evaporation. The irrigation efficiency can be increased by reducing the losses of water.

1. Irrigation Efficiency (Ex) : Irrigation efficiency is ratio, usually expressed as the percent of amount of irrigation, water transpired by plant, evaporated from soil and that necessary to regulate salt concentration in the soil solution and that of consumptive use to the total volume of water delivered or pumped for irrigation.

$$\frac{Wt - Ws - Er}{Wsr}$$

where,

Ex = Irrigation efficiency (in percentage)

Wt. = Amount of irrigation water per unit area of land lost in evapotranspiration including that necessary for the preparation of land and nurser.

Ws = Amount of irrigation water per unit area of land to regulate salt concentration in soil solution.

Er = Effective rainfall.

Wsr = Amount of water per unit area of land that is stored in reservoir or divertual for irrigation.

Irrigation efficiency indicated how effectively the available irrigation water is used for the production of the crop. It is as low as about 10–20 per cent in Indian Irrigation projects.

2. Water Conveyance Efficiency : Water conveyance efficiency (EC) indicates the losses of water, that occur during conveyance from the source of supply to the field. It is the ratio between total quantity of water delivered at source and water delivered to the irrigual plot.

$$EC = \frac{WF}{WS} \times 100$$

where,

EC = Water conveyance efficiency (in percentage)

WF = Water delivered to the field.

WS = Water delivered at the source.

3. Water Application Efficiency : Water application efficiency is the ration between the quantity of water stored in root zone and the water delivered to the fields.

$$Ea = \frac{WS}{WF} \times 100$$

where,

Ea = Water application efficiency (in percentage)

WS = Water stored in hot zone.

WF = Water delivered to the field.

4. Water Distribution Efficiency : Water distribution efficiency is a measure of water distribution within the field. It indicated uniformity in distribution of water over the entire hot zone.

$$Ed = \left[1 - \frac{Y}{d}\right] \times 100$$

where,

Ed = Water distribution efficiency (in percentage)

Y = Average numerical, deviation in depth of water stored from average depth stored during irrigation.

d = Average depth of water stored during irrigation

5. Water Storage Efficiency : Water storage efficiency is the ration between water stored in root zone and water needed in the root zone, prior to irrigation.

$$ES = \frac{WS}{Wn} \times 100$$

where,

ES = Water storage efficiency (in percentage)

WS = Water stored in root zone during irrigation.

Wn = Water needed in the root zone.

6. Water Use Efficiency : Water use efficiency is the ratio between crop yield and water used in evapotranspiration.

$$WUE = \frac{Y}{ET}$$

where,

WUE = Water use efficiency.

Y = Marizetable yield (ks/ha)

ET = Evapotranspiration (mm.)

Factors Influencing Water Use Efficiency : There are some factors influencing water use effective as follows :

(*i*) ***Nature of the Plant :*** WUE varies according to the nature of the plant.

Table 7.9 : Water Use Efficiency of Different Crops

Crop	*Water requirement (mm.)*	*Grain yield kg/ha.*	*WUE Ckg (Ha mm.)*
Rice	2000	6,000	3.0
Wheat	260	3,534	12.6
Maize	625	5,000	8.00
Sorghum	500	4,000 rev	8.0
Groundnut	506	4,680	9.2

(*ii*) ***Climatic Condition :*** Climatic condition effects both yield and evapotranspiration. It is not possible to manipulate the climate condition. However, evapotranspiration can be reduced to a limited extent by mulating and use of antitranspirants. But it may not be economical or practical. It is most effective means of reducing evapotranspiration losses by controlling weeds and increasing the amount of water available to plants, thereby increasing water use efficiency.

(*iii*) ***Soil Moisture :*** Inadequate and excess moisture in the soil have an adverse effects on the growth and productivity of the crop. Proper scheduling of irrigation will increases water use efficiency.

(*iv*) ***Plant Population :*** The yield of a crop depends on the plant population per unit areas. Proper scheduling of irrigation have a effect on plant population.

(*v*) ***Application of Fertilizers :*** The nutricats availability is highest for most crops when water tension is low. Under suitable moisture content, fertilization generally increased yield considerably with a relatively small increase in evapotranspiration, and therefore marzelly impores water use efficiency.

7.5.11 Demerits of Irrigation

It is possible to increase the crop intensity by administering proper irrigation schedule. But excessive irrigation creates problems as follows :

1. Decreasing Soil Fertility : Irrigation aids in cultivating the crop in the year round. But excessive irrigation water causes leaching loss of plant nutrient from the soil. The profitable crop such as potato, jute, sugarcane etc. is possible to cultivate extensively. As a result of which the fertility of soil decreases simultaneously. Excessive irrigation water increases the leaching losses of nutrient from the soil.

2. Problem of Stagnant Water : Heavy amount of irrigation water in the field causes damage of the crop, especially the water sensitive crop like wheat. As for example, it is found that wheat crop turns yellow and dies if the field is irrigated heavily and water remains stagnant in the field for some days. Land becomes a marsh, and unfit for cultivation. Salt accumulates on the soil surface due to lack of sufficient percolation and drainage. Crop diseases may occur in the water logged condition.

3. Increasing Salinity in the Field : The irrigation water of tubewell (*i.e.* shallow tube well, deep tubewell, submersible etc.) contains salt and salt deposits on the soil. As a result, the fertility of soil decreases. Irrigation water having no quality standard causes damage to the soil.

4. Creation of Unhealthy Environment : Multipurpose river valley project has been started in our country. Irrigation facility has been provided in large sector of our country through canal. Canal water remains stagnant in the water reservoir and acts as a breeding place of mosquito. The mosquitoes act as a carrier of some diseases like malaria, filaria, encephalitis etc. and these effect on the health of the farmers, crop diseases may occur in water logged condition.

7.5.12 System and Methods of Irrigation

Followings are the four principal methods or irrigation generally adopted in our country.

1. Surface Irrigation Methods : In surface methods of irrigation, water is applied directly to the soil surface from a channel located at the upper reach of the field. Followings are the general requirements of prime importance to obtain high efficiency in surface methods of irrigation.

(*i*) Properly constructed water distribution systems to provide adequate control of water to the field.

(*ii*) Proper land preparation to permit uniform distribution of water over the field. The land should be levelled well to achieve a higher water application efficiency.

There are different methods of surface irrigation as follows:

(a) Free Flooding Method : This is a common method of irrigation of wet land rice where water may be applied as a continuous flood, rotational flood or an intermittent flood (De, 1979). This is a suitable method for an area where cheap and abundant supply of water is available. Flooding consist of opening a water channel in a plot or field. So that water can flow freely in all directions and cover the surface of the land in a continuous sheet.

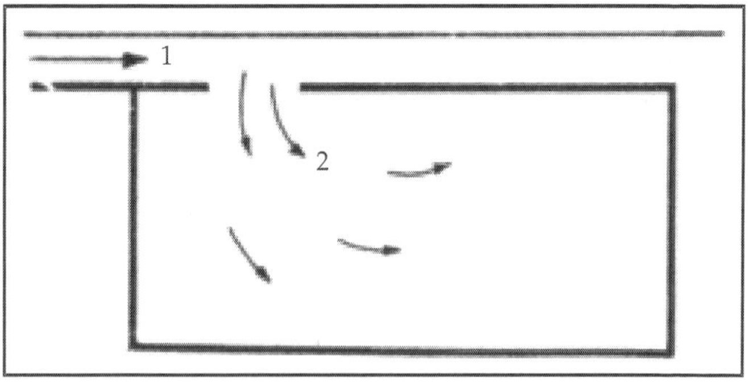

Fig. 7.1 : Free flooding method of irrigation

Advantages

(*i*) No other field layout becomes necessary except the levelling and preparation of impervious boundary bunds.

(*ii*) No land area is utilized for water distribution. As a result wastage of land becomes minimum.

(*iii*) Less supervision is required for application of water in the field.

(*iv*) Labour requirement is minimum.

Disadvantages

(*i*) It is the most inefficient method of irrigation as only about 20 per cent of water is actually used by plants and the rest is being lost as run-off, seepage and evaporation (Prasad 1967).

(*ii*) Levelling of land increases cost of cultivation.

(*iii*) This method is unsuitable for crops that are sensitive to water logging and uneven land.

(*iv*) Crop growth is not uniform as the water distribution by this method is very uneven.

(*v*) There is a possibility of soil erosion and loss of nutrients by leaching.

(*b*) *Furrow Methods* : This method is suitable for row crops like Potato, Cabbage, Cauliflower, Brinjal, Chilli, Sugarcane, Cotton, Tobacco, Tomato etc. Water is applied in the furrows which are developed between crop rows in the planting and cultivating processes. There is usually one furrow between each row of crop. In orchards and widely spread vegetable crops, several furrows may be opened between the rows of the crop and the water is led simultaneously in all furrows. Furrows are laid down along the slope when the slope of the land is about three per cent. The furrows are laid on graded contour when the slope exceeds three per cent and it upto 15 per cent. The length of furrows varies with soil type, slope of the land and the quantity of water to be applied and may vary from 10 metres to 100 metres.

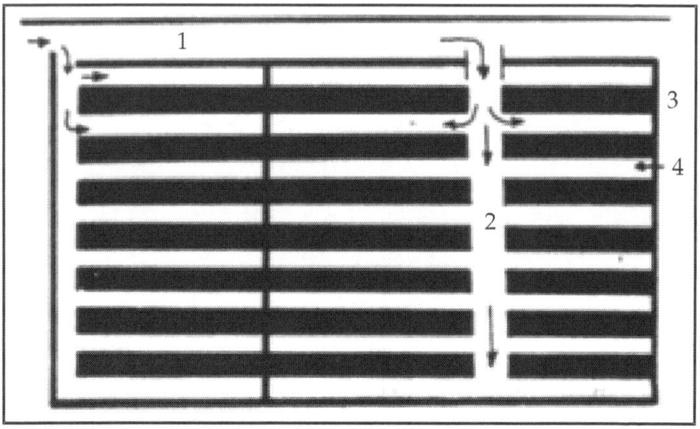

Fig. 7.2 : Furrow Method of Irrigation

Advantages

(*i*) There is a great advantages in the use of water which is of great consideration in lift irrigation.

(*ii*) This method is suitable for crops which are sensitive to saturated soil condition in the root zone.

(*iii*) Water in the furrows contracts only one fifth to one half of the land surface, thereby reducing puddling, crusting of the soil and evaporation losses.

(*iv*) Water is applied uniformly by this method.

(*v*) It is especially suitable to crops like maize that are injured by contact with water as water does not come in contact with the plant stem.

(*vi*) The method reduces labour requirements in land preparation and irrigation.

Disadvantages

(*i*) More cost is involved in the preparation of furrow.

(*ii*) This method is not suitable for all areas, specially canal areas.

(*iii*) Uniform distribution of water in all furrows is not possible.

Following are the different types of furrow irrigations :

(*i*) *Straight Furrow Method of Irrigation :* This method is practised in flat lands.

(*ii*) *Contour Furrow Method of Irrigation :* This method is practised in sloping land in which the direction of flow is across a slope rather than down slope. The furrows are laid out with a gentle uniform grade to carry the irrigation streams.

(*iii*) *Corrugated Furrow Irrigation :* Corrugations are small furrow which are used for close growing crops such as Beet, Cabbage, Cauliflower, Turnip etc. In this system more and smaller furrows are utilised and crop rows are not necessarily related to the irrigation furrows. The size and spacing of corrugation vary with the soil. The corrugation should be more closely spaced, generally 45 to 90 cm.

(c) *Check and Furrow System :* It is a modified form of furrow irrigation. This method is suitable for a land which is quite level. The water is lead in first furrow and the flow of water is continued to second furrow through a breach in the ridge at the terminal end of the run. After completion of the run in the second furrows, the terminal end of the flow is continued to the third furrow through a breach. A number of furrows are irrigated with a single run of water that moves in a zig-zag way through furrows (De, 1989).

Advantages

(*i*) Water is applied uniformly and the wastage of water is minimum. There is less evaporative loss from furrows.

(*ii*) Soil gets sufficient time to absorb water.

(*iii*) It is a very suitable method for irrigating row crops in level land.

(*iv*) Soil erosion is checked.

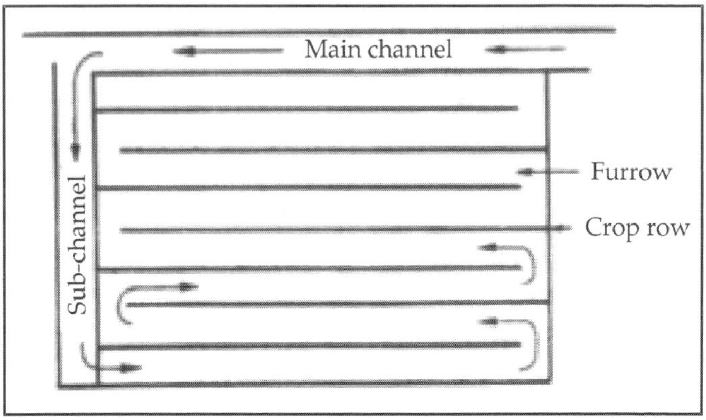

Fig. 7.3 : Check and furrow method of irrigation

Disadvantages

(*i*) Uneven distribution of water in the different furrows.

(*ii*) This method is only suitable for row-crops.

(*iii*) Irrigation cost is higher as it involves much labour.

(*iv*) More time is required for irrigation.

(d) Boarder or Check Basin Irrigation : The field is divided into smaller unit areas or beds measuring 10 to 100 sq. m. or even more. Bunds or ridges are constructed around the areas forming basin within which irrigation water may be controlled. The basins are levelled in both directions. Small irrigation channels are provided between two adjacent rows of beds. Water is conveyed to the field by a stream of supply channel and lateral field channel. The supply channel is aligned in the upper side of the area and there is usually one lateral for every two rows of check basins. Water from the lateral channel is turned in the beds and is cut off when sufficient water has been admitted to the basin. The basins are filled to the desired depth and the water is retained until it infiltrates the soil. This method is especially suitable for irrigating grain and fodder crops in heavy soil where water is absorbed very slowly.

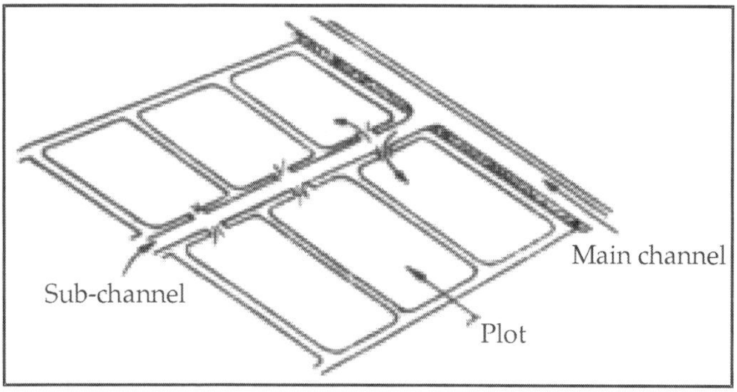

Fig. 7.4 : Boarder or check basin method of irrigation

Advantages

(*i*) It is more efficient method of irrigation and ensures uniform application of water.

(*ii*) It has good control on irrigation water and high water application efficiency.

(*iii*) Though the initial cost is high, it requires less labour and has low maintenance cost.

(*iv*) Wastage of water and soil erosion is reduced as there is bund around the basin.

Disadvantages

(*i*) There are two many ridges which occupy the larger areas of land. It involves the wastage of land.

(*ii*) Lay out requirement in land preparation and irrigation are much higher in check basin irrigation as compared to other methods.

(*iii*) Repairing of ridges and careful supervision during irrigation are needed.

(*iv*) Uneven distribution of water within the plot.

(*v*) This method impede surface drainage, precise land grading and shaping are required.

(e) Border Strip Method of Irrigation : The field is divided into a number of long parallel (3 – 10 m × 50 – 300 m.) strips which is called border. The border is separated with small parallel low ridges on the sides. The individual strips are levelled perfectly and each strip is irrigated independently by turning a stream of water at the upper elevation. The water can flow down the slope with nearly uniform depth. This method is suitable to irrigated all close growing crops like wheat, barley, fodder crops and legumes. But this method is not suitable to irrigated rice which requires standing water during most part of their growing season. This method is suitable to soils having moderately low to moderately high infiltration rate.

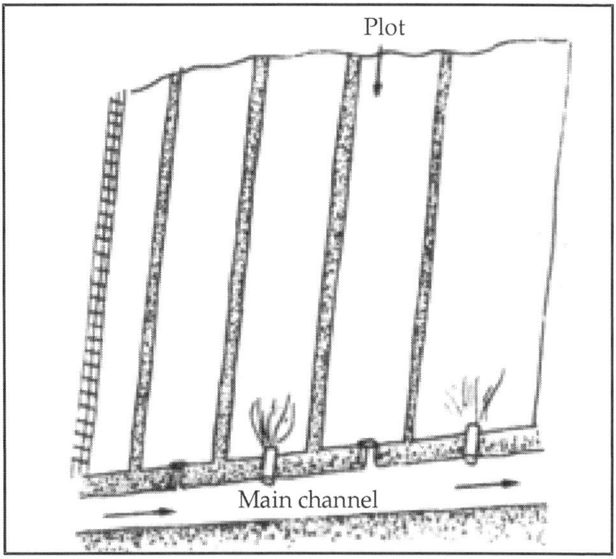

Fig. 7.5 : Boarder strip method of irrigation

Advantages

(*i*) It is a best method to irrigated close growing crops.

(*ii*) Uniform distribution and high water application efficiencies are possible if the system is properly designed.

(*iii*) Labour requirement is less to irrigated the field.

(*iv*) Operation of this system is simple and easy.

(*v*) Excess rainwater are drained out if outlets are available.

Disadvantages

(*i*) More labour is required for levelling the field.

(*ii*) Ridges cut down the neat cropped area.

(*iii*) Large irrigation streams are required.

(*iv*) Repair of ridges and supervision during irrigation are needed.

(f) ***Basin Method :*** Basin irrigation is suitable for orchards and other crops such as pumkin, gourd, ashgourd etc. A basin is made around the stump of the plant at a distance of about one to two feet which may be square, rectangular or circular in shape. The individual plant or plant grown in pits is irrigated by this method. The soil around the plant is soaked with irrigation water and not the entire land. Generally ring basin is used for irrigating fruit trees. The basin is connected with irrigation channels for receiving water.

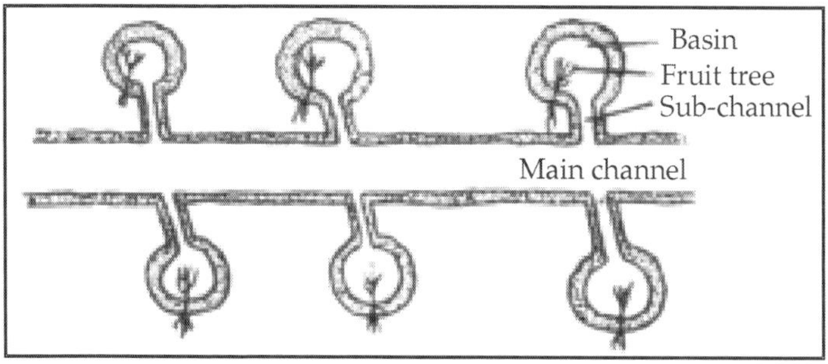

Fig. 7.6 : Basin method of irrigation

Advantages

(*i*) It is a good method for irrigating orchard plant and also for all types of soil.

(*ii*) It is an efficient method in the use of water.

(*iii*) This method economise the water considerably as there is no loss of water by run-off.

Disadvantages

(*i*) Initial cost for preparing basin is high.

(*ii*) This method is suitable for irrigating field crops and some orchard plant such as Banana, Papaya, Pine apple etc.

(*iii*) All plants do not get same amount of water.

2. Sub-Irrigation or Sub-Surface Irrigation System : Sub-irrigation or Sub-surface irrigation may be natural or artificial. Natural sub-surface irrigation is possible where an impervious layer exists below the root zone. Water is led into a series of ditches dug up to the impervious layer which then moves laterely and then vertically through capillaries and saturates the root zone. In artificial sub-surface irrigation, perforated or porous pipes are laid underground below the root zone and water is led into the pipes by suitable means. In either case, the idea is to raise the water table so that plant root can get water by capillary movement. (Talwar and Prasad, 1967). This method is suited to soils having reasonably uniform texture and permeable enough for water to move rapidly both horizontally and vertically with and for some distance below the crop root zone. This system is practised at present in Gujarat and Kashmir for growing cash crops on sandy loam soil (Dastane, 1980).

Advantages

(*i*) It is a permanent and efficient method of irrigation.

(*ii*) Evaporation loss of irrigation water is minimum.

(*iii*) Sub-irrigation do not interfere with the agrotechnical process.

(*iv*) No wastage of land.

(*v*) The pipes may be used for drainage in artificial sub–surface irrigation.

(*vi*) No possibility of accumulation of salts crust formation on surface soil.

(*vii*) No possibility of soil erosion. Loss of nutrient by leaching is minimum.

Disadvantages

(*i*) Initial cost is high.

(*ii*) The neighbouring land gets damaged due to seepage.

(*iii*) The use of this method is very limited as it requires special soil condition.

3. Sprinkler or Overhead Irrigation : This method of irrigation is generally adopted in an area having uneven topography and where erosion hazards are great and the surface irrigation is not congenial. In this system, water is conveyed under pressure through pipes to the area to be irrigated where it is passed out through nozzles or sprinklers. The system comprise four parts, such as power generator, pump, pipe lines and sprinkler.

Classification of Sprinkler Irrigation

(*i*) *Permanent System :* This is generally adopted in nurseries and orchards. In this system, the pipes (both main and lateral) are burried underground permanently and the area to be irrigated is served by rotating wide–radius sprinklers. The pipes burried underground do not interfere the tillage operation. Installation cost is much higher, but labour and maintenance costs are less than that of semi-permanent or portable system.

(*ii*) *Semi-Permanent System :* In this system, the main lines are burried and laterals are portable. The installation cost is some what less than the permanent system, but labour and maintenance cost is more.

(iii) Portable System : In this system, the main and lateral lines are portable. So it can be used in different fields as needed. The installation cost is less, but the labour and maintenance cost is high.

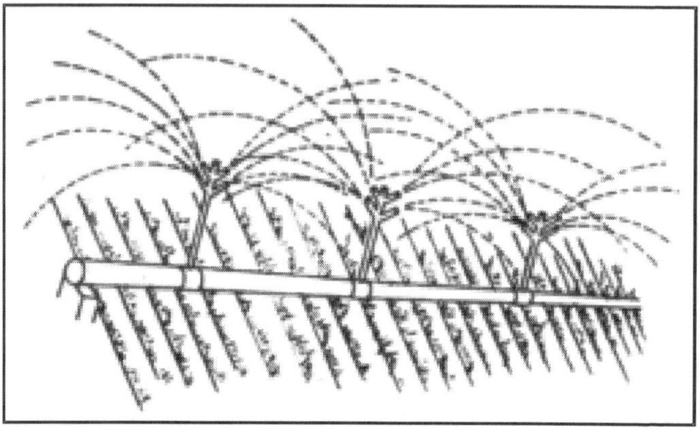

Fig. 7.7 : Sprinkler or overhead irrigation

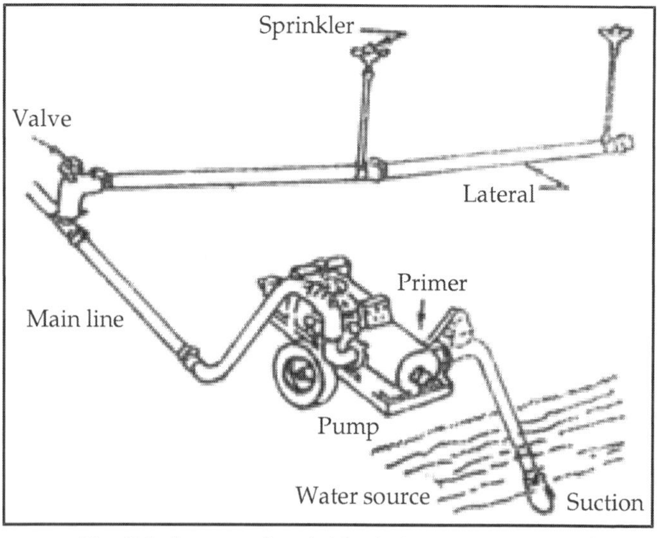

Fig. 7.8 : Layout of sprinkler irrigation system with a centrifugal pump

(iv) Semi-Portable System : In this system, the pumping device is kept permanent and the others are portable.

Advantages

(*i*) This method can be used effectively in an area where surface irrigation is not feasible.

(*ii*) This is a very suitable method for irrigating shallow sandy soils of uneven topography where levelling is not *practicable* for surface irrigation.

(*iii*) Control of irrigation and water management is possible.

(*iv*) This is a suitable method in all area where water and labour are scarce.

(*v*) Water soluble fertilisers can be very effectively applied through sprinklers.

(*vi*) Irrigation water is applied on plants and soil. Soil gets sufficient time to absorb the water.

(*vii*) It is a good method for irrigating young plants and in the lawn.

(*viii*) No wastage of land as it does not involve in the preparation of channels.

(*ix*) It is also used for cooling the plants during high temperature and for frost protection during freezing temperature (Dastane, 1980).

Disadvantages

(*i*) The original cost along with operation and management cost is high.

(*ii*) Application of water is effected if wind velocity is high.

(*iii*) Evaporation loss of water specially in clay soil is high and the soil do get time to absorb water.

Drip or Trickle Irrigation : Drip irrigation is one of the most recent development in irrigation. It was originally developed in Israel and it is becoming popular in areas having water scarcity and salt problem. This system involves slow and frequent application of water drop by drop to the root zone of a crop. Water is lifted and distributed through overhead pipelines which are fitted with nozzles to drop water in the desired site. Under medium of low pressure, required quantity of water is given daily so that plant may not suffer from water stress.

Advantages

(*i*) Water is used economically by this method. Since it minimizes the loss of water by deep percolation or by evaporation from the soil surface and water saving to the extent of 30–50 per cent.

(*ii*) Drip irrigation gives higher yield with poor quality water.

(*iii*) This method is gaining popularity in arid region of the world.

(*iv*) This method can be used for applying fertilizers in solution.

(*v*) This method reduces weed growth.

(*vi*) This method reduces operational labour and salinity hazard to plant.

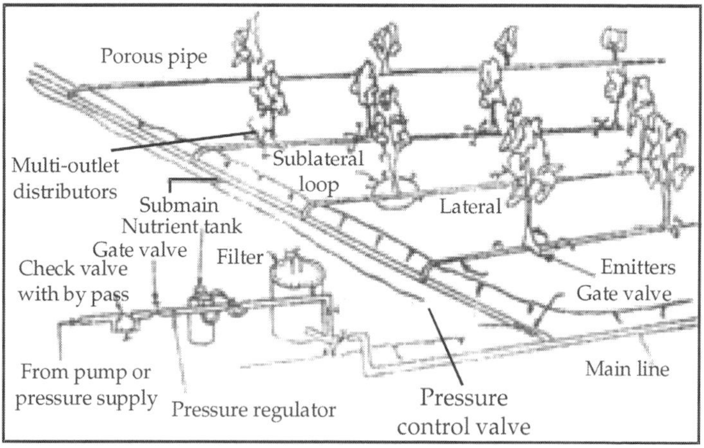

Fig. 7.9 : A typical layout of drip irrigation system

Disadvantages

(*i*) The initial cost of equipment is high.

(*ii*) Maintenance cost is also high.

(*iii*) Salt accumulates near the root zone of the plant.

(*iv*) Soil water distribution and plant root development are restricted.

(*v*) This method is not suitable for irrigating all crops. Maize, Sugarcane ,Tomato, Tobacco, Orange, Apple etc. are only irrigated by this method.

7.5.13 Irrigation Practices for Major Crops

1. Paddy *(Oryza sativa* L.)

(a) Aus Paddy : Life cycle of H.Y.V. of paddy is 90–100 days. Broad casted Aus paddy is generally cultivated depending on monsoon rains. But it is essential to give irrigation regularly to get higher yield. The transplanted paddy is also cultivated depending on rainfall. Where irrigation facility exist, it is necessary to maintain 1.25 to 2.5 cm. water upto 35–45 days of sowing or transplanting of seedlings. The water requirement of Aus Paddy is 30–35 hectare centrimetre. Irrigation should be provided @ 3.0 – 3.5 hectare centimetre and an interval of 6–7 days.

(*b*) *Aman Paddy* : The life cycle of early high yielding of Aman Paddy is 110–120 days. The transplanting of seedlings should be done in a Puddled land having 2 cm. water. It is necessary to maintain 2.5 cm of water upto maturing stage of paddy from the transplanting of seedling if possible. More water at hillering stage hampers in the hillering of paddy. The total water requirement of Aman Paddy is 40–50 hectare centimetre and this amount of water should be provided through 8–10 irrigations, if there is no regular rainfall.

(*c*) *Boro Paddy* : The average life cycle of high yielding varieties of Boro paddy is 120–150 days. Boro paddy is cultivated as irrigated crop. Maintenance of water in boro paddy helps in getting higher yield as per following schedule.

(*i*) 2 cm at transplanting, 5 cm upto 30 days after planting, 2 cm upto panicle initiation. 5 cm upto 21 days after flowering followed by gradual withholding of irrigation.

(*ii*) Draining of excess water at late tillering for three days and irrigation is given again after top dressing of fertilizer.

The water requirement of Boro paddy is 60.00–70.0 hectare centimetre and this amount of water should be provided through irrigation at an interval of 4–5 days.

2. Wheat (*Triticum* sp) : The life cycle of high yielding varieties of wheat is 90–100 days. Irrigation plays a key role in getting best yield of dwarf wheat. Wheat needs to be irrigated at the right time with adequate quantity of water. Tall wheats needs three or four irrigation, giving the first one while preparing land for sowing, the second when the crop is four to five weeks old, the third at 9–10 weeks stage and the last one when the crop is the third at 12–13 weeks old. Irrigation should not be given at grain development stage as the crop lodges.

Dwarf wheat needs about six timely irrigations to get maximum yield. These exclude the one that is given during land preparation and sowing as presowing irrigation for germination of seed and stand establishment. Four irrigations should be given to get higher yields as follows :

(*i*) At crown root initiation stage *i.e.* 18–21 days after sowing of seed.

(*ii*) At late littering stage *i.e.* 40–45 days after sowing of seed.

(*iii*) At also flowering stage *i.e.* 70–75 days after sowing of seed.

(*iv*) At dough stage of grain *i.e.* 90–95 days after sowing of seed.

Irrigation should be given according to the availability of water as per the following schedule :

(*i*) In case of water available for one irrigation, it should be given at crown root initiation stage which comes around 18–22 days of sowing.

(*ii*) In case of water available for two irrigation, the first and second irrigation should be given at crown root initiation stage and at late jointing stage (*i.e.* 65 DAS) respectively.

(*iii*) In case of water available for three irrigations, irrigation should be given as per following :

(*a*) First at crown root initiation stage.

(*b*) Second at late jointing stage.

(*c*) Third at flowering stage.

(*iv*) In case of water available for four irrigations, irrigation should be given as per following :

(*a*) First at crown root initiation stage.

(*b*) Second at late tillering stage (*i.e.* 45 DAS).

(*c*) Third at flowering stage.

(*d*) Fourth at milk stage of the grain.

(*v*) In case of water available for five irrigations, irrigation should be given as per following :

(*a*) First at crown root initiation stage.

(*b*) Second at late tillering stage.

(*c*) Third at late jointing stage.

(*d*) Fourth at flowering stage.

(*e*) Fifth at milk stage of the grain.

(*vi*) In case of water available for six irrigations, the irrigation should be given as per followings :

(*a*) First at crown root initiation stage.

(*b*) Second at late tillering stage.

(*c*) Third at late jointing stage.

(*d*) Fourth at flowering stage.

(*e*) Fifth at milk stage.

(*f*) Sixth at dough stage of the grain.

Experiments at Indian Agricultural Research Institute, New Delhi, and elsewhere showing that irrigation given at 25, 45, 65, 85, 105 and 125 days after sowing (DAS) are highly beneficial of the crop.

Table 7.10 : Schedule of Irrigation Depending on the Availability of Water

Supply of water available for number of irrigation	*Days from sowing for giving irrigation*
6	25, 45, 65, 85, 105, 125
5	25, 45, 85, 105
4	25, 45, 85, 105 or 25, 65, 85, 105
3	25, 65, 105
2	25, 65
1	25.

The water requirement of wheat is 14.0–15.0 hectare centrimetre and this amount of water should be provided through 5–6 irrigations.

3. Maize (*Zea mays* L.) : The life cycle of hybrid maize is about 100 days. Maize does stand drought and standing water. Irrigation is to be given to provide proper soil moisture for the growth of Maize plant. In kharif, one or two irrigations is to be given if needed. In all four to five irrigations are needed to get a good crop in rabi seasons at an interval of 10–12 days. In addition to this, one presowing or post sowing irrigation is to be given for good seed germination and stand establishment. The need of water is higher at the tasseling stage (tassel is the male flowering structure of maize plant) and seed development stage. The ripening crop should be irrigated less frequently. The water do not allowed to stagnate in the field for more than 6 to 8 hours. So the excess water should be drained out.

4. Jute (*Corchorus* sp) : The jute is normally cultivated as rainfed crop. For timely sowing of jute, one presowing irrigation is helpful and thereafter two to three post sowing irrigations may be required depending on rainfall. Jute requires plenty of water and humid condition for its growth. But the growth of the crop is adversely affected under water logged conditions.

The water requirement of jute is 20.0–24.0 hectare centimetre and this amount of water should be provided through 6–7 irrigations.

5. Potato (*Solanum tuberosum* L.) : The life cycle of potato is 85–100 days. In general, potato is summer crop in the hills and the winter crop in the plains. In the plains, irrigation is absolutely necessary for successful potato cultivation. Potato responds well to timely irrigation. It requires 25.0–26.0 hectare centimetre of water which should be spread through out the growth period. Ten to twelve irrigations are necessary during growing session, particularly in the plains. The critical stages of irrigation are pre-emergence stage, stolonization stage, tuberization stage and bulking stage. Irrigation scheduling depends on moisture retention capacity of soil, stage of crop growth, atmospheric factors and rooting pattern of the cultivar. Preplanting irrigation is usually applied for land preparation and one to two pre-emergence light irrigations are also applied to ensure quick plant emergence depending on initial moisture condition of soil (Grewal *et. al.,* 1990). The success of potato cultivation is primarily dependent on timely irrigation. First irrigation should be light and given 4–7 days after planting or earlier depending upon soil moisture. The subsequent irrigation should be given at an interval of 5–7 days in September–October, 7–10 days in November, 10–15 days in December and 5–7 days in February–March and onward. The most critical period of irrigation was observed at stolon elongation stage *i.e.* from 20–40 days (Singh *et. al.,* 1968 and Hukkeri *et. al.,* 1970). The irrigation should be stopped about a week before harvesting. In hills, it is necessary to give supplemental irrigation in dry periods from April to June.

6. Sugarcane (*Saccharum officinarum* L.) : The life cycle of sugarcane is about 365 days. In sugarcane, the maintenance of optimum soil moisture during all stages of growth is one of the essential requisites for higher yield. First irrigation should be given at 40–50 days of planting after the completion of germination. But in case of moisture is insufficient for germination and attack of termites, irrigation must be given before germination. Subsequent irrigations may be given at 10–15 days interval till maturity during post monsoon season. In areas where frost occur, irrigation is given to save the crop from frost. Irrigations should be stopped at one month before harvesting of the crop. The water requirement of sugarcane is 95 hectare centimetre and this amount of water should be provided through irrigation as per requirement.

7. Mustard (*Brassica sp.*) : The average life cycle of mustard is 85–90 days. Mustard is grown as rainfed and irrigated crop. Toria (*Brassica campestris var Toria*) is absolutely a rainfed crop in West Bengal. One pre-sowing irrigation is needed in a land having no sufficient moisture for germination of seed. For irrigated crop of mustard, two irrigations under normal condition would be sufficient. Irrigation at the time of flowering and pod filling stage have a significant effect in increasing yield. The water requirement of mustard is 10.0–12.0 hectare centrimetre and this amount of water should be provided through 2–3 irrigations.

8. Sesamum (*Sesamum indicum* L.) : Under West Bengal condition, sesamum can be cultivated three times in a year as summer crop, Rain crop and Autumn crop. Summer crop of sesamum is usually grown as irrigated crop and 2–3 irrigations required depending on the moisture and rainfall. First irrigation is to be given 21–30 days after sowing, second at full flowering stage and third at Pod development stage. Rainy season crop does not require irrigation.

9. Ground Nut (*Arachis hypogaea* L.) : Ground nut is an irrigated crop during rabi and to certain extent in summer season.

Two or three irrigations are required in areas with a rainfall of 50–70 cm. First irrigation should be given at the start of flowering and subsequent irrigation whenever required to keep the soil moist during the flowering period which last from 45 days to 90 days after sowing to facilitate the peg penetration in the soil. Sometime, pre-sowing irrigation is to be given to facilitate the field preparation and good germination of seed. In low rainfall area, 3 or 4 irrigations will be required to raise a good crop. In areas where water is much limited and is available for single irrigation, the crop should be irrigated during flowering stage and where two irrigations are possible, the crop should be irrigated first time at the commencement of flowering and second during pod formation. The later irrigation will facilitate the penetration of pegs. The last irrigation given about 9 week before the harvesting of the crop will help in full recovery of the pods and will also be useful in sowing the succeeding crops. Frequency of irrigation is very important, otherwise a high proportion of immature and shrivelled (under developed) nuts will be formed. On an average bunch type–requires 6–8 irrigations and spreading type requires 10–12 irrigations. The water requirement of groundnut is 26 hectare centrimetre and this amount of water should be provided through requisite number of irrigation.

10. Sun flower (*Helianthus annuus* Linn) : Sunflower is an irrigated crop during rabi and kharif season. Soil, deficient in moisture, requires one pre-soaking irrigation for the good germination of seed. Thereafter, irrigation must be given according to the need of the crop. If there is sufficient timely rain in kharif season, no irrigation is needed. Two or three irrigations are needed for a good crop of rabi season and kharif crop should be irrigated at an interval of 15–20 days before flowering and twenty days after flowering is the critical period for water requirement.

11. Gram (*Cicer arietinum* L.) : Gram is a cold or winter crop in India and grown as dry crop in rabi season. Sometimes it is also grown as a regularly or partially irrigated crop. In times of acute drought, one to two irrigation at the time of secondary branching stage and pod filling stage have been found beneficial. Irrigation should not be given at the flowering time of the grain crop.

12. World Pea (*Pisum* sp) : Pea is popular pulse crop throughout the word and it is grown as vegetable and pulse crop.

In the moisture deficient soil, pre-sowing irrigation is beneficial for good germination of seed. Irrigation, there after is given whenever required. Irrigation at the time of flowering and pod filling stage has been found beneficial.

13. Red Gram or Pigeon Pea (*Cajanus cajan* Millsp. *C. millcus* Spreng) : Red gram is a hardy plant often cultivated as an annual crop. It is generally grown as rainfed crop. In dry season, irrigation should be given according to the need of the crop. The crop is sometimes irrigated to protect it from frost in late December or early January.

14. Soyabean (*Glycine max* Linn) : Soyabean is an important pulse crop rich in food value. The kharif crop being grown during rainy season usually does not require any irrigation. However, if monsoon fails or is irregular, irrigation should be given as and when necessary. The field should be kept sufficiently wet during germination, flowering, podding and pod filling stage. Moisture stress during any of these periods affects the yield. Pre-sowing irrigation is found beneficial for good germination of seed.

The spring or summer crop of soyabean requires assured irrigation facilities. Irrigation at 8–10 days interval is required, during April and May.

7.5.14 Water Lifting Devices Used for Irrigation

There is a large number of devices used for lifting water from the different sources of irrigation water which was in vague in the past and are also in functioning at present in different places as follows :

1. Don : Don is one of the most important and popular water lifting device in our country. It looks like a boat having one end closed. It is 12–18 feet (3.6 –5.4 metre) long, one feet (30 cm.) broad at middle and 7 inches (17.5 cm.) deep. It is suitable for lifting of water upto 4–6 feet (120–180 cm.) The quantity of water lifted varies from 1000–2000 gallon per hour depending on the efficiency of the labour.

Fig. 10 : Don

Fig. 7.11 : Swing Basket

2. Swing Basket : It is a triangular shaped water lifting device made of tin. This can easily be made at home. Four ropes are attached to this device, two are attached at the top and two are attached at bottom. Two men are engaged in operating this device. The quantity of water lifted varies from 350–400 gallon per hour.

3. Mhote : Mhote is one of the most important water lifting device meant for lifting water from the well. Bucket, made of leather or iron, is tied in one end of a rope and the other end is tied to the bullock's yoke which move over a wooder pulley suitably fixed on the well. The bucket is hauled up and down the well by one pair of bullock walking on a sloping ramp constructed near the well. The quantity of water lifted varies from 1000–1500 gallon per hour over a lift of 30 feet (9 metre).

4. Persian Wheel : The buckets that holds about one gallon water, are fixed on a endless belt or spoket chain in series which moves on a vertical wheel or drum. The bullocks walking in a circle turn the drum through a set of bevel gears and shafts, which in turn moves the endless belt. As a result, the buckets full of water are raised on one side which are emptied as they come

Fig. 7.12 : Persian Wheel

to the top and become filled with water when they come to the bottom of the well. The discharge is about 2,500 gallons per hour for a lift of 25 to 30 feet (7.5–9.0 metres).

5. Picota : A wooden beam is mounted on a vertical piller fixed near a well in which a bucket with long rope is tied at one end and the other end is counter balanced with heavy stone. The quantity of water lifted varies from 1500–2000 gallons per hour over a lift of 15 feet (4.6 metres) from well.

6. Egyptian Screw : It is hollow pipe like device having 12–16 feet (3.60 – 4.80 metre) length and 15 inches (37.5 cm) diameter. A auger like projection is inserted inside the hollowpipe in which a handle is attached at one end. The one end of this devices is emersed in water and the handle is rotated at clockwise. The water is lifted and emptied at the other end. The quantity of water lifted varies from 2000–3000 gallon per hour for a lift of 3–4 feet (90–120cm.)

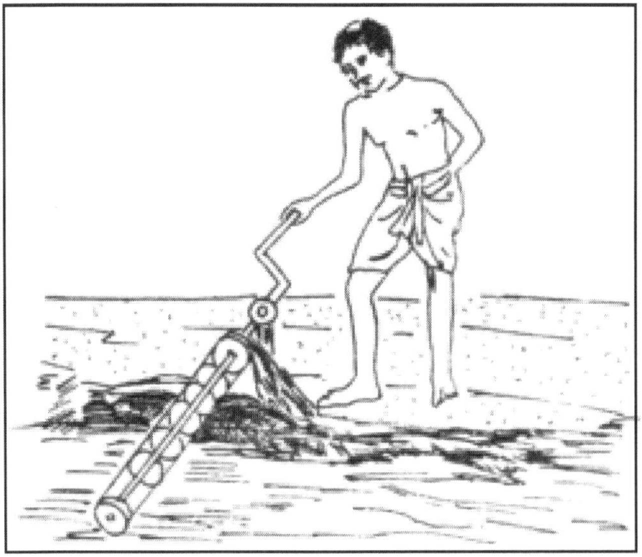

Fig. 7.13 : Egyptian Screw

7. Wind Mill : It is a contrivance in which wind is employed as a source of power. A wind velocity of not less than

6 m.p.h. is necessary to run a wind mill. The operation of the wind mill depends on wind and it operate well where wind velocity is high. A 12 feet (3.6 metre) wind mill mounted on a 70 feet (21 metres) tower raises on a average about 30,000 gallon of water per day.

8. Pump : Pump is a mechanically operated appliance used extensively at present for irrigation. There are three types of pump as follows :

(*i*) Piston pumps

(*ii*) Centrifugal pumps

(*iii*) Turbine pumps

Out of these, centrifugal pump is extensively used for irrigation, pumps are operated by engines or electric motors. Diesel, Kerosine or Petrol are used for operating the engines. The water lifting capacity of pump depends on its horse power (H.P.) A pump having 5 H.P. can discharge 45,000 to 70,000 litres of water for a lift of 20–25 feet (60–75 metres).

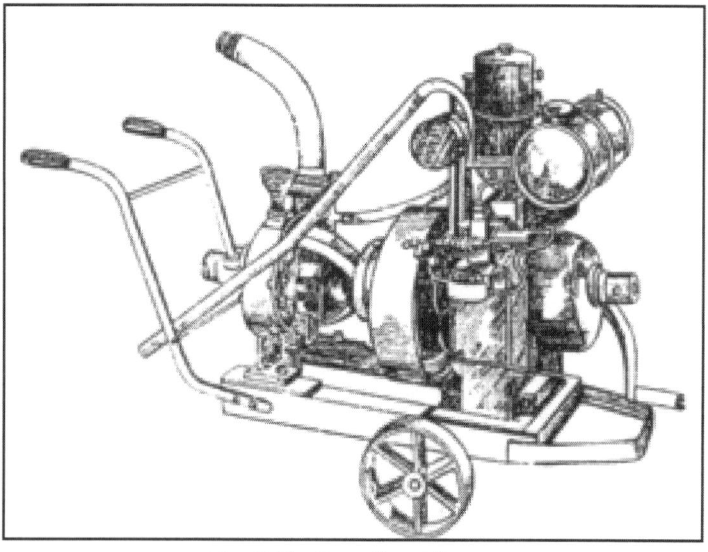

Fig. 7.14 : Centrifugal Pump

7.6 DRAINAGE

Agricultural drainage is the removal of excess water, known as free water or gravitational water, from the surface or below the farm so as to create favourable soil condition for plant growth. The lowering down of the ground water table (G.W.T.) below the root zone of the crops to improve the plant growth or to reduce the accumulation of salt is included in drainage.

Irrigation and drainage are equally important for successful crop production. Irrigation and drainage cannot be separated from one another and they should 80 hand to hand. Excess water in the soil even for a short period may cause severe damage to crop and soil. Irrigation provides sufficient moisture in the soil for satisfactory growth of plant and drainage is important for avoiding excess moisture in the root zone. The excess moisture may originated from excess rainfall, over irrigation, Seepage from canal or reservoirs or ditches. A land having high water table or water stands on the *land* surface for a long period, excessive soil moisture content, humid or superhumid regions with continuous or intermittent rainfall or flat land with fine textured soil may need drainage for high agricultural productivity.

Following are the two types of drainage :

(*i*) Surface Drainage : The process of removing the excess water from the land surface is known as surface drainage. Adequate arrangement for surface drainage in heavy rainfall areas is essential for speedy disposal of water. Rainfall, snowmelt, waste run off, seepage from adjoining higher land, overflow from stream channel etc. are the sources of surface water. Surface drainage system may be required in either humid or in irrigated areas which is usually an integral part of irrigation system on slowly permeable soils or in areas of high precipitation rates. It is therefore told that irrigation and drainage are like the two faces of a coin.

(*ii*) Sub-Surface or Internal Drainages : The process of removing the water from sub-surface or soil profile is known as sub-surface or internal drainage. There should be sub-surface drainage in high water table areas.

In either case, the system is conveniently divided into the following three functional parts.

(*i*) Collection System : Budding, surface field ditches, row ditches or diversion ditches are the integral part of the system that first picks up water from the land.

(*ii*) Disposal System : This is a part of the system that receives water from the collection system and conveys it, usually in an open channel to the outlet.

(*iii*) Outlet : This is the end point of the drainage system. The first step in the design of any drainage system is to locate the outlet, because most of the drainage systems fail due to failure of outlet. The most economic outlet is a natural water way, such as rivers, streams, ditches, municipal drains etc.

Sub-surface drains are underground artificial channels through which excess water of a land may flow to a particular outlet.

Role of Drainage in Crop Production

Drainage and irrigation are equally important for successful crop production. Drainage, if not properly administered, causes various troubles in crop production, as follows :

(*i*) Soil aeration is hampered which ultimately effect the respiration process of plant root. As a result, the plant may die due to the damage of plant root.

(*ii*) Soil becomes deficient in oxygen. The soil organism fails to survive in such soil. As a result, some poisonous substances are developed due to improper decomposition of organic matter which hamper the growth of plants.

(*iii*) Loss of nutrient may occur due to leaching and run–off of excess water.

(*iv*) Except some crops, most of the crops are very sensitive to water logging condition.

(*v*) The accumulated drainage water may be used for irrigation purposes by lifting them with suitable water lifting divice.

(*vi*) Soil erosion may occur.

(*vii*) Sometimes drainage becomes a great problem than irrigation, specially in low lying areas.

(*viii*) Surface salt accumulation may be the result of improper drainage.

(*ix*) Drainage water of one field may be used as irrigation water with proper planning and designing.

Objective of Drainage

There are some objects of drainage :

(*i*) Removal of excess water from the soil.

(*ii*) Arrangement for good aeration in the soil.

(*iii*) To control the loss of plant nutrients.

(*iv*) To enhance the activity of soil organism.

(*v*) To improve the soil structure.

(*vi*) To control soil erosion.

(*vii*) To prevent the salt accumulation of surface soil.

(*viii*) To develop favourable condition for the growth of plant root.

(*ix*) To lower down the water level below the root zone.

Method of Drainage

Drainage system depends on the following factors :

(*i*) Topography and gradient of the land.

(*ii*) Types of the soil.

(*iii*) Permeability characteristics of the soil.

(*iv*) Crop and its water tolerance capacity.

(*v*) The quantity of water to be disposed and its speed.

There are generally two methods that are employed for drainage of crop fields :

1. Open Drainage : The open drains are easy to construct and maintain. The open drains are generally employed for drainage of surface water in our country. Open drains are being increasingly uses. Open drains are very suitable for an area where rainfall is heavy and drainage of more water is needed within a short period or time. Drainage channels are laid on the lowest contours of the land and they are utilised for draining of excess water from the land. The soils, having impervious lower layer, are best suited for open drainage. But this system have the following disadvantages :

(*i*) Wastage of land is the main disadvantage of this method as the drains occupy considerable land.

(*ii*) Drains within the land hamper the cultivation practices.

(*iii*) Soil erosion is more.

(*iv*) The drains need repairing every year. As a result, cost of cultivation is increased.

(*v*) Weeds grow on the side of drainage channels and their seeds are disposed of every where.

(*vi*) Drainage of subsurface water is not possible.

2. Closed Drainage : The closed drains are laid underground. The closed drains are desirable for a place where the land is costly. The closed drains save the land, but it is very costly. Generally tile drains, mole drains, bush stone drains etc. are used as sub-surface drains.

(a) Tile Drains : The tile made of clay or concrete pipe or perforated steel pipes are mostly used for tile drains. The tile is about 30 to 50 cm. in length and 7 to 12cm. in diametre. The tiles are installed end to end in the field after digging trenches with 2–3 millimetre spacing in the joint. The drains are usually spaced 50–150 feet apart according to the type of soil. Excess water enters the system through the space in between the two tiles and conveyed along the gradient (Dastane, 1980).

The tiles are placed in underground soil in the following two methods :

(*i*) *Herring bone or Fishbone Method :* Herring bone pattern is used if the above submain is laid in a depression. The laterals join from each side alternatively and collects the water from eitherside.

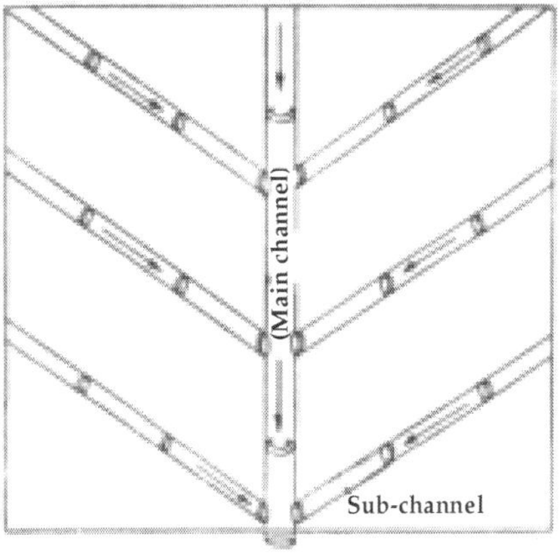

Fig. 7.15 : Herring bone method of drainage

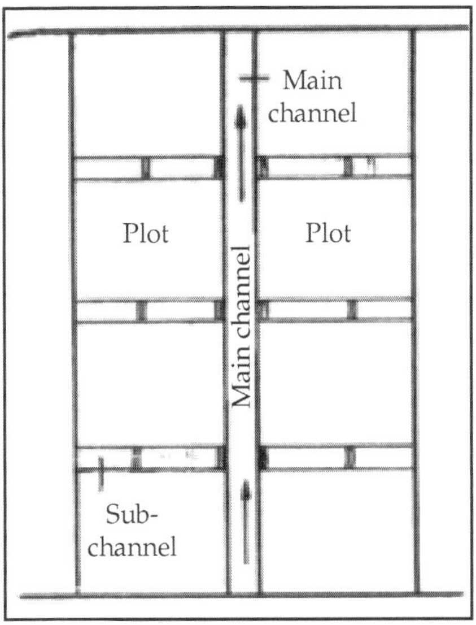

Fig. 7.16 : Grid iron method of drainage

(*ii*) *Grid Iron Method :* The grid iron method must be used where the land is practically level. It is a good method when the entire area is to be drained. Laterals enter the submain one side only to minimize double drainage.

The distance and depth of drains vary with the types of soil as follows :

Soil type	*Distance of the drains*	*Depth of the drains*
Sandy-loam soil	9–12 metre	90–120 cm.
Loamy soil	6–9 metre	75–90 cm.
Clay soil	3.5–6 metre	60–75 cm.
Peat soil	5.5–6.3 metre	100–120 cm.

(*b*) *Mole Drains :* Mole drains are unlined, circular earthen channels in the subsoil made with a tractor drawn mole plough. The channels are made 3.0–3.5 metres apart. The channels receive and discharge the water to drainage ditches at the edge of the field. Mole drains are suitable only on heavy clay soils.

(*c*) *Bush and Stone Drains :* Trenches are made in the field and they are filled up with bush and stone. Water may pass through the open space of bush and stone. Afterwards, the bushes are dried and decomposed and made into a fine channel. Drainage is continued until the drains are closed with soil. There is every possibility of closing the drains with soil.

REFERENCES

Ashawa G.H. Irrigation Engineering (1999), New Age International Publishers, New Delhi.

Balasubramaniyan P. and S.P. Palaniappan Principles and Practices of Agronomy (2004), Agrobios (India), Jodhpur.

Hiran K.S. Jaspal Singh and Acharya M.S. Irrigation Scheduling (1990), C.B.S Publishers and Distributors, New Delhi.

Lenka D., Irrigation and Drainage (1991), Kalyani Publishers, Ludhiana.

Michel A.M. Irrigation Theory and Practice (1978), Vikas Publishing House Pvt. Ltd., New Delhi.

Misra R.D. and Ahmed M. Manual on Irrigation Agronomy (1987), Oxford and IBH Publishers Co., New Delhi.

Murty V.V.W. Land and Water Management Engineering (1985), Kalyani Publishers, Ludhiana.

Pande S.C. Principles and Practices of Water Management (2003), Agrobios (India), Jodhpur.

Parihar S.S. and Sadhu B.S. Irrigation of Field Crop : Principles and Practices, Indian Council of Agricultural Research, New Delhi.

Sivanappan R.L. and K.R. Karan Gowder. Irrigation and Drainage, Agrobios (India), Jodhpur.

8

WEED AND WEED CONTROL

8.1 INTRODUCTION

India is an agricultural based country. Agriculture is a main source to meet our basic need. Crop cultivation is an important aspect of agriculture to meet the demand of growing population. Crop cultivation suffers due to attacks of pest. Pest are of four types as follows :

1. Weed,
2. Insect,
3. Diseases, and
4. Other Pests.

e.g. Nematode, store grain pest, birds, bat, rat etc. Damages caused by different pests are as follows (in percentage) :

Weeds : 45

Diseases : 30

Insects : 20

Other Pests : 5

8.2 DEFINITION OF WEED

Weed are defined in different ways by different scientists as follows :

(*i*) Weed is a plant growing where it is not desired.

–I.S.A. (1987)

(*ii*) Weed is defined as a herbaceous plant not valued for use or beauty, growing wild and rank and regarded as cumbering the ground or hindering the superior vegetation. *–Onions (1956)*

(*iii*) A weed is a plant growing out of place where it is not wanted, either because of its inherent disagreeable or poisonous character or because it is taking the place reserved for something else.

–Encyclopedia Americana (1962)

(*iv*) Weed is a general term for troublesome or otherwise undesirable plant usually introduced, grows without intentional cultivation. *–Herbert (1962)*

(*v*) Weed is used by gardeners in the sense of a plant growing without human encouragement, compound names of lower plants/animals. *–Peter (1967)*

(*vi*) Weeds are plants growing in places and at times when men wanted either some other plants to grow or no plants at all.

–NAS (1971), Gupta and Lamba (1978), Rao (1983)

(*vii*) Weeds are abnoxious, unwanted, ugly and persistent plants which are injurious for the crop, interfere with agricultural operation, increasing cost of labour and finally reducing the crop yield. *–Robinson (1973)*

Weeds may be of two types :

(*i*) *Absolute Weeds* : There are non-useful, unwanted, persistent and harmful in potentialities. *Argemone mexicana, Cyperus rotundus, Mikania scandens, Phalaris minor* etc. are the examples of this weed.

(*ii*) *Relative Weeds* : These are also known as Rouges. Rouges are off type plants in the cropfields. They are economical plants but not desirable by the cultivators as this type of plant grows in unwanted place. The jute plant, an economical crops, if grown in paddy field should be roughed out to get the yield of the principal crop. Even in an area where Ratna is cultivated, off type variety of Pankaj or IR–50 is found to grow, they should be uprooted to maintain the purity of seed and these unwanted plants are termed as weed.

8.3 CLASSIFICATION OF WEEDS

Weds are classified in different ways as is done with the economic crops. But the following two methods of classification of weed are considered very useful.

1. Classification of Weeds According to Ontogeny (Life Cycle) : Weeds are classified into three broad groups as follows:

(*a*) Annuals : The weeds that complete their life cycle within a year or season are called annual weeds. They are associated with annual crops and propagated mainly through seeds. Annuals are subdivided according to the season of prevalence.

(*i*) ***Kharif Seasonal Annuals of Kharif Weeds :*** The weeds that grow during the summer and attains maturity with the onset of winter and finally dies are called summer annuals. They complete their life cycles during the warm wet season (*i.e.* from June to September).

Example : Echinochloa colonum, Echinochloa crusgalli, Heliotropium indicum, Digera arvensis Cyanotis axillaris etc.

(*ii*) ***Winter Seasonal Annuals or Rabi Weeds :*** The weeds that germinate and grow during winter and die at the onset of the summer are called winter annuals. They complete their life cycle during dry season (*i.e.* from October to February).

Example : Chenopodium album, Vicia hirsuta, Avena fatua, Phalaris minor, Asphodelus tenuifolius Melilotus alba etc.

(*iii*) ***Summer Seasonal Annuals or Zaid Weeds :*** The weeds that grow and complete their life cycle during hot dry season (*i.e.* February to June) are called summer annuals.

Examples : Solanum nigrum, Argemone mexicana, Physalis minima, Portulaca oleracea etc.

(b) Biennials : Biennials weeds complete their life cycle in two years or in two seasons. They grow vegetatively during first year and starts flowering and sets fruits during second year. So they need two distinct seasons, one for vegetative and another for reproductive growth. Biennial weeds are less troublesome than annual weeds as they are harvested along with the crops before flowering and fruiting.

Example : Melilotus alba, Eichorium intybus, Alternanthera achinata etc.

(c) Perennials : Perennials weeds live for more than two years and may live almost indefinitely. These weeds may propagate by seeds and vegetative parts like roots, rhizomes, tubers, bulbs etc. or both. Even if the aerial parts of these weeds die or are damaged, the underground parts may continue to live and starts new growth forming flowering shoots and producing seeds year after year. The perennials weeds may be classified into three groups on the basis of their vegetative reproduction.

(*i*) ***Simple Perennials :*** These perennials may reproduce solely by seeds. Vegetative reproduction occurs only when roots and stems are mechanically cut or damaged. The each cut piece may grow individually forming separate roots and stems.

Examples : Lantana camara, Ipomoea carnea, Dactyloctenium egyptium etc.

(*ii*) ***Bulbous Perennials :*** This perennial propagates through bulbs, bulblets or tuber as well as by seeds.

Examples : Cyperus rotundus, Allium vineale, Asphodelus tenuifolius etc.

(*iii*) ***Creeping Perennials :*** These perennials may propagate by lateral extension of stem along the soil surface, roots or by seeds. In some creeping perennials new shoots may develop either from roots or rhizomes.

Examples : Cynodon dactylon, Convolvulus arvensis, Imperata cylindrica, Marsilea quadrifolia, Sorghum halipense, Oxalis corniculata etc.

2. Classification of Weeds According to the Type of Habitat : Weeds are of different types according to the type of habitat :

(*a*) Terrestrial Weeds : The weeds that grow on crop field are known as terrestrial weeds.

Examples : Chenopodium album, Cyperus rotundus, Melilotus alba, Solanum nigrum, Cynodon dactylon etc.

(*b*) Acquatic Weeds : Acquatic weeds are water inhabitant plants. They are classified into the following three types.

(*i*) ***Floating :*** This weed float freely either on the water or in a fixed area. They are or may be free floating (*e.g. Eichhornia crassipes, Pistia stratiotes* etc.) and rooted floating (*e.g. Ludwigia adscendens, Trapa bispinosa* etc.).

(*ii*) ***Submerged :*** These weeds are anchored to the bottom of the ditch and grow entirely beneath the surface of water.

Examples : Vallisneria spiralis, Hydrilla verticillata, Myriophyllum heterophyllum etc.

(*iii*) ***Emerged :*** This weeds have their roots beneath the surface of water but the leaves and stems peep above the water line.

Examples : Typha elephantina, Sagittaria sagittifolia, Nelumbo lutea, Monochoria vaginalis etc.

8.4 CHARACTERISTICS OF WEEDS

Weeds have the following characteristics :

(*i*) Weeds are hardy and can grow under adverse edaphic, climatic and biotic condition *e.g. Lantana camara, Eupatorium odoratum* etc.

(*ii*) Weeds are in general prolific with abundant seed production potentialities *e.g. Amaranthus sp. Chenopodium album etc. as well as by seeds.*

(*iii*) Weeds propagate vegetatively by rhizomes (*e.g. Cyperus rotundus*), bulbs (*e.g. Allium vineale*) stem (*e.g. Cynodon dactylon, Paspalum distichum* etc.) as well as by seeds.

(*iv*) The seeds or many weeds species are specially designed in such a way that they are easily dispersed by water (*e.g. Amaranthus viridis, Amaranthus spinosus, Cyperus cephalodes* etc.) wind (*e.g. Blumea lacera, Ageratam conyzoides, Vernonia cinerea* etc.) and by animal or human beings (*e.g. Xanthium strumonium, Andropogon aciculatus, Achyranthes aspera* etc.) some weeds have explosive mechanism for seed dispersal (*e.g. Ruellia prostrata* etc.)

(*v*) Weed seeds remain dormant and viable for 30 to 40 years *e.g. Chenopodium album Cypers iria* etc. Dormancy acts as a safely mechanism against adverse environmental conditions.

(*vi*) The weeds have disagreable flavour and taste (*e.g. Parthenium hysterophorus, Calotropis gigantea* etc.) or equipt with spines or thorns (*e.g. Solanum xanthocarpum, Solanum khasianum* etc.) which is not accepted by cattle and as a result of which these weeds grow luxuriantly under any situation irrespective of soil reaction.

(*vii*) The seeds of some weeds are very similar in shapes and size to the crop seeds and their separation is very difficult *e.g. Argemone mexicana and mustrad* (*Brassica* sp.), *Asphodelus* sp and onion (*Allium cepa*), *Avena fatua* and oat, barley or wheat etc.

(*viii*) Weeds are euryoecious (wide range of tolerance) compared to crop plants which are more stenoecious in nature. Some weeds are able to survive and multiply under such climatic and soil condition where the crop plants can not survive *e.g. Eupatorium odoratum, Glycosmis pentaphylla, Mikania scandens* etc.

(*ix*) The seeds of weed do not germinate at time. But they germinate in succession and create problem in crop field *e.g. Chenopodium album* in potato field.

8.5 DISPERSAL OF WEEDS

Dispersal means the spreading of weeds from one place to

another place. Dispersal of weeds generally takes place through seeds and vegetative parts.

1. Dispersal of Weeds through Seeds : Seeds are protectors and propagators of their kind. Weed seeds remain viable for a long time and their dispersal mainly takes place through natural and human agencies.

(A) *Natural Agencies :* The following are some of the devices through which the natural agencies work.

(*i*) *Through Wind :* The weed seeds having parachute like structure called pappess are generally dispersed through wind. The seeds of most plant belonging to the compositae family are dispersed by winds from one place to another to invade new areas. The following weeds are generally dispersed through wind. *Ageratum conyzoides, Blumea lacera, Sonchus arvensis, Calotropis gigantea, Vernonia cineria* etc.

(*ii*) *Through Water :* The aquatic weeds are generally dispersed through water. The weed seeds are carried through irrigation and drainage channels and invade new areas. The following weeds are generally dispersed through water : *Amaranthus viridis, Amaranthus spinosus, Cyperus cephalotes*. The seed of *C. cephalotes* having corky outgrowth in the body floats in water and thus carried to a long distance.

(*iii*) *By Animals :* Animals, birds etc. have a key role in the dispersal of weed seeds. The weed seeds having hooks (*e.g. Xanthium strumonium*), scaly bracts (*e.g. Achyranthes aspera*), spines (*Chorkanta*) etc. are dispersed by animals.

(*iv*) *By Explosive Fruits :* Fruits of some of the weeds (*e.g. Argemone mexicana, Ruellia prostrata* etc.) burst open as soon as the seed attains maturity and shaken out from the pores of the fruits. This type of explosive mechanism helps in the dispersal of weed seeds.

(B) *Human Agencies*

(*i*) *Sowing of Impure Seeds :* The sowing of crop seed

admixtured with weed seed (*e.g. Phalaris minor* in wheat, *Argemone mexicana* in mustard, *Echinochloa colonum in paddy, Eichorium intybus* in *berseem*) is one of the most important means of dispersal of weeds. This is occurred if the cultivator is not alert about the purity of crop seed.

(*ii*) *Through Implements :* The seeds or vegetative parts of weeds is carried from one land to another through the agricultural implements (*e.g.* Country Plough, M.B. plough, Tractor, Power tiller, Wheel hoe etc.) to cause infestation in new areas *e.g. Dicanthium annulatum, Chenopodium album* etc.

(*iii*) Through F.Y.M., compost etc. the seeds of some weeds (*e.g. Chenopodium album, Amaranthus viridis, Amaranthus spinosus* etc.) are dispersed through organic manuers. The weed seeds may remain viable in the organic manure and carried to new land where its infestation was not there.

2. Dispersal of Weeds Through Vegetative Parts : Most of perennial weeds and few annuals are reproduced and spread vegetatively in addition to seeds propagation. The following weeds spread vegetatively through the following vegetative parts :

(*i*) *Rhizome : Sorghum halipense*

(*ii*) *Tuber : Cyperus rotundus, Cyperus iria* etc.

(*iii*) *Runner : Cynodon dactylon*

(*iv*) *Bulb : Asphodelus tenuifolius*

(*v*) *Creeper : Convolvulus arvensis.*

8.6 CROP-WEED ASSOCIATION

There are some weeds which prefer to grow in crop field and some of them have a distinct association with some specific crops. Following weeds are almost always associated with the respective crop plants.

(*i*) **Rice :** *Echinochloa colonum, Echinochloa crusgalli, Cynodon dactylon, Marsilea sp. Ipomoea reptans* etc.

(*ii*) **Wheat :** *Cynodon dactylon, Chenopodium album, Phalaris minor, Argemone mexicana, Melilotus sp. Leucas aspera, Enagallis arvensis* etc.

(*iii*) **Maize, Jowar Bajara and Groundnut :** *Cynodon dactylon, Cyperus rotundus Echinochloa colonum, Eclipta alba, Eleusine indica, Croton sparsiflorum* etc.

(*iv*) **Barley :** *Chenopodium album, Phalaris minor, Avena fatua* etc.

(*v*) **Potato :** *Chenopodium album, Cynodon dactylon, Echinochloa colonum, Melilotus sp. Cyperus rotundus. Argemone mexicana, Anagallis arvensis* etc.

(*vi*) **Jute :** *Cyperus rotundus, Cynodon dactylon, Echinochloa colonum, Euphorbia hirta, Eleusine indica, Digera arvensis* etc.

(*vii*) **Cotton :** *Cynodon dactylon, Cyperus rotundus, Amaranthus* sp, *Eleusine* sp., *Celosia argentea, Digera arvensis, Trianthema monogyna, Sorghum halipense* etc.

(*viii*) **Sugarcane :** *Cyperus rotundus, Cynodon dactylon, Chenopodium album, Euphorbia hirta, Digera arvensis, Sorghum halipense, Commelina benghalensis, Croton sparsiflorum,* etc.

(*ix*) **Rape and Mustard and Linseed :** *Chenopodium album, Spergula arvensis, Cyperus rotundus, Vicia* sp., *Fumaria parviflora, Anagallis arvensis, Melilotus* sp. etc.

(*x*) **Gram, Lentil, Pea :** *Chenopodium album, Cyperus rotundus, Melilotus* sp., *Vicia* sp., *Asphodelus teneuifolius, Anagallis arvensis, Fumaria parviflora,* etc.

(*xi*) **Green Gram (Moog), Black Gram (Urd), Cow-Pea and Soybean :** *Cyperus rotundus, Sorghum halipense, Trianthema monogyna, Digera arvensis, Amaranthus sp., Phyllanthus niruri, Commelina benghalensis* etc.

(*xii*) **Sunflower :** *Cyperus rotundus, Cynodon dactylon, Digera arvensis, Euphorbia hirta, Trianthema monogyna, Commelina benghalensis, Phyllanthus niruri, Eleusine* sp. etc.

8.7 CROP-WEED COMPETITION

Plant competition is a powerful force for the eliminator of some weaker plants. Plants require nutrients, water, light, oxygen and carbon-dioxide for their growth and development. Weeds compete with the crop plants for those growth factors and the competition is maximum at the early growth stages of the crop. But the competiting ability of the weed becomes practically nil as soon the crop attains maturity. Weed needs more nutrient and water than crop plants. The weeds in general grow faster and shade the crop plant if they are not controlled at proper time and afterwards interfere in the growth of crop

Table 8.1 : Allelopathic Effects of Weed on Crop Plant

S.No.	*Weeds*	*Crop Plant*	*Effect*
1.	*Avena fatua* (Seeds)	Wheat	Poor germination.
2.	*Parthenium sp* (leaves)	Wheat, Maize	Central growth is retarded. Affect germination and seedling growth.
3.	*Cyperus rotundus* (Rhizomes)	Wheat, Chick pea	Dry matter production is retarded.
4.	*Amaranthus spinosus* (Leaves and inflorescence)	Wheat	Vegetative and reproductive growth is retarded.
5.	*Argemone mexicana*	Wheat, Finger millet.	Affects germination and seedling growth.
6.	*Parthenium* sp (Root)	Wheat, Maize and Sorghum	Affects germination and seedling growth.
7.	*Solanum* sp (Root)	Wheat, Maize and Sorghum	Affect germination and growth.
8.	*Cyperus esculentus*	Maize, Soyabean	Affects dry matter production.
9.	*Cyperus rotundus* (Tuber and plant residues).	Chick pea, Pearl millet	Affects tillering and dry matter production.
10.	*Celosia argentea*	Pearl millet	Affects germination and growth.
11.	*Datura* sp.	Sunflower	Affects germination and growth.

Source : All about weeds. Subramanian, Ali and Kumar, 1993, Kalyani Publishers, New Delhi.

plant. Weeds have allelopathic effect on crop growth. Some weeds excrete certain chemicals into the soil which inhibit the germination and growth of other plants in their vicinity and the phenomenon is called allelopathy. Weed competition can result in complete crop failure.

Table 8.2 : Allelopathic Effect of Crop Plants on Weed

S.No.	*Weeds*	*Crop Plant*	*Effect*
1.	Wheat, Oat, Pea, Buckwheat	*Amaranthus* Sp *Chenopodium album*	Suppresses growth and accumulation of biomass and leaf area.
2.	Wheat (Cold water extracts of straw)	*Abutilon* Sp and *Ipomoea* Sp	Affects germination and growth.
3.	Corn (Root exudation)	*Amaranthus* Sp and *Chenopodium album*	Inhibits growth.

Source : Principles and practices of Agronomy. S.S. Singh, 2001, Kalyani Publishers, New Delhi.

Result of Competition : Competition between crop and weeds leads to the reduction in growth and yield of the crop.

1. *Reduction in Crop Growth* : Weeds compete with crop plants for nutrients, water, light, space, oxygen and carbon dioxide and causes heavy reduction in the growth of crop. But the extent of reduction depend on the following :

(*i*) Readiness and uniformity *of seed germination.*

(*ii*) Rapid development of foliage at the seedling stage and rate of seedling growth.

(*iii*) Presence of a large Photosynthetic surface.

(*iv*) Rapid development of root system.

(*v*) Method of cultivation practices followed.

The early growth phases of the plants are found to be more critical in respect to competition with weeds and the competition is most acute when and where available resources become limiting to plant growth.

Table 8.3 : Loss of Yield of Different Crops Due to Weed (Bhan, 1994)

Crop	*Yield reduction (%)*
Paddy (Drilled)	70–80
Paddy (Transplanted)	30–40
Maize	50–70
Sorghum	50–70
Pulses (Kharif)	40–60
Pulses (Rabi)	20–30
Groundnut	40–50
Mustard	10–15
Soyabean	40–60
Cotton	40–50
Sugarcane	20–30

2. *Reduction in Crop Yield* : Yield is associated with the proper growth of the crop. The weed interferes in the growth of crop and the yield is reduced. Different weeds compete with the crop differently. For examples *Brassica arvensis*, if grown in association with wheat reduces the number of tiller and fertile shoot of the wheat crop and *Raphanus raphanistnum* causes heavy reduction of ear head size in addition to the reduction of number of tiller and fertile shoots.

8.8 ECOLOGY OF WEEDS

Ecology of weeds means, the inter-relationship of weeds with climate, soil, crops and other biotic factors.

1. Weeds and Climate : The growth and development of weeds are primarily governed by the climate and the distribution of weeds is related to climate in the same way as the economic crops. Weeds differ in their adaptability and grow better in certain climatic condition as follows:

(*i*) ***Desert land*** : *Opuntia dillenii.*

(*ii*) ***Humid areas :*** *Canada thisttle* (*Cilrsium arvens*).

(*iii*) ***Dried regions :*** Field bind weed. (*Convolvulus arvensis*) quack grass (*Agroplyron repens*).

(*iv*) ***Mesophytic condition :*** *Amaranthus viridis.*

(*v*) ***Aquatic :*** *Eichhornia crassipes, Ipomoea* Sp.

2. Weed and Soil : Weeds in general grow in any type of soil. But few weeds species are directly related to any particular soil. The following weeds are equally distributed in all types of soils.

Chenopodium album, Raphanus raphanistrum, Cirsium arvens etc.

There are some weeds that grow well in different soil reaction. *e.g.*

(*i*) ***Saline Soils :*** *Suaeda maritima, Salsola kali (saltwort)* etc.

(*ii*) ***Alkali Soils :*** *Distichlis spicata, Sporobolus* aeroides, *Sporobolus diander, Cressa ereceta* etc.

(*iii*) ***Acid Soils :*** *Pteridium equilinum, Rumex acetosella* etc.

3. Weeds and Crop : Weeds have definite association with certain crop. The crop weed association in certain cases become so close that it is very difficult to differentiate the weed and crop seedling at the early growth stage. As for example *Echinochloa colonum* and *Phalaris minor* resembles paddy and wheat seedling respectively. This phenomenon is known as MIMICRY.

8.9 HARMFUL EFFECTS OF WEEDS

Weeds interfere with man's utilization of land and water in various ways. The extent of losses due to weeds depends on the intensity of infestation, time of occurrence and type of weeds.

1. Competition Between Weeds and Crops for Growth Factors : Weeds compete with crop plants for water, air space, light, mineral nutrient and other factors in the micro environment. Weeds are capable of absorbing more amount of nutrients from soil than the crop plants. In dry land, weeds exhaust the soil moisture and put the crop under water stress earlier and thus reduce the yield (De, 1990). The Parasitic weeds may even kill the crop plants.

2. Increase the Cost of Cultivation : Frequent tillage is needed for controlling of weeds and more than 50 per cent tillage cost is incurred for the same. Weeds interfere with agricultural operations and harvesting is difficult and delayed due to presence of weeds. This increases the total cost of cultivation.

3. Need for More Labour in Agriculture : More labour are required for controlling weeds. Weeds increase the cost of labour which ultimately increase the cost of cultivation.

4. Increase the Cost of Agricultural Implements : The weed infested crop fields needs frequent weedbug to get crop. The wear and tear of farm implement may be the result of frequent weeding, which ultimately increases the cost of cultivation. Weed reduces the efficiency of farm implements.

5. Increase the Irrigation Requirements : Weeds absorb more water from soil than crop plants and thus increase the frequency and depth of irrigation water which is very often expensive and difficult to oftain. The transpiration ratio of weed is high compared to crops, the amount of water required to produce unit amount of dry matter is known as transpiration ratio. The transpiration ratio of *Cynodon dactylon* is around 813 compered to 430 for pearl millet and 430 for Sorghum.

6. Reduce the Selling Price of Crops : There is very possibility of admixture of weed seed with agricultural produce in the weed infested Plots and the quality of the produce becomes Poor. Agricultural produce mixed with seeds and trashes of weeds are poorly accepted by the customers and fetches a lower price, because weeds may lower the market value of the crop by reducing quality.

7. Decreases the Selling Price of Land : The selling price of land infested with perennial and problematic weed (*e.g. Saccharum* sp., *Phichea* sp., *Lantana camara, Parthenium hysterophorus* etc. becomes low as they incur heavy expenditure to clean or to eradicate the weeds.

8. Weeds Harbour Insectpests and Disease : Weeds present in off-season on field, bunds, wastelands, irrigation channel etc.

provide shelter to insects pest. They act also as alternate host of number of disease-producing organism. Weeds harbour the pest and diseases which attack the crop sown subsequently.

9. Weeds Causes Harm to the Living Beings : There are some weeds that cause ailment to live stock as follows :

(*i*) *Datura fastuosa, Datura stramonium, Datura metel* etc.

(*ii*) *Solanum nigrum.*

(*iii*) *Abrus precatorius.*

(*iv*) *Withania somnifera.*

The seeds (*e.g. Datura* sp., *Abrus precatorius* etc.) and berry (*Withania somnifera*) of some weeds are poisonous to both animals and human beings.

Table 8.4 : List of Weeds that Acts as Host of Insect Pest and Diseases

Name of Weed	*Crop*	*Insectspest/Diseases*
1. *Ageratum conyzoides* (kukur sonka)	Vegetables Tobacco Potato other up land crop	Virus, Wilt, Mosaic. Leaf curl, wilt. Wilt Virus, Nematode
2. *Agrophyron repens*	Wheat	Black rust
3. *Amaranthus viridis* (Note) *A. spinosus* (kantanote)	Upland crops Redgram Tobacco, Groundnut	Nematodes, Virus, Rosette. Gram caterpillar Mosaic, Blight, Tikka disease, Nematode.
4. *Brachiaria platyphylla* (kara grass)	Paddy	Tungro virus, Rice leaf hopper.
5. *Capsella bursapastoris* (Ban sarson)	Vegetable	Mosaic, (Cucumber), Ring spot (Beet), Black ring spot (Cabbage) Mosaic (Cauliflower).
6. *Chenopodium album* (*Bathua*)	Potato	Yellow dwarf, Blight, Mosaic, Wilt Vegetable (Beet, Radish, Turnip etc.) Mosaic, Stripe.
7. *Commelina benghalensis* (Kansira)	Groundnut Vegetables	Nematodes, Rosette, spotted wilt etc.
8. *Cynodon dactylon* (Durba grass)	Paddy	Bacterial leaf blight, Stripe disease Nematodes.

Name of Weed	*Crop*	*Insectspest/Diseases*
9. *Cyperus rotundus*	Several crops.	Mosaic, Nematodes, Virus.
10. *Digitaria sanguinalis*	Wheat	Nematodes, Streak
Echino-chloa colonum (Jangli dhan)	Paddy	Rice bug, virus disease, stem borer
	Sugarcane	Mosaic
11. *Eleusine indica* (Kedai)	Paddy	Leafcurl, Nematodes, Virus
	Maize	Dwarf mosaic, Leafgall
	Groundnut	Rosette
	Sugarcane	Streak Disease.
12. *Echinochloa crusgalli* (Shyama grass)	Paddy	Tungro Virus, Stemborer
	Maize	Dwarf mosaic
	Barley	Stripe mosaic
	Sugarcane	Mosaic
13. *Euphorbia hirta*	Groundnut,	Nematodes, Rust, Nematodes
	Tobacco	Leaf curl
	Wheat	Nematodes, Rust.
14. *Fimbristylis liftoralis* (Bara javani)	Paddy	Nemotodes, Tungro virus
15. *Tschaemum rugosum* (Moraro grass)	Paddy	Gallfly, Tungro virus
16. *Leersia hexandra* (Para grass	Paddy	Brown spot disease, caseworm.
17. *Paspalum congatum* (hilograss)	Paddy	Yellow dwarf,
	Sugarcane	Blight, Mosaic
18. *Portulaca oleracea* (Punarnova)	Vegetables	Brown ring, Curl disease.
19. *Solanum nigrum* (Banbegum)	Potato	Leaf curl
	Tobacco	Leaf roll, Ring spot
	Vegetables (Cucumber chilli, Tomato etc.).	Bunchytop, Mosaic, Nematodes
20. *Sorghum halipense*	Paddy	Gall fly
21. *Saccharum spontaneum*	Maize	Downy mildew.

10. Decrease the Yield of Crop : Weeds compete with crop plant for the growth factors and affect the growth of standing crop. Weeds reduces the yield of crop by shareing the growth factors. The average losses due to weed are 15-20 per cent in cereals and 30-40 per cent in potato, maize, vegetables, fodder and root crops. Yield losses due to weeds were higher in broadcast paddy (27.3 per cent) than the transplanted paddy (21.6 per cent). So weeds have a direct effect on yield of the crop.

11. Reduces the Quality and Quantity of Farm and Animal Products : The milk produces undesirable flavour if the milch animals eat some weed such as *Asphodelus tenuifolius, Cleome viscosa, Paederia foetida* etc. When poisonous weed (*e.g. Datura stramonium* etc.) is eaten by animals causes death, disorder or permanent deformity or disability. The fruits (*e.g. Xanthium strumarium, Achyranthes aspera* etc.) hooks of some weed (*e.g. Martynia annua*) entangle with the wool of sheep and impare the quality which fetches lower price.

12. Causes Health Problems to Animals and Men : Weeds harbour mosquitoes that cause or transmit diseases like malaria, encephalitis, filaria, dengue fever, yellow fever etc. Some weeds are poisonous and other produce Pollen which causes allergies to human when they come in contact with them.

Table 8.5 : Health Problem Caused By Weeds

Name of weed	*Health problem caused by respective weeds*
(*i*) *Parthenium hysterophorus* and *Ambrosia sp.*	Dermotitis (Skin allergy)
(*ii*) Pollen of *Ambrosia sp.* and *Franseria sp.*	Hay fever and Asthma.
(*iii*) *Urtica sp.*	Itching and inflammation.
(*iv*) Brush weed	African sleeping sickness.

13. Impair the Purity of Varieties : Cross pollination with compatible weed plants impair genetic purity of crop seeds.

The aquatic weeds are very harmful due to the following reasons :

(*i*) Weeds that grow at the bottom or that float on the moving water impede water flow in canals, channels, river etc. and the irrigation may be delayed resulting in poor irrigation efficiency.

(*ii*) Weeds that grow in catchment areas impede drainage resulting in damage to crops.

(*iii*) Weeds are a menace to fisheries. They are harmful to fishes in various ways such as :

(*a*) They affect the growth of phytoplankton which is the major food constituents of fishes.

(*b*) They produce toxic gasses and increase the acidity of water which is harmful to fish.

(*c*) They liberate carbondioxide which when dissolve in water affects the respiration of fishes.

(*d*) They affects fish breeding and netting.

(*e*) They affect the rearing of ducks.

(*f*) Weeds share in the food of fishes and they take up nutrients that are applied for fishes. They causes the deficiency of food for fishes as they take up major portion of the applied nutrients.

(*g*) The shortage of light and oxygen occurs in water reservoirs when water hyacinth, algae etc. grow luxuriantly. The fishes suffer most due to shortage of oxygen and they sometime die.

(*h*) Weeds cause the curtailment of species favourable for the growth of fishes as they do not move freely.

(*i*) Weeds harbour predator, insectpests etc. and they do harm to the fishes.

(*iv*) Weeds impair the cultivation of aquatic crops such as deep water and flooded paddy, lotus, cork plants, *Ipomoea sp*. water chestnut, reeds etc.

(*v*) Weeds prevent or spoil the recreational value of water bodies. Weeds impair also angling, boating and water sports.

(*vi*) Weed on decomposition pollute the water and make the water unfit for irrigation, drinking and washing.

(*vii*) Irrigation water from weed infested water bodies carries seed or seeds of different species (*e.g. Chara* sp., *Pistia* sp., *Nitella* sp., *Hydrolea zeylanica, Cyperus difformis, Eichhornia crassipes* etc.) to the crop field when they grow profusely (De, 1990).

8.10 MECHANISM OF SURVIVAL OF WEED

1. Seed Production : Most of weeds propagates through viable seed. The survival of weeds depends on the production of viable seeds. They are in general prolific with abundant seed production potentialities. *e.g. Amaranthus* sp., *Chenopodium album* etc. The seeds of the weed growing together with crops shed in the land before harvesting of the crop. They germinate and do competition with the succeed crop.

2. Dispersal of Seed : The seeds of weeds disperse through wind, water, animals, human beings etc. due to having some characteristics. The weeds seeds may be spread through crop seeds that are admixed with weed seed.

3. Germination of Seed : The weed seeds remain viable even in adverse environment. It is found that weeds grow in a field remaining submerged for long time when the land dries up. Weeds do not loose their viability even when they remain under submerged condition. Under such condition, weed seeds remain in dormant condition. Weeds seeds do not germinate at a time, but they germinate by instalments.

As a result, small weeds are seen coming up after duly uprooting all grown up weeds.

4. Dormancy of Seeds : Seeds do not germinate even under suitable condition for germination. This condition is called 'Dormancy of seed'. Dormancy is a mechanism by which seeds remain viable for longer times. Weed seeds have the dormancy and this condition is favourable for their survival.

5. Vegetative Propagation : Weeds propagate vegetatively by different vegetative organs as well as seed as follows :

Rhizome : *Cyperus rotundus, Sorghum halipense* etc.

Stem : *Cynodon dactylon* etc.

Bulb : *Allium vineale, Asphodelus tenuifolius* etc.

8.11 BENEFICIAL EFFECTS OF WEEDS

No plant in the earth is completely worthless. Weeds have

also some beneficial effect although they do much harm in different ways. Following are the beneficial effects of weeds.

(*i*) Most of the weeds absorb nutrient from the deeper layer of soil. When such weeds are incorporated in the soil, the fertility of plough layer from where plants absorb most of their nutrients *is enriched.*

(*ii*) Weeds help in controlling soil erosion by the soil binding effect of their roots and by forming a living mulch of which resist the beating action of rain drops and sweeping action of wind.

(*iii*) Weeds when incorporated in the soil increases the organic matter content of the soil. Organic matter is the essential constituent of arable soil.

(*iv*) Some weeds such as *Echinochloa colonum, Cynodon dactylon* etc. are used for green manuring and some aquatic weeds such as water hyacinth *Eichhornia crassipes*), *Ipomoea reptans* etc. are used for composting.

(*v*) Some weeds such as *Echinochloa colonum, Cynodon dactylon, Melilotus indica, Avena fatua* etc. are used as fodder for cattle and some weeds such *as Chenopodium album, Amaranthus* sp., *Portulaca oleracea, Ipomoea reptans* etc. are used for vegetable purposes by human beings.

(*vi*) Weeds belonging to legume family such as algae, fern (*Azolla* sp.) can fix atmospheric nitrogen biochemically in their bodies and their incorporation increases the soil fertility.

(*vii*) Some weeds have medicinal value. As for example *Cynodon dactylon, Croton sparsiflorum* etc. are used in wounds, *Leucas aspera* is used against snake bite and *Argemone mexicana* seeds are used to control skin disease. The nuts of *Cyperus rotundus* are used in making scents etc.

(*viii*) A large number of weeds are used for different economic utilization. As for example, plant of *Saccharum* sp., *Imperata cylindrica, Typha elephantina* etc. are used for making thatches, *Cyperus rotundus* nuts are used for making scents, *Eulaliopsis binata, Sansevieria roxburghiana* etc. are used for making ropes and strings, *Andropogon squarrosus (vetiver), Phragmites karka* etc. are used for making mats and screens.

(*ix*) Some weeds such as cactus, saccharum etc. are used for fencing of crop fields.

(*x*) Some weeds are used for various religions and ritual purposes *e.g. Cynodon dactylon, Ocimum sanctum, Saccharum spontaneum* etc.

(*xi*) Some weeds such as *Cynodon dactylon, Argemone mexicana* etc. is incorporated in the soil help to reclaim the alkalinity of soils.

(*xii*) A large number of weeds that are sod, turf and bush types of weeds are valued for protecting different types of bunds *e.g. Cynodon dactylon, Saccharum squarrosus, Dichanthium annulatum* etc.

(*xiii*) Some weeds are used for making paper pulp *e.g. Parthenium hysterophorus, Imperata cylindrica, Eichhornia crassipes* etc.

Names of Weeds	*Composition (in Percentage)*		
	N	P_2O_5	K_2O
1	2	3	4
Achyranthes aspera	2.21	1.63	1.32
Amaranthus viridis	3.16	0.06	4.51
Cassia occidentalis	3.08	1.56	2.31
Chenopodium album	2.59	0.37	4.34
Cleome viscosa	1.96	1.53	5.81
Dactyloctenium aegyptium	2.76	0.24	1.65
Digitaria sanguinallis	2.00	3.36	3.48
Echinochloa colonum	2.98	0.40	2.96
Melilotus alba	2.45	1.53	1.85
Portilaca quadrifida	2.40	0.09	4.57

Source : O.P. Gupta, Weed Management (1998), Agro Botanica, Bikaner.

(*xiv*) The floating weed such as *Eichhornia crassipes* reduces the evaporation losses of water from different water reservoirs.

(*xv*) Some aquatic weeds are used for feeds for fishes.

(*xvi*) There are some weeds that possess very high quantity of Nitrogen (*N*), Phosphorus (P_2O_5) and Potassium and they may be used as good composting materials as shown above.

8.12 METHODS OF WEED CONTROL

Weed control methods may be classified into the following two groups :

(*A*) Preventive method (*B*) Curative or remedial measures, Curative method is subdivided into eradication and control.

A. Preventive Method : Prevention is the measures taken to prevent the introduction or spread of weed to new areas and to check the infestation prior to the germination of weed seed. To achieve this, a wise step is to be follow the principle "Prevention is better than cure."

The following methods can be adopted to prevent the introduction and infestation of weeds in crop fields.

(*i*) The weeds should never be allowed to form seed which will minimize the infestation of weeds in the following year. It has been reported that most of the weed seeds germinate even after ten years. Hence the term "one years" seeding seven years weeding is quite appropriate.

(*ii*) The crop seed free from weed seed should be used for sowing.

(*iii*) Weeds having viable seed should never be fed to cattle. Otherwise weeds will spread from dung or manure.

(*iv*) The manures free from weed seed should be used for manuring the crop. The fresh or partially decomposed organic manures containing viable weed seed should not be used.

(*v*) The seed bed should be kept free from weeds. Otherwise, the weed get carried to the main field with the seedlings of the crop.

(*vi*) The irrigation and drainage channels should be kept free from weed to avoid the spreading of weeds. The weed of such area, if present, may be prevented from seeding.

(*viii*) The irrigation water should not be contaminated with weed seeds and other propagules of weeds.

(*ix*) The threshing yard, granary and compost pit should be kept free from weeds.

(*x*) The cattle should not be allowed to move from weed infested area to clean or cultivated areas as they carry weed propagules through their fur, fleece, hoops and droppings.

(*xi*) The farm fences, roads, bunds etc. should be kept clean or free from weeds.

(*xii*) The weeds having viable seeds should not be used for organic *mulching*.

(*xiii*) The farm implements and machinery should be cleaned properly after their use in weed infested fields and before using them in clean fields.

(*xiv*) The soil from weeds infested areas should not be used in clean or cultivated areas.

(*xv*) The weeds should be destroyed as soon as they are seen to emerge.

B. Curative or Remedial Measures : These methods include control and eradication of weeds from field.

(a) Eradication Methods : Weeds must be destroyed immediately before its multiplication, dispersal and acclimatisation as and when they are seen to emerge. Eradication, that is, complete kills of weeds is considered to be a very difficult task due to the reason that they not only propagate through seed but also propagated vegetatively. Eradication of weeds at an early growth stage have been found effective.

(b) Control Methods : Control means the check of weed to some extent or reduce the number of weeds to the minimum, so that they do not affect the crop yield. The control methods may be broadly classified as follows :

1. Cultural methods.
2. Mechanical methods.
3. Biological methods.
4. Chemical methods.

8.12.1 Cultural Methods : The following cultural methods are employed for controlling the weeds.

(*i*) ***Crop Rotation :*** Rotation of crops is an efficient way to reduce weed growth which means growing of different kinds of crops on the same land in proper and well planned sequence. Weeds thrive best in a particular crop. Weeds become increasingly aggressive each succeeding year if that particular crop is grown. Crop rotation reduces the weed problem.

(*ii*) ***Cultivation of Cover Crops :*** Crops such as Kalai, Cowpea, Pea, Sweet Potato etc. is cultivated, cover the land completely within a short period of time. As a result, the weed can not get the chance to grow or if grows, may die due to want of light and air.

(*iii*) ***Changing the Seed Rate and Date of Sowing :*** Crop competition is one of the most useful and cheapest method of weed control. The crop plant density becomes more if the crop is sown at higher seed rate. The dense crops have a relatively higher competitive ability to suppress weeds or to escape competition generated by weeds.

Weeds already present in the soil starts germinating within two or three days of rainfall or irrigation. The land is ploughed to get good tilth for sowing of seed which is helpful for uprooting the germinating seed. The crop is sown afterward which faces the less weed problem.

(*iv*) ***Catch Cropping :*** Cultivation of short growing crop in between the two main crops in the field is known as catch cropping. The weed fails to grow due to presence of crops in the field.

(*v*) ***Bare Fallowing and Bastard Fallowing :*** Fallowing is one of the important tool in controlling weeds but one year's fallow should be followed by a row crop in the next year. Keeping the land completely fallow *i.e.* without cultivating any crops, the repeated tillage is given with a view to control the weeds which is known as Bare fallowing. Taking one short duration crop, the land is kept fallow and repeated ploughing is given in the rest of the period with a view to control weeds which is known as Bare fallowing.

(*vi*) ***Intercropping :*** The process of growing a short duration companion crops in between the spaces of widely spaced principal or main crop is known as inter-cropping. Intercropping suppresses weeds better than sole crop. Cultivation of short duration pulse crop (*i.e.* cowpea, green gram, soyabean etc.) as intercrop effectively smother weeds reduces weeding cost and gives bonus yield of intercrop.

(*vii*) Crop rotation should include Jute, Potato etc. The land of these crop should be kept clean and for this repeated weeding is to be given in the field. As a result, the weeds are controlled automatically.

8.12.2 Mechanical Methods : Mechanical methods include the use of differents implements and power to reduce the infestation and growth of weeds. The following mechanical methods are employed for controlling the weeds.

(*i*) ***Tillage :*** Tillage is a practical and economical methods for controlling of all types of weeds. Tillage encourages the germination of weed seeds and the subsequent tillage followed by harrowing destroys the germinating weds. Shallow ploughing is given to control annual weed and deep ploughing is given to control perennial and abnoxious weed such as *Saccharum spontaneum*, *Sorghum halipense* etc. But for controlling shallow rooted perennial weeds such as *Cynodon dactylon* etc. surface tillage should be employed.

(*ii*) ***Hand Weeding :*** Hand weeding is the most efficient and easiest method of controlling weeds in cropped fields. Pulling out of weeds by hand or uprooting weeds by using small hand tools is known as hand weeding. Hand weeding is generally done 2–4 times at an interval of 15–20 days for most of field crops. 3–4 hand weedings are necessary for long duration crops like sugarcane, cotton and upland rice. The number of hand weeding to be done depends on the growth of crop and weeds, at the critical period of crop weed competition.

Hand weeding is a time consuming, costly and back–breaking operation. This method of weed control is not economical where labour is scarce and expensive.

(*iii*) ***Hoeing :*** Hoeing includes the use of some agricultural implements such as spade, nirani, khurpi harrow, cultivator, wheel hoe etc. Wheel hoe is used to control the weeds in the row crops. The weeds are generally uprooted and removed. This method is less time consuming and less costly than hand weeding. But this method cannot always be recommended due to the fact that it can not kill the weed within the row and there is every possibility of regeneration of weed and sometimes damaging of the roots of crop plants. This method is considered useful because it improves the physical condition of soil in addition to the removal of weed.

(*iv*) ***Mowing :*** Mowing include the use of common mowing tools such as sickle, lawn-mower, Scythe etc. Mowing tools are used for cutting of weeds to ground level in non-cropped areas, lawns and gardens to improve the aesthetic value. Mowing helps to prevent multiplication of weed seed as it is employed to remove the top of the weeds before flowering. Repeated mowing can also weaken the underground parts of perennial weeds.

(*v*) ***Puddling :*** Puddling is an efficient method of controlling weeds in a Aman field. Puddling operation incorporates all weeds in the soil and they are decomposed in course of time. Puddling is an important method, of controlling some weeds such as *Cynodon dactylon, Cyperus rotundus* etc.

(*vi*) ***Digging :*** Some weeds such as *Cynodon dactylon, Cyperus rotundus* etc. can be effectively controlled by digging them with the help of pick axes or crow bars. The underground storage organs are removed and destroyed.

(*vii*) ***Cutting :*** The weeds are cut above the ground surface with the help of sickles, axes and saws. The road side weeds such as brush and trees are controlled by this method.

(*viii*) ***Flooding :*** At first ploughing is given in the weed infested field and then the filed is submerged with 15–20 cm of water for 3–8 weeks. Flooding cause the death of weeds by excluding the air when they are completely submerged. A number of weeds such as *Croton sparsiflorum Clerodendrum infortunatum, Tephrosia purpurea* etc. can be controlled effectively by flooding.

(*ix*) ***Burning and Flaming :*** This method is generally employed to control weeds in hilly areas. The weeds such as *Saccharum sp.* from irrigation and drainage channels and also on road side are controlled by burning them. The flames that emanated from burning liquid petroleum are used towards weeds in between the rows with a hood cover. Flaming kills the weeds by coagulating their cell protoplasm. The thermal death point for most of the plant lies between 45°C to 55°C (113°F – 131°F). It is quite obvious that the higher temperature kills the plant at a shorter time.

(*x*) ***Smoothering :*** Generally some non-living materials such as straws, old gunny bags, paper, plastic film and rice hulls are used as artificial mulch. Mulches when applied to the soil exclude completely the light and at the same time prevent germination of seed and top growth of weed. Mulch should be thicker for perennial weeds. A number of weeds such as *Cynodon dactylon, Cyperus rotundus, Sorghum halipense* etc. are easily controlled by this method. This method is generally employed for controlling weeds in small areas. (*i.e.* kitchen garden, flower garden etc.).

(*xi*) ***Solarization :*** The plastic sheet is placed on moist soil to trap the solar radiation. In this method, the soil temperature is increase to the lethal level for weed. Due to increase of soil temperation, the dormancy of seed is broken and the seeds start germinating. The possible mechanism of weed control by this process is direct killing of germinating seed.

8.12.3 Biological Methods : In biological weed control, a "Natural enemy" of the plant is used which is harmless to desired plants. Insects or disease organisms are the usual natural enemies. Besides this, the parasitic plants, selective grazing by livestocks and highly competitive replacement plants are other forms of biological control (Moracham, 1986). First attempt to biological weed control was carried out in the Hawai Island on *Lantana camara* in the year 1902 through the use of insects like *Agromyza lantanae, Thecla echion, T. bagochi, Crocidiana lantana* etc., particularly to avoid weedicide pollution.

Following are the outstanding and feasible examples of biological weed control :

(*i*) ***Insects :*** Prickly pear weed (*Opuntia inermis*) was controlled by *Cactoblastis eactorum* in Australia and by *Dactylopius tomentosus* in India. *Lantana camara* was controlled by *Agromyza lantanae* in Hawai and by two beetles *viz. Octotoma scabripennis* and *Uroplata giraldi* in Australia. St. Johns wort (*Hypericum perforatum*), a common roadside plant, was effectively controlled by

Chrysollia gamellata and *Zeunideplosis giorde* in Hawai and Europe respectively. *Eichhornia crassipes* can be controlled by *Gasonula punctifrons*.

(*ii*) ***Geese :*** The geese are used to control the Johnson grass (*Sorghum halepense*) in cotton in some areas in South-West America. They selectively feed on Johnson grass without causing any damage to the cotton crop. Ducks help to destroy the floating aquatic weeds such as *Lemna minor, Pistia stratiotes* etc.

(*iii*) ***Fish :*** The herbivorous fishes like grass carp (*Ctenopharygodon idella*) common carp (*Cyprinus caropio*), tellapia (*Tellapia mossambica*) etc. are generally employed for destruction of aquatic vegetation in India and abroad. Grass carp is being employed to control the submerged aquatic algal weeds like *Hydrilla verticillata, Chara fragilis, Najas mino* etc.

(*iv*) ***Animals :*** A number of herbivorous animals such as cattle, goat, camel, sheep, pigs, deer, horse etc. feed on weeds selectively. They remove the regenerated organs and top of weeds by grazing and as a result of which the reproduction and multiplication of weeds are reduced. Pigs are efficient for controlling the *Cyperus rotundus* weed.

(*v*) ***Snails :*** The submerged weeds like coontail, algae etc. can be controlled by *Marisa sp.* and other fresh water snails.

(*vi*) ***Plants :*** Cowpea sown in between the rows of sorghum reduces the growth of weeds. Kans (*Saccharum spontaneum*) and *Parthenium hysterophorus* may be controlled by growing basket grass and *Cassia tora* as their root is supposed to excrete some toxins causing death of kans and parthenium respectively.

(*vii*) ***Mites :*** Prickly pear is controlled by spider mite (*Tetranychus* sp.).

(*viii*) ***Fungi :*** A number of fungi is found to be useful for controlling some weeds. Such as *Alternaria eiccorniae* on

Eichhornia crassipes, Rhizoctinia on water hyacinth, *Fusarium* sp. on *Opuntia* sp. *Puccinia canaliculata* in *Cyperus esculentus, Puccinia lagenophorae* on *Senecio vulgaris* etc.

Table 8.6 : Commercial Mycoherbicides used in Abroad

Product	*Content*	*Target Weed*
1. Devine	A liquid Suspension of fungal spores of *Phytopthrora palmivora* which causes root in the target weed.	Strangle vine (*Morrenia odorata*) in citrus orchards
2. Collego	Wettable powder containing fungal spores of *Colletotrichum, Gloeosporoides* Sp.	Joint ventch (*Aeschynomone*) Sp. which causes stem and leaf-blight in the weed.
3. Bipolaris	A suspension of fungal spores, *Bipolaris sorghicila*	*Sorghum halepense* (Johnson grass).
4. Biolophos	A microbial toxin product as fermentation product of *Streptomyces hygroscopicus.*	

Source : O.P. Gupta, Weed Management–Principles and Practices, 1998, Agro-Botanica, Bikaner.

8.12.4 Chemical Method : In this method, chemicals are used for controlling weeds very effectively. The chemicals that are used to kill weeds are called herbicides. The use of chemicals for weed control has developed rapidly since 1944.

8.12.4.1 Classification of Herbicides

A. Classification of Herbicides According to Use : Herbicides are classified according to their use *i.e.,* methods of application and mechanism of action are as follows.

1. Selective Herbicides : The herbicides that are more toxic to one plant than another plant are called selective herbicides. They affect plant species differently by killing plants of one species, while plants of another species survive. The selectivity is not absolute, but is governed by the amount of chemical applied, the way it is applied, the degree of wetting of foliage, the amount of rainfall following, the tolerance of different plants to a specific chemical and the growth habit of the crop and the weed.

(*a*) *Foliar Application :* These treatments are made to the leaves of growing plants usually as sprays, but in few cases as dust application.

(*i*) ***Contact Herbicides :*** The plant or portions of plant that are contacted by the chemicals are killed. The contact herbicides do not move in the plant very far beyond the point of contact.

Examples : Gramoxone (Paraquate), Stamp F–34 (Propanil), Sulphuric acid, Potassium cyanate, Dicryl, Sodium arsenite, Calcium cyanamide, Minoseb etc.

(*ii*) ***Translocated Herbicides :*** The herbicides that move within the plant and destroy them afterwards are called translocated herbicides. They may be effective in destroying roots of perennial plant and a low volume applications is possible. Systematic herbicides is also a term employed for a herbicide which is translocated. The selectivity depends primarily on the physiological differences among plants.

Examples : 2,4 D, 2, 3,6 TBA, 2, 4–5 TES, CIPC, Propham, M.S.M.A. (Aasar–529), M.C.P.B. (Tropotox), 2, 4–DB, 2, 4, 5–T, Silvex, Barban, MCPA, Dalapon, Simazine (Prinzep), Atrazine (Primatol), MCPA, TCA etc.

(*b*) *Soil Application :* In this method, the herbicides are applied in the soil as per-emergence (*i.e.* before emergence of crop) and post-emergence (*i.e.* after emergence of crop) treatment. The soil must have sufficient moisture for the efficient use of herbicides. The selectivity depends upon the plant tolerance, location of herbicides in the soil and the difference in the growth habit of crop and weed.

Examples : Simazine, Atrazine, Propazine, Propham, Eptam, Avadex, Diuron, Amizen, TCA etc.

2. Non-Selective Herbicides : The herbicides that kill all vegetation in which they come in contact with irrespective of weeds and plants are called non-selective herbicides.

(*a*) *Foliar Application*

(*i*) ***Contact Herbicides :*** Through coverage of all vegetation with sufficient amount of toxic is needed to kill annual weeds, perennial weeds require retreatment (Mani, 1967).

Examples : Paraquate, PCP (Penta chloro phenol), Diquat, Ammate, Sodium arsenite, Sodium chlorate, Sulphuric acid, minerals oils etc.

(*ii*) ***Translocated Herbicides :*** The relatively few chemicals in this group affect a wide range of plants but there is as yet no truly non-selective transplanted herbicide.

Example : Dalapon, MH, 2–4–D, 2, 4, 5–T, Sodium arsenite, Amitole etc.

(*b*) Soil Application : In this method the herbicides are applied in the soil where it is desired to remove all plant or to keep areas free of plant.

(*i*) ***Soil Fumigants :*** Soil fumigants are volatile compound and they function as a gas or vapour that diffuses through soil. They are applied for killing all plant growth before seeding of more desirable species. This type of chemical is effective at warmer temperature (over 18.5°C) on soil that are well aerated by tillage and in soils with moisture levels favourable for biological activity. The chemicals that are used as soil fumigants have relatively short life in the soil. The treated area may be replanted within a month or less with no toxicity remaining.

Examples : Carbon disulphide, Vapam, Methyl bromide, Chloropicrin, Cyanamide etc.

(*ii*) ***Soil Sterilants :*** The chemicals that prevent the growth of green plant when present in the soil considered a soil sterilant. They used where it is desired to prevent all plant growth for periods varying from months to years. The length of time the soil remains sterile will depend on the Chemical used, the amount applied, rainfall, soil type, organic matter content etc.

Example : Simazine, Atrazine, Triazines, Diuron, Monouron, Fenuron, Meburon, Borates, Sodium arsenites, Sodium chlorates etc.

(*c*) Acquatic Application : A number of chemicals are used for control of submerged aquatic weeds. The chemicals are used in different water reservoirs after dissolving or emulsifying them in water with a view to control aquatic weeds.

Examples : 2.4–D, Copper sulphate, Sodium arsenites, Aqualin, Chlorinated benzenes, Aromatic solvents, Fenac, Endothal etc.

B. Classification of Herbicides According to the Time of Application : Herbicides are classified according to the time of application as follows :

(*i*) ***Pre-Planting Herbicides :*** The herbicides that are applied before planting of the crop in the field are called pre-planting herbicide. This herbicides need to be applied into soil and incorporated into soil as they are volatile in nature. This types of herbicide affect the emerging weed seedlings and they act as contact killers.

Examples : Fluchloralin, Triflurialin, Nitrofon, etc.

(*ii*) *Pre-Emergence Herbicides* : The herbicides that are applied in the soil after sowing of crop, but before a crop or weed has emerged or germinated are called pre-emergence herbicides. It is pre-emergence to crops in case of annual crops and post-emergence to weeds in case of perennial crops. Most pre-emergence chemicals are absorbed in the surface layer of the soil where they can poison seedlings as they emergence through the soil crust. If the chemicals last a suitable length of time without breakdown, a single spray can control the weeds for entire season. (Morachan, 1986). The herbicides having too long life may interfere in the germination of the next crop sown in the area. Pre-emergence herbicides provide an almost ideal weed control chemicals.

Examples : Alachlor, Fluchloralin, Methabenzthia, Zuron, Paraquat etc.

Table 8.7 : Chemical Classification of Herbicides

Chemical Class	*Chemical Subclass*	*Name of Herbicide*	*Application rate time (kg. or L per ha)*	*Crops Recommended for*	*Weeds controlled*
1	*2*	*3*	*4*	*5*	*6*
Carbamic acid esters	Carbamates	Rowmate (Sirmate)	4–8 pre and post emergence	Mustard, cotton, groundnut, potato, soybean, sunflower and tobacco	It controls both dicot and monocot weeds.
		Barban (Carbynbe) emergence	0.25–1.0 post-emergence	Wheat, barley pear and sugarbeet	Monocot (*Avena fatua*).
	Thiocarbamates	Diallate (Avadex)	1.5–3.0. pre-emergence	Sugarbeet, maize, linseed, potato, pea, urd, mung	Dicot and monocot plus *Avena fatua*
		Triallate (Avadex BW)	1–3.0 pre-emergence	Wheat, barley, pea, potato	Dicot and monocot plus *Avena fatua.,*
		EPTC (Eptam)	2–5.0, pre and post-emergence	Maize, sugarcane, potato, rice	Dicot and monocot *Echinochloa spp., Digitaria spp., Setaria spp., Panicum spp. Cyperus spp.*
Nitrogen compouds	Amnnes	Trifluralin (Treflan)	0.5–2.0 pre-emergence P.P.I	Cotton, sunflower, soybean and pea	Dicot and monocot *Echinochloa spp., Digitaria spp., Setaria spp., Panicum spp.*

1	2	3	4	5	6
	Ureas		5–20, 1–0.5 pre and post-emergence	Sugarcane, cotton	Total Dicot and monocot
		Fluometuroa (cotoran)	0.5–.0, pre and post-emergence	cotton maize, sugarcane, wheat	Dicot and monocot
		Metoxuron Dosanex)	3–5.0 pre and post-emergence	Wheat	Dicot and monocot
		Duron (Karmex)	0.5–5.0 pre and post-emergence	Cotton, sugarcane, jowar, maize, soybean	Dicot and monocot
		Linuron (Aftan, Lorox)	0.5–2.5 pre and post-emergence	Soybean, maize, potato	Dicot and monocot
Oxygen compounds, ethers		Nitrofen (Tok E-25)	4–6.0 pre and post-emergence	Rice, pea, cucurbits, groundnut, potato, cotton, sunflower	Dicot and monocot plus *Echinochloa, spp.*
		Fluorodifen (preforan)	2–5.0 pre and post-emergence	Rice, cotton, soya-bean, groundnut	Dicot and monocot plus *Digitaria spp., Cyperus spp., Porachiaraia spp.*
Hyterocycles	Dipyridyls	Paraquat (Gramaxone)	0.5–2.0 post-emergence	—	Total killer
		Diquat (Reglone)	post-emergence	—	Total killer
	Triazines	Simazine (prinzep)	0.5–3.0 pre and post-emergence	Maize, potato, sugar-cane, wheat, barley sugarbeet	Dicot and monocot and acts as total killer at 5-10 kg/ha.

1	2	3	4	5	6
		Atrazine (Primatol)	0.5-3.0 pre and post-emergence	Maize, Jowar, Sugarcane, linseed	
		Prometryn (Caparol) Gesagard, Primetol Q)	0.5-2.5 pre and post-emergence	Cotton, wheat, barley, sugarbeet	Dicot and monocot *Echinochloa* spp., *Digi*-taria *spp., Setaria* spp.
	Miscellaneous	BAS 3480 H	1-3.0 post-emergence	Maize, wheat, jowar, barley	Dicot and *Cyperus* spp.
		Bentazon (Basagran)	1-3.0 post-emergence	Rice, maize, jowar, wheat, barley	Dicot and *Echinochloa* spp.
		Basalin	1-3.0 pre-plant and pre-emergence	Rice, wheat, cotton, groundnut, soy-bean, tomato	Dicot and monocot.
	Dalapon (Dowpon)	1-5.0, pre and post-emergence	Sugarbeet, sugar-cane, potato, maize	Monocot but acts as total killer when applied at 5-20 kg/ha	
	Ansar 138 (Phytar 138)	1-3.0 post-emergence (directed)	Cotton	Total killer	
	DSHA (Ansar 184)	2-4.0 post-emergence (directed	Cotton	Total killer	
	Tricamba (Banvel T)	0.5-3.0, pre and post-emergence	Wheat, barley	Dicot and monocot	
	Fenac (Fenatrol)	0-8.0, pre post-emergence	Sugarcane	Dicot and moncot but acts as total killer when applied at 4-40 kg/ha	

1	2	3	4	5	6
Phenoxy fatty acids. Acid derivatives	2, 4-D (Weedon 638) Acids	1-1.5 post-emergence TCA (Nata)	Wheat, barley 5-15.0 pre and post-emergence	Dicot	Monocot has acts as total killer when applied at higher doses of 15-30 kg/ha.
	MCPA (Mathoxon) 2,	1-1.5, post-emergence	Wheat, barley	Dicot	
	4, 5-T	0.25-30 post-emergence	Wheat barley	Dicot	
	2, 4-DB	1-2.0 post-emergence	Wheat, barley	Dicot	
	MCPB (Tropotox)	2-3.0 post-emergence	Pea	Dicot	
Acid derivatives	Easters	Chlorothal (Decthal)	4–10.0 pre-emergence	—	Dicot and monocot (sedges).
	Nitriles	Dichlobenil (Casoron)	1–5.0 pre and post-emergence	Rice, wheat	Dicot and monocot but acts as total killer at higher doses of 10–30 kg/ha.
	Amides	RH 315 (Kerb)	0.5–2.0 pre and post-emergence	Cotton, Soybean, Rice, Jowar, Potato	Dicot and monocot (*Avena fatua*).
		Propanil (Rogue, Stam F–34)	3–4.0 pre and post emergence	Rice	Dicot and monocot (sedges).
		Anachior (Lasso)	0.5–4.0 pre-emergence	Maize, Cotton, Soybean, Tomato, Groundnut, sugarcane, Sunflower	Dicot and monocot (sedges).

Source : Herbicides 1971 by Dr. A Fischer.

Table 8.8 : Classification of Herbicides Based on Their Mode of Action, Method of Application and Chemical Affinity

Chemical Group	Herbicides	Soil Applied Systemic	Foliage Applied	
			Systemic	Contact
1	2	3	4	5
1. Aliphatics	Acrolxin Allyl Alcohol	Allyl alcohol, TCA	Dalapon, Acrolein,	
2. Amides and Acet-amides	Alchlor, Butachlor, Carbetamide,CDAA, Cyprazole, Delachlor, Dicryl, Diphenamid, Metachlor, Napropae-mide, Naptal and Per-fluidone, Pron-amide, Prynachor, Solan, Ter-tuchlor Xylachlor.	Alachlor, Butachlor Carbetamide, CDAA, Cyprazole, Delachior. Dicryl, Diphenamid, Metolachi or, Napropar-mide, Naptalam, Per-fluidone, Pronamide. Propachlor, Prynachlor, terbuchlor, Xylachlor.	Alachlor, Butachlor, Carbetamide, Delachlor, Proamide, Propachlor, propanil	Solan
3. Arsenicals	ANA, Cacodylic acid, CMA, DSMA, Hesaflurate, MAA, MAMA, MSMA	—	AMA, CMA, DSMAm Hexaflurate, MAA, MAMA, MSMA	Cacodylic acid.
4. Denzoics and Phenylacetates	Bifenox Chloramben, Dicamba, Fenac, PBA, 2,3,6–TBA, TIBA	Bifenox, Chloramben, Fenac, FBA, 2, 3, 6–TBA	Dicamba, Fenac, PBA, 2,3,6–TBA	—
5. Bipyridilliums	Cyperquat, Diquat, Paraquat	—	—	Cyperquat, Diquat, Paraquat.

1	2	3	4	5
6. Carbamates (Carbanilates)	Asulam, Barban, Chlorbupham. Chlorpropham, Cycloate, Desmediham, Ethiolate, Karbultilate, Phenmedipham, Propham, Swep, Terbulol.	Chlorbupham, Chlorpropham, Cycoloate, Ethiolate, Karbutilate, Propham, Swep, Terbulol	Asulam, Barban, Desmedipham, Phenmedipham, Propham.	
7. Dinitroanilines (Nitroanilines)	Benefin, Butralin, Dinitramine, Ethalfluraline, Fluchloralin, Isopropalin, Methalpropalin, Nitralin, Nitrofluralin, Cryzalin, Pendimethalin, Profluralin, Prosulfalin, Trifluralin	Benefin, Butralin, Dinitramine, Ethalfuralin, Fluchloralin, Isopropalin Methalpropalin, Nitralin, Nitrofluralin, Oryzalin, Pendimethalian, Profluralin, Proulfalin, Trifluralin		
8. Diphenylethers	Chlomitrofen, Flurodifen, Fluoronitrofen, Nitrofen, Nitroflurfen, Oxyflurfen	Chlorinitrofen, Flurodifen, Flouronitrofen, Nitrofen, Nitrofluofen, Oxyflourfen, Dichlobenil	Flurodifen	—
9. Nitriles	Bromoxynil, Dichlobenil Ioxynil	Dichlobenil	Bromoxynil Ioxynil	—
10. Phenols	Dinoseb, DNOC, PCP	Denoseb	—	Denoseb DNOC, PC.
11. Phenosyacids	Dichlorprop, Diclofop, 2, 4-D, 4-DB, 2, 4-DEP, Erbon, MCPA, MCPB, Mecoprop, Sesonex, Silvex, 2, 4, 5-T	2, 4-D, 2, 4-DEP, Sesone	Dichlorprop., Diclofop 2, 4-D 2, 4-DB, Erbon, MCPA, MCPB, Mecoprop, Silvex, 2, 4, 5-T	—

1	2	3	4	5
12. Pyridaziones	Norflurazon, Oxapyrazon, Pyrazon	Norflurazon, Oxapyrazon, Pyrazon	Pyrazon	—
13. Thiocar-bamates	Benthiocarb, Butylate, CDEC	EPTC, Ethiolate, Merharn, Molinate, Pepulate, Tnallate, Vermolate.	Bentylate CDEC, Dialiate EPTC, Mertham, Molinate,	
14. Triazines	Ametryn, Atraton, Atrazine, Cyazine, Cyprazine, Desmetryne, Dipropetryne, Hexazinone, Ipazine, Metribuzine, Procyazine, Prometryn	Propazine, Secbutneton Simazine, Simeton, Simetryn, Terbuthyla-zine, Terbutryn, Trietazine	All of them may be used	Atrazine (with oil), Cyprazine, Hexazi-none, Procyazine, Prometon, Prome-tryn, Terbuthy-lazine, Terbetryn.
15. Triazoles	Amitrole, Amitrole-T	—	Amitrole, Ami-trole-T	—
16. Uracils	Bromacil, Isocil, Lenacil, Terbacil	Bromacil, Isocil, Lenancil, Terbacil	—	—
17. Ureas	Benthiazuron, Chlorbro-muron, Chloroxuron, Cycluron, Difenoxuron, Diuron, Fenuron, TCA, Fluometuron, Karbutilate, Linuron, Methabezthiazuron, Metobromuron, Monolinuron, Monuron, Monuron-TCA, Neburon, Siduron, Terbuthiuron, Tetraflouron, Trimeturon.	All may be used	Chlorbromuron Fluometuron, Linuron, Monuron.	—

1	*2*	*3*	*4*	*5*
18. Unclassified	Bensulide, Bentazon, Benzadox, Cypromid, Dazomet, DCPA, Endot hall, Ethofumesate, MH, Methazole, Oxadiazon, Perfluidone Picloram, Pyrichlor.	Bensulide, Cypromid, Dazomet, DCPA, Endothall, Ethofumesate, Methazole, Oxadiazon, Perfluidone, Pyriclor	Benzadox, Endothall, Fosamine, Glyphosate, MH, Methazole, Picloram	Bentazon.

Source : Rao, V.S. 1983, Principles of Weed Science, Oxford and IBH Publishing Co., New Delhi.

(iii) Post-Emergence Herbicides : The herbicides that are applied after the emergence of crop and weed on the soil surface are termed as post-emergence herbicides. This herbicides is applied on crops at the seedling stage generally at about 30–40 days after sowing.

Examples : 2, 4–D, Sodium salts, Propanil etc.

C. Classification of Herbicides According to the Structure and Mode of Action : The herbicides are classified according to the chemical structure and mode of action as follows : (Table 8.7).

8.12.4.2 Formulation of Herbicides

Herbicides are available in the market in the following formulations.

1. Soluble Powder (SP) : These herbicides are formulated as water soluble powder and they form a homogenous solution as and when dissolved in water which can be applied to the crop by spraying.

Examples : Dalapon, sodium salt of 2,4–D, TCA. etc.

2. Wettable Powder (WP) : These herbicides are made of finely divided solid particles of less than three microns in size for suspension in water.

Examples : Atrazine 80% WP, Diuron–80% WP, Simazine – 50% WP etc.

3. Soluble concentrates (WSC) : These herbicides are formulated in the form of Soluble liquid which are called water soluble concentrates (WSC).

Examples : Amine Salt of 2, 4–D, 2,4,5 T, Silver (2,4,5–TP), Diaquat, Paraquat etc.

4. Emulsifiable Concentrate (EC) : The active ingredient (*a.i.*) of this herbicides is dissolved in solvent and an emulsifier is added to it. These materials from stable emulsion, when diluted with water become suitable for spraying on the plant.

Examples : Alachlor, Dialate, Nitrofen, 2,4–D ester etc.

Table 8.9 : Critical Period of Weed Control and Extent of Reduction in Crop Yield, if Competition Continues.

Name of crops	*Critical period for control of weeds*	*Extent of reduction in crop yield if weeds are not controlled (per cent)*	*Remarks*
1	2	3	4
Wheat and Barley	3-5 leaves stage or wheat of 25–31 DAYS	10 – 25	If 2, 4–D is applied at higher concentration or later than the periods specified, the crop suffers from ear–cockling.
Rice (Upland)	15 – 40 DAYS	50 – 60	
Rice (Transplanted)	25 – 30 DAYS	15 – 20	
Maize	20 – 30 DAYS	20 – 30	
Jowar and Bajra	20 – 25 DAYS	20 – 30	
Gram	20 – 40 DAYS	10 – 15	
Mung and Urd	15 – 25 DAYS	10 – 20	
Peas	30 DAYS	20 – 25	
Groundnut	20 – 25 DAYS	50 in erect and 20 in spreading types.	
Sunflower	20 – 40 DAYS	30 – 50	The extent of reduction in yield.
Sugarcane	30 – 50 DAYS	20 – 70 depending on weed flora	Mainly depends upon the types of weed present in the field for all the crops.
Cotton	20 – 40 DAYS	30 – 30	
Jute	20 – 35 DAYS	30 – 70	

5. Liquid Suspension (LS) : The active ingredient of this herbicides is solubilized in organic solvent which when added to water for spraying forms a liquid suspension. The main drawback of this herbicides is that it requires constant agitation during spraying to avoid settling.

Examples : Atrazine, Cyprazine, Nitralin etc.

8.13 SALIENT FEATURES OF SOME WEEDS

Table 8.10 : List of Common Weeds

Scientific Name	*Local Name*	*Season of Growth*	*Habitat*	*Propagation Method*	*Life Period*
1	2	3	4	5	6
Abutilon indicum	Kanghi	Kharif	Wasteland	By seed	Annual
Achyranthes aspera	Latjeera,Chitchita	Kharif	Wayside, wasteland	By seed	Annual
Ageratum conyzoides	Mahkua	Kharif	Cultivated fields	By seed	Annual
Alhagi camelonum	lawasa	Rabi, summer	Wasteland	By underground parts	Perennial
Amaranthus polygamus	—	Rabi	Cultivated fields and garden	By seed	Annual
Amarnathus spinosus	Katili-chaulai	All season gardens	Cutivated fields and	By seed	Annual
Amarnathus viridis	Jungli chaulai	Monsoon and Rabi	Fields and wasteland	By seed	Annual
Ammannia baccifera	Dadonari	Monsoon	Rice fields	By seed	Annual
Anagallis arvensis	Krishnaneel	Rabi	Cultivated fields	By seed	Annual
Andropogon annualatus	Jowar	Mosoon	Fields and gardens	By seeds and underground parts	Perennial
Argemone mexicana	Bharghar, Satyanasi	Rabi	Wasteland	By seed	Annual
Asphodelus tenuifolius	Ban piajee	Rabi	Cultivated fields	By seed	Annual
Avena fatua	Jangli jayee	Rabi	Wheat field	By seed	Annual
Blumea lacera	Kukur-aundha	All seasons	Wasteland	By seed	Annual

1	2	3	4	5	6
Boerhavia diffusa	Khapra, Puneranwa, Patharchatta	Monsoon	Wasteland	By seed	Perennial
Calotropis gigantea	Madar, Akand	All seasons	Wasteland	By seed	Perennial
Commelina benghalensis	Kankaua	Monsoon	Cultivated fields	By seed	Annual
Cannabis sativa	Ganja	Rabi	Cultivated and wasteland	By seed	Annual
Celosia argentea	Safed Murga ka phool	Monsoon	Wasteland and cultivated fields	By seed	Annual
Chenopodium album	Bathua	Rabi	Cultivated fields	By seed	Annual
Chenopodium murale	Khartua	Rabi	Cutivated fields	By seed	Annual
Chicorium intybus	Kasani	Rabi	Berseem fields	By seed	Annual
Cleome viscosa	hul–hul	Monsoon	Wasteland	By seed	Annual
Convolvulus arvensis	Hiran khuri	All seasons	Cultivated fields	By seed and rhizomes	Perennial
Commelina benghalensis	Isansia	All seasons	Up and moist land	By seed	Perennial
Corchorus acutangul-aris	Jangli jute chench sag	Monsoon	Wayside and cultivated field	By seed and rhizones	Perennial
Croton sparsiflorum	Ban mircha	All seasons	Way side and cultivated field	By seed	Perennial
Cucumis trigonus	Goma	Monsoon	Cultivated fields and wastelands	By seed	Perennial

1	2	3	4	5	6
Cuscuta reflexa	Amar bail	All seasons	Plant parasite (complete)	By seed and vegetative parts	Perennial
Cynodon dactylon	Doob grass	All seasons	Wasteland and cultivated fields	By seed and creeping rootstock	Perennial
Cyperus rotundus	Motha purple Nut sedge	All seasons	All over under abundance of water	By seed and nut	Perennial
Cyperus iria	Jalmutha	All seasons	Low land rice field	By seed	Annual
Cyanotis axillaris	Baghmula	All seasons	Roadside and wetland	By seed	Perennial
Datura fastuosa	Dhatura	Monsoon	Wasteland	By seed	Perennial
Dactyloctenium aegyptium	Stargrass	Kharif	Upland	By seed	Annual
Desmodium trifolium	Tipatia	Monsoon	Pasture and banks	By seed	Perennial
Digera arvensis	Lahasua	Monsoon	Wasteland and cultivated fields	By seed	Annual
Digitaria sanguinalis	—	Monsoon	Cultivated fields	By seed	Annual
Eclipta alba	Bhagara	Monsoon and Rabi	Wasteland and cultivated fields	By seed	Annual
Euphorbia thymifolia	Chotti dudhi	Monsoon	Wasteland	By seed	Perennial

1	2	3	4	5	6
Euphorbia hirta	Bari dudhi	Monsoon	Wasteland	By seed	Perennial
Echinochloa colonum	Jangli dan	All season	Ponds and water reservoirs	By seed	Annual
Echinochloa crusgalli	Sawan	Monsoon	Paddy fields	By seed	Annual
Eichhornia crassipes	Jalkumbhi	All season	Ponds and water reservoirs	By floating rhizomes	Perennial
Eleusine indica	—	Monsoon	Wasteland and cultivated fields	By seed	Annual
Eragrostis cynosuroides	—	All seasons	Cultivated fields	By seed and rootlets	Annual
Fumaria parviflora	Pitpara, Ban soya	Rabi	Cultivated fields	By seed	Annual
Gynandropsis pentaphylla	Hulhul	Monsoon	Wasteland and cultivated fields	By seed	Annual
Indigofera enneaphylla	—	Monsoon	Wasteland and pastures	By seed	Annual
Ischaemum rugosum	Moraroghas	Monsoon	Wet and swampy areas	By seed	Annual
Ipomoea pestigridis	Tigerfoot, morning glory	Monsoon	Cultivated	By seed	Perennial
Ipomoea sp.	Behaya	All seasons	Fence plants	By plants parts	Perennial
Ipomoea reptans	Karami sag	Kharif	Ponds or moist places	By seed	Annual
Launaea asplenifolia	Jangli gobhi	Rabi	Cultivated fields	By seeds	Annual

1	2	3	4	5	6
Leucas aspera	Gumas	Monsoon	Wasteland and cultivated fields	By seed	Annual
Lathyrus aphaca	Banmatar, chapra	Rabi	Cultivated fields	By seeds	Annual
Lippia nodiflora	Jal pimpri	Rabi	Ponds or moist places	By seeds	Annual
Laggera aurita	Kukroundha	Rabi	Wasteland and fields	By seeds	Annual
Ludwigia parviflora	Banlabanga	Kharif	Lowland	By seeds	Annual
Melilotus alba	Safed senji	Rabi	Cultivated fields	By seeds	Annual
Melilotus indica	Pili senji	Rabi	Cultivated fields	By seeds	Annual
Mimosa pudica	Lajawanti	All seasons	Roadside and wasteland	By seeds	Perennial
Mollugo lotodies	Gokhula	Kharif and Rabi	Roadside and wasteland	By seeds	Perennial
Nicotiana plumbag-inifolia	Jangali tambakoo	Kharif	Wasteland	By seeds	Perennial
Orobanche aegyptiaca	Vakumba	Kharif and Rabi	Root parasite of mustard, brinjal and tobacco	By seeds	Annual
Ocimum canum	Ram Tulsi	Kharif and Rabi	Wasteland and garden	By seeds	Perennial
Ocimum sanctum	Tulsi	Rabi	Garden	By seeds	Perennial
Oxalis corniculata	Tin patia	Rabi	Garden	By seeds	Perennial
Phyllanthus niruri	Hazar dena	Kharif	Wasteland and fields	By seeds	Perennial

1	2	3	4	5	6
Physalis minima	Ban makorya	Kharif	Wasteland	By seeds	Perennial
Pennisetum cenchroides	Anjan	Kharif	Wasteland	By seeds	Perennial
Paspalum sanguinale	Crab grass	Kharif	Wasteland and cultivated field	By seeds	Perennial
Panicum crusgalli	Barnyard grass	Kharif	In or near water or paddy fields	By seeds	Perennial
Portulaca oleracea	Kulfa	Kharif and summer	Cultivated fields	By seeds	Perennial
Portulaca quadrifida	Noniya	Kharif and summer	Cultivated fields	By seeds	Perennial
Pluchea lanceolata	Baisuri	Kharif	Wasteland and cultivated fields	By seeds	Perennial
Rhynchosia minima	Tusel	Rabi	Cultivated field	By seeds	Annual
Saccharum spontaneum	Kans	All seasons	Wasteland	By seeds, by root	Perennial
Saccharum arundinaceum	Sarpat, munj	All seasons	Wasteland	By seeds, By root	Perennial
Spergula arvensis	Sathgathia	Rabi	Cultivated fields	By seed	Annual
Setaria glauca	Banara–Banari	Kharif	Rice, maize and sorghum fields	By seed	Annual
Solanum nigrum	Makhoi	Rabi, Kharif	Wasteland and cultivated fields	By seed	Annual

1	2	3	4	5	6
Solanum xanthoca-rpum	Bhatkataiya	Kharif and Rabi	Cultivated fields	By seed	Annual
Sorghum halopense	Baru grass	Kharif and Rabi	By seed and under-ground parts	Perennial	Annual
Striga luteaauphrasiodes	Tovati Agia	Kharif	Root parasite on jowar	By seed	Annual
Saponaria vaccaria	Soap wort	Rabi	Cultivated fields	By seed	Annual
Sesbania acuileata	Dhaincha	Kharif	Wasteland and cultivated fields	By seed	Annual
Scoparia dulcis	Bondhoney	Rabi, Kharif	Cultivated fields	By seed	Annual
Trianthema monogyna	Pathari	Kharif and Summer	Cultivated fields	By seed	Annual
Tridax procumbens	phulani	–	Roadside	By seed	Annual
Tribulus terrestris	Gukhru	Kharif, Rabi and Summer	Waste and cultivated field	By seed	Perennial
Vicia hirsuta	Ankari	Rabi	Wasteland and cultivated fields	By seed	Annual
Vicia sativa	Ankara	Rabi	Wasteland and cultivated fields	By seed	Annual
Vernonia cinerea	Kala jeera	Rabi and Kharif	Cultivated fields	By seed	Perennial
Volutarella divaricata	–	–	–	–	–
Vicoa vestita	Jangali poster	–	–	–	Perennial

1	2	3	4	5	6
Xanthium strumarium	Chhota gokharu	Kharif	Bunds, Wasteland and Roadside	By seed	Perennial
Ziziphus rotundifolia	Jharberi	Kharif	Wasteland and bunds	By seed	Perennial
Aquatic Weeds					
A. True acquatic Weeds					
1. *Submersed Weed*			**2. *Surface weeds* (Free floating)**		
Chara spp.	Musk grass		*Salvia auriculata*		Water fern
Nitella spp.	Stone wort		*Azolla caroliniana*		Water velvert
Potamogeton crispus	Curly leaf pond weed				
Potomogeton nodosus	Long leaf pond weed				
Hydrilla verticillata	Hydrilla				
Elodea densa					
Scirpus acicularis					
			3. Emersed Weeds and Marginal Weeds		
Eichhornia crassipes	Water hyacinth		*Pontederia sagittaria*		Arrowhead
Pistia stratiotes	Water lettuce		*Typha spp.*		Cattails
Trapa natans	Water chestnut		*Paspalum spp.*		
Lemna minor	Common duckweed		*Erianthus spp.*		
Nymphaea odorata	Fragrant water–lily		*Panicum spp.*		
Luphar advena	Yellow water–lily		*Cladium spp.*		

1	2	3	4	5	6
Nelumbo lutea	Water–nut			*Narenga spp.*	
				Scirpus	Club rushes
				Cyperus grossus spp.	Flat sedge
				Cosex spp.	Sedges

(*Source* : Singh S.S. 2001. Principles and Practices of Agronomy, Kalyani Publishers, New Delhi).

Table 8.11 : Weed Control Practices for Important Crops (An integrated Approach)

Name of Crop	*Name of Associated Weeds*	*Mechanical Methods of Controlling Weeds*	*Name of the herbicides found selective*	*Rate of application of herbicide*	*Time of application of herbicide*
1	*2*	*3*	*4*	*5*	*6*
Wheat and Barley	*Fumaria parviflora*, *Melilotus* spp., *Chenopodium album*, *Vicia* spp. *Asphodelus tenuifolius*, *Spergula arvensis*, *Convolvulus arvensis*, *Avena fatua*, *Phalaris minor etc.*	1. Alternate winter cropping with wheat and berseem keeps down the population of wild oats *Phalaris minor* and kasani. 2. Hand weeding at 18-20 and 35-40 DAYS.	1. For wild oats : Triallate or Diallate. 2. For *Phalaris* spp. Nitrofen, Isoproturon, Pendemethalin, Metoxuron, Methabenzthiozuron. 3. For broad leaved	1.25 lires a i/ha, 2.0 kg ai/ha, 1.0 kg ai/ha, 1–1.5kg/ha, 1–2kg/ga, 0.5kg ai/ha	Pre–plant incorporated. Pre-emergence. Post–emergence 30–35 DAS Post-emergence
		2. Hoeing between 4–6 weeks after sowing checks the weeds.	Picloram, Dinoseb, Dicamba + 2,4–D	0.5kg ai/ha., 0.25–0.5 kg/ha., 0.5–0.75 kg/ha., 0.25+0.5 kg/ha.	30–32 DAS
Rice	*Echinochloa* spp. *Paspalum* spp. *Panicum* spp.	1. Puddling 2. Use of rotary weeder or paddy weeder 3. Use of weed free seeds	1. In upland rice, 2. 4–D and Propanil, 3. Direct seeded puddled rice : Beathiocarb/	0.75 kg ai/ha, 3.5 kg ai/ha, 3.5 g.ha, 0.75 kg/ha	7DAS Post-emergence 20–05 DAS Post-emergence 20–25 DAS.

1	2	3	4	5	6
	Cyperus spp. *Ammania baccifera*, *Eclipta* spp. etc.	4. Summer ploughing	Butachlor +2,4–D IPE 3. Transplated rice propanil + 2,4–D IPE Benthiocarb/Butachlor		
Maize	*Cyperus* spp., *Digera arvensis*, *Phyllanthus*, *niruri*, *Solanum nigrum*, *Cynodon dactylon* etc.	1. Two shallow hoeings at 15 days interval after sowing	1. Atrazine or Simazine, Metribuzin, Terbutrya or Butylate 2,4–D, Dinoseb, dicamba, Atrazine + Phytobland oil	2.5 kg/ha 1.0 kg ai/ha 1–2 kg/ha 1–2 kg.ha, 0.75–1 kg.ha, 0.5–1 kg/ha 1.25 kg ai/ha 1–2 kg	Pre-emergence Pre-emergence Post-emergence at 2–4 leaf stage at 2–4 leaf stage at 2–4 leaf stage
Jowar and Bajra	Same as of maize	Two hoeings 1st at 30–45 DAS and 2nd at 60–70 DAS	Atrazine, Propazine, Noruron propachlor Isoproturon and	1.25 kg ai/ha, 1–2 kg, 1.0 kg ai/ha	Pre-emergence Post-emergence 40DAS
	For Striga		2,4–D Dicamba Linuron and monolinuron Vernalate + Benefin	2–3 kg	Pre-plant
Groundnut	Same as of maize	Two seedings between 3–5 weeks after sowing	Flurodifen/Lasso	1.5 kg ai/ha, 2.0 kg ai/ha	Pre-emergence Pre-emergence
			Prometryne, Ambiben, Dinoseb	1.0 kg ai/ha, 3.0kg/ha, 1.5 kg/hs	Pre-emergence Post-emergence Post-emergence

1	2	3	4	5	6
Sugarcane	*Cyperus rotundus* *Cynodon dactylon*, *Euphorbia hirta* *Commelina benghalensis*,	1. Hot weather cultivation. 2. Repeated and hoeings weedings.	1. Atrazine of Nitrofen Metoxuron, Diuron Diachlor.	1.5–2.0 kg/ha, 3.0–5.0 kg. ai/ha, 4–6 kg., 1.5–25 kg. 1.5–25 kg,	Pre-emergence Pre-emergence Pre-emergence Pre-emergence Post-emergence
	Digera arvensis *Phyllanthus niruri* *Chenopodium album*, *Sorghum helopense*, *Croton-sparsiflorum*, *Corchorus* spp.	3. Mulching., 4. Companion cropping	2. 2,4–D amine, Glyphosate, Asulam., 3. Paraquat or Dalapon	1.0 kg ai/ha, 0.8–1.6 kg, 3–4kg/ha, 0.4–0.5kg ai/ha, 3.0–5.0kg ai/ha	Post-emergence directed
Cotton	*Cyperus rotundus* *Cynodon dactylon*, *Eleusine* spp.,	1. Crop rotation	1. Cotoron or Nitrofen Trifluralin + Diuron Trifluralin+Prometryn,	1.5 kg ai/ha, 1.5 kg ai/ha, 1 + 1kg,	Pre-emergence Pre-emergence Pre-emergence
	Celosia argentea *Sorghum halipense* *Amaranthus* spp. *Corchorus* spp.	2. Intercropping, 3. Band application of fertilizers, 4. Weeding and hoeing	2. Ansar 529 (MSMA) (DSMA), Dalapon	1 + 1 kg, 5 litres/ha, 2–3 kg, 3–4 kg.	Pre-emergence Post-emergence
	Digera arvensis *Trianthema monogyna* etc.	5. Earthing, 6. Skip row planting of cotton			Post-emergence Post-emergence

1	2	3	4	5	6
Jute and Mesta	*Echinochloa* spp. *Eleusine* spp.	Two weedings 1st at 3 weeks after sowing and 2nd at 5 weeks after sowing	1. Tetrapion/TPP Sodium, Butachlor	2 kg ai/ha 1-2 kg 1-2 kg 1-2 kg	Pre-plant application
	Cyperus rotundus, Phyllanthus niruri, Digitaria spp., Euphorbia hirta, Eclipta alba., Paspalum spp., *Panicum* spp., *Desmodium triflorum Scoparia dulcis* etc.		2. Ansar 529 ai/ha and (MSMA) and Dalapoe	3-4 kg ai/ha 3-5 kg ai/ha	Post-emergence
Pigeon pea	*Cyperus rotundus* *Amaranthus spp.* *Celosia argentea* *Sorghum halipense* *Convolvulus arvensis, Solanum nigrum, Digera arvensis, Trianthema monogyna, Phyllanthus nirurl*, etc.	Two hoeings at 20 days interval after sowing	1. Triflurialin EPTC, 2. Linuron or Alachor (Lasso), Terbutryn, Choroxuron, Nitrofen.	1–1.5kg/ha, 3.0kg ai/ha, 0.5–1.0 kg/ha 0.75–1 kg/ha, 1.0 kg ai/ha.	Pre-plant incorporated Pre-emergence Pre-emergence Pre-emergence Pre-emergence Pre-emergence Pre-emergence Pre-emergence Pre-emergence

1	*2*	*3*	*4*	*5*	*6*
Green gram (Moong) and Black gram (urd)	*Cyperus rotundus, Sorghum halopense Amaranthus* spp. *Phyllanthus niruri, Digera arvensis, Commelina benghalensis, Trianthema monogyna* etc.	2. Weedings at 15 days interval after sowing	1. Trifluralin, Nitrofen, 2. Alachlor or Nitralin	1–1.5kg/ha, 1–1.5kg/ha, 20kg ai/ha, 1.0 kg ai/ha.	Pre-plant incorporated Pre-emergence Pre-emergence
Cowpea	Same as above	Same as above	1. Trifluralin or Amiben	1.0 kg a.i/ha 2.0 kg ai/ha.	Pre-emergence Pre-emergence.
Soybean	Same as above	Same as above	1. Treflan, Linuron or Nitrofen	1.0 litre ai/ha, 0.5 kg ai/ha, 2.0 litres ai/ha	Pre-plant application Pre-emergence.
Pea	*Chenopodium album, Asphodelus tenuifolius, Fumaria parviflora, Anagallis arvensis, Cyperus rotundus, Melilolus* spp. *Vicia spp.* etc	One weeding at 30 DAS	1. Linuron prometryne, 2. MCPB	0.5 kg ai/ha, 0.5 kg ai/ha, 1.0kg ai/ha	Pre-emergence Pre-emergence
Gram (Bengal gram) and Massor	Same as above	Two weeding or 60 DAS	1. Trifluralin or prometryne	1.5 kg ai/ha or 1.0kg ai/ha	Pre-emergence

1	2	3	4	5	6
Castor	Kharif season weeds	1. Intercultivation, 2. Intercropping	1. Eptam, 2. Prometryne	— 10kg ai/ha	Pre-plant incorporated Pre-emergence.
Sesame	Same as above	Two weedings at 15 days imterval after sowing	1. Monsoon with Eptam	—	Pre-emergence.
Sunflower	*Cyperus rotundus, Cynodon dactylon, Euphorbia hirta, Eleusine spp. Trianthema monogyna, Phyllanthus niruri, Digera arvensis, Commelina benghalensis* etc.	Inter-row cultivation 2 times at 15 days interval after sowing	1. Eptam or Treflan, 2. Linuron of Alachlor, 3. Dalapoa	1. Litre ai/ha, 0.5 kg ai/ha, 4.0 litre ai/ha	Pre-plant Pre-emergence Post-emergence
Mustard Linseed and Safflower	*Chenopodium album, Fumaria parviflora, Spergula arvensis, Anagallis arvensis, Cyperus rotundus, Vicia spp., Melilotus* spp. etc.	1. Weeding at 15 DAS	1. Eptam, 2. Alachlor	2.0 ai/ha	Pre-plant Pre-emergence

1	2	3	4	5	6
Potato	Same as above	1. 2 wedings at 20 days interval after sowing, 2. Earthing 35–40 DAS	1. Nitrofen, 2. Liauron	2.5 kg ai/ha, 0.25 kg ai/ha	Pre-plant Pre-emergence
Deep water crops	Aquatic weeds	1. Hand cutting, 2. Chaining,	1. Fena, 2. Acroteim or xlyene or Hydro-	4–8.0 kg ai/ha, 3.5–kg ai/ha,	In drained bottom canals. In running water.
		3. Desilting, 4. Drying of canals, 5. Burning, 6. Use of weed cutter.	thal 191 (Granular), 3. 2. 4–D or 2, 4, 5–T, 4. $CuSO_4$ or Simazine	0.5–1,0 ppm or 0.2–2.0 ppm	For floating weeds like water hyacinth, Ipomoea spp. For algae.
Orchard crops	*Brush viz Lantana* spp. *Rosa moshata*	Digging or uprooting	2,4, 5–T + 2, 4–D	0.5% solution (5-6 litres a.i./ ha)	Smearing of cut stumps at 5–7 cm above ground level.

Table 8.12 : Selective Herbicides Control of Some Specific Problem Weeds

Local name	*Scientific name*	*Selective herbicides*	*Remarks (rate and time of application)*
1	*2*	*3*	*4*
Goat weed	*Ageratum conyzoides*	Simazine, Atrazine, Diuron, Oxadiazon Oxyfluarfen, Methazole, Metribuzin any of them and 2, 4–D	Pre-emergence Post-emergence.
Amarnath	*Amaranthus* spp.	Simazine, Atrazine, Trifluralin, Proflralin, Fluchloralin, Dinitramine, Butralin, Chloramben, Diuron, Osyfluorfen, Terbail, Alachlor (as Pre-emergence) and 2, 4–D, MCPA, paraquat, as	Post-emergence.
Pimpernel	*Anagallis arvensis* L.	Triazine, Ureas, Carbamates as soil applied and 2,4–D, MCPA, Diclofop, Mecoprop, Bentazon as foliage applied	Pre-plant, Pre-emerg.
Prickly of Mexican poppy	*Argemone mexicana* L.	2,4–D and MCPA as Post–emergence at seedling stage of mixture of MCPA and 2,3,6–TBA or dinoseb and paraquat	Weed germinates througoughout Post-emergence.
Wild oats	*Avena fatua* L.	Triallate and trifluralin as	Pre-plant soil incorporated.
		Barban, Asulem, Diclofop methyl, Difensoquat, Benzoylpropethyl as Post-emergence as 1–3 leaf stage. Difenarogan mixed with 2,4–D or MCPA controls broad leaved weeds also.	

1	2	3	4
Field bindweed	*Convolvulus arrensis*	Dicamba, HCPA 2,4–D Glyphosate as Post-emergence.	
Carpet or Savannah grass	*Axonopus compressus*	Paraquat + diuron (0.4 + 0.5 or 1.0 kg/ha) paraquat MSMA (0.4 + 1.0 kg/ha), dalapon + MSMA (3 + kg/ha) as Post-emergence. (Glyphosate (0.8–1/2 kg/ha) herbicide.	Stolons help in its spread
Hairy *beggarticks black jack or* cobblers peg	*Bidens pilosa* L.	Simazine, Atrazine, Metribuzin, Diuron, as pre-emergence : 2, 4–D as post–emergence at 2–4 leaf stage.	
Button weed	*Borreria hispida* L. *B. articularis (Lf) Wild*	Simazine, Diuron, Oxyfluorfen, Oxadiazon, Metribuzine, methozole as pre-emergence and 2, 4D as post-emergence at 3 leaf stage. Triazines, ureas as pre-emergence and phenoxy acids as post-emergence.	Common weed in tea gardens.
Cocks' comb	*Celosia argentea* L.	Post-emergence.	
Lamb's quaters	*Chenopodium album* L.	Triazines, Ureas as soil applied, Metolachlor (2–3 kg.ha) + Metribuzin (0.5 kg/ha) or Alachlor (2–3kg/ha) + Metribuzin (0.5 kg/ha); Propachlor (4 kg/ha) Chloramben (0.5 kg/ha), Linuron = Nitralin (1 + 1) kg/ha) and Chlorprofan = Propyzamide (1 + 1kg/ha) as pre-emergence.	

1	2	3	4
Day flower	*Commelina benghalensis* L.	Simazine, Metribuzin, Bentazon, Linuron, Diuron, as pre-emergence and Glyphosate as	
Wandering jew		post-emergence.	
Morning glory (*Ipomoea purpurea*) or *Ipomoea* spp.		Dicamba, MCPA 2, 4–D Glyphosate as post-emergence.	
Dodder	*Cuscuta chinesis* Dank	Chlorpropham (4 kg/ha) as pre-emergence of host plants or pronamide (1.5kg/ha) 15 days after germination or 2 kg/ha 20 days after germination.	
Bermuda or star grass	*Cynodon dactylon* (L) Pers	Dalapon of TCA or Glyphosate (1.2–2kg./ha), Bromacil (10/kg/ha), Dalapon +2, 4-D (1–1.5 + 3–4.5 kg/ha) or Terbacil followed by Dalapon as post-emergence EPTC (4.5 kg/ha) preplant incorporated or Diuron as pre-emergence.	
Nutgrass or purple nutsedge	*Cyperus rotundus* L.	Atrazine (4 kg/ha), Perfluidone (3–4 kg/ha) as Pre-emergence soil applied, EPTC soil applied incorporated, Pre-plant, Alachlor or alachlor + Atrazine or Dichlobenil as Pre-emergence.	
Crab grass	*Digitaria sanguinalis* (L) Scop.	Diuron, Trifluralin, EPTC, Vemolate, Metribuzin, Oxyfluorfen, Oxadiazon etc, may kill the emerging seedlings. Butylate + Atrazin, Alachlor + Linuron, Alachlor + Chloramben as Pre-emergence soil incorporated paraquat or 4 applications of Methazole + MSMA (0.14 + 22 kg/ha each time).	

1	2	3	4
Barnyard grass of water grass	*Echinochloa crusgalli* (L) Beauv.	Alachor, Chloramben, Atrazine, Simazine, Diuron, Linuron, Fenurofr, Metolachlor, EPTC, Pebulate, Vernolate as Pre-emergence. Benthlicarb and Butachlor 3 days after rice transplanting.	
Wild rice (Shyama)	*Echinochloa colonum* (L)	Link, Alachlor, EPTC, Trifluralin. Benefin, Oxadiazon, Diuron, Methabenzthiazuron, Triallate as Pre-emergence and Paraquat, Glyphosate, Dalapon as Post-emergence.	
Goose grass wild fingermillet	*Eleusine indica* (L) Gaertn. Eleusine	Simazine, Atrazine, Monuron, Diuron, Oxyflourfen, Linuron, Metribuzin, Alachlor as Pre-emergence and Glyphosate, MSMA + Paraquat or two applications of Napropamide (2.2 kg/ha) each time or Methazole – MSMA (0.28 + 2.2kg/ha) each time as Post-emergence.	
Broomrape	*Orobanche cernua Loeff.*	No Chemical except growing trap crops like sun-flower, sesame, cotton, soybean, finger millet, etc. in rotation with tobacco takes care.	
Torpedo grass	*Panicum reptans* L.	Glyphosate, Dalapon and Paraquat as Post-emergence.	
Wild carrot weed or congress grass	*Parthenium hysterophorus* L.	Simazine, Atrazine, Alachlor, Metribuzin, Diuron, Butachlor, Terbutryne, Nitrofen etc. as Pre-emergence may prevent its emergence upto 5 month. Glyphosate, 2, 4–D MSMA, MSMA + Paraquat as Post-emergence.	

1	2	3	4
Sourgrass, Hilograss	*Paspalum conjugarun* Berg.	Diuron, Oxythrolen, Methazole as pre-emergence and Glyphosate, Paraquat + Diuron, Paraquat + MSMA as post-emergence.	
	Pennisetum pedicellatum Trin.	Paraquat, Dalapon (repeated sprayings) or Glyphosate as post-emergence application.	
Canary grass	*Phalaris minor* Retz	Methabenzthiazuroh (0.75 to 1.5 kg/ha), Metoxuron (1–2 kg/ha). Isoproturon (0.75 to 1.5kg/ha) and Nitrofen (2–4 kg per ha) as Post-emergence 25 days after sowing wheat crop. Methabenzthiazuron Pendimethalin (1–1.5 kg/ha). Nitrofen and Terbutryn (both 1–1.5 kg/ha). as Pre-emergence.	
Common purslane	*Portulaca oleracea* L.	Simazine, Atrazine, Metolachlor, Diuron, Monuron, Linuron, Nor flurazon. Chlorproham, Metribuzine (0.5 kg/ha) Metribuzine (0.5 kg/ha) + (2–3 kg/ha) or all as Pre-emergence.	
Kansgrass	*Saccharum spontaneum* L.	Paraquat and Glyphosate as Post-emergence.	
Black night shade	*Solanum nigrum* L.	2, 4–D, MCPA, Diquat, Dinoseb, as Post-emergence and Chlorproham (1kg/ha) + Propyzamide (1 kg/ha), Chlorpropham + Metolachlor (1 + 1.5 kg/ha), Butralin (2 kg/ha), Linuron (1 kg/ha), Chloramben (2kg/ha), Alachlor (4 kg/ha) as Pre-emergence.	

1	2	3	4
Johnson grass	*Sorghum halepense* L.	Atrazine + EPTC, Alachlor, Butylate, Metolachlor, Butralin, Diuron, Fluchloralin, Profluralin, Vernolate, Trifluralin, Pendimethalin as Pre-emergence, three Post-emergence applications at 20 days interval of Dalapon (3 kg/ha) or Asulan (3 kg/ha).	
Witchweed	*Striga lutea*	Fenac or 2,3,6–TBA as pre-plant soil incorporated, Simazine, Atrazine, pro-pazin as Pre-emergence or 2,4–D MPA, Paraquat or diesel oil as Post-emergence. Growing trap crop like soybean, cotton, cowpea, groundnut, bengal gram, pedgram etc. with catch crops like sorghum, maize and millets control the *Striga lutea*.	

Source : Rao. V.S. (1983) Principles of Weed Science, Oxford and IBM Publishing Co., New Delhi.

Table 8.13 : Chemical Weed Control for Aquatic Weeds

Name	*Scientific Name*	*Chemicals found Effective*
Water hyacinth	*Eichhornia crassipes*	2, 4–D (4 kg/ha). Paraquat (0.5 kg), Duqyat (1–1.5 kg). Amitrole–T) 0.5–1.5 kg), Ametryn (3–4kg).
Cat tails	*Typha augustata*	Dalapon (25–30 kg), Amitrole–T (8 kg/h), Paraquat (1.2–1.6 kg), Dalapon + Amitrofe–T (15 + 3 kg), Amitrole + TCA (5 + 10 kg). Use of simazine or diuron as Pre-emergence in ditch or drainage channel prevents seed germination followed by Dalpon + Amitrole T as post-emergence controls the weed effectively.
Pond-weed	*Potamogeton* spp.	Pre-emergence herbicides like Smiazine, Diuron and Monuron or Post-emergence herbicides like 2, 4–D and 2, 4, 5–T.
Hydrilla	*Hydrilla verticillata* casp.	2, 4–D (2,5 ppmw). Diuron (0.5 ppmw), Paraquat (1 ppmw), $CuSO_4$ (4ppmw), Endothall (2 ppmw) and Acrolein (4 ppmw)
Water lettuce	*Pistia straliotes* L.	Diquat (1–1.5kg), Paraquat 1.5 to 1.75 kg)
Salvinia	*Salvinia molesta Mitchell*	2, 4–D + Paraquat (4kg + 10kg)
Swamp morning glory	*Ipomoea aquatica* Forsk.	2, 4–D + Paraquat (4kg + 0.5 kg)
Other weeds		2, 4–D Silvex, Dalapon, TCA Amitrole–T as post-emergence or simazine, monuron as pre-emergence.

6. Granules (G) : The active ingredient of this herbicides is mixed with granular inert materials of suitable particle size. Granules are generally made, with inert clay. The granules mixed with the solution of toxicant are dried and then paced for subsequent use.

Examples : Butachlor–G, 2, 4–DEPG etc.

Table 8.14 : Classification of Herbicides Based on Chemical Nature

Group	*Common Herbicides*	*Time of Application*	*Rate*	*Crops/Weed Species*
1	*2*	*3*	*4*	*5*
Acetamides	Acetochlor	PPI or PRE	1.5-3.0kg/ha	Soybean, maize
	Alachlor	PPI, or PRE, or E. Post	2.0-4.0 kg/ha	Maize, sorghum, groundnut, soybean, cotton
	Butachlor	PRE	1.0-3.0 kg/ha	Rice, groundnut, soybean
	Metolachlor	E–PPS, PPI, PRE, or E-Post	1.5-4.0 kg/ha	Maize, cotton, groundnut
	Popanil	POST	3.0–5.0 kg/ha	Rice, Wheat
Aliphatics	Acrolein	—	15 g/100 li	Floating weeds
Arsenicals	DSMA	POST	2.5–5.0 kg/ha	Cotton, non–crop areas
	MSMA	POST	2.5 kg/ha	Cotton, non–crop areas
Benzamides	Isoxaben	PRE, POST or PPI	1.5–2.0kg/ha	Tobacco
Benzoics	Dicamba	PPS, PRE or POST	0.25–0.50	Maize, sorghum, wheat
Bipyridiliums	Diquat	POST	2–0.4 kg/ha	Aquatic weeds and desiccant in cotton
	Paraquat	POST	0.25–1.0 kf/ha	Non crop areas and plantation crops
Carbamates	Asulam	POST	1.0–3.0 kg/ha	Sugarcane, non–crop areas
	Desmedipham	POST	0.8–1.5 kg/ha	Sugarbeet
Cineoles	Cinmethylin	PPI or PRE	0.5–1.5 kg/ha	Soybean, cotton, groundnut

1	2	3	4	5
Cyclohexanediones (Cyclohexenones)	Sethoxydim	POST	0.10–0.5 kg/ha	Cotton, sunflower
Dinitroanilines	Fluchloralin	PRE	1.0–4.0 kg/ha	Groundnut, pulse crops
	Pendimethalin	PRE or E-POST	0.5–2.0 kg/ha	Maize, sorghum, wheat
	Trifluralin	PPI	0.5–1.0 kg/ha	Maize, soybean, sunflower, castor, safflower
Diphenylethers	Oxyfluorfen	POST	0.25–2.0 kg/ha	Soybean, groundnut
Imidazolidinones	Buthidazole	PRE or POST	–	Maize, wheat
Imidazolinones	Imazethaphyr	POST	50–70 g/ha	Soybean
	Imazamethabenz	POST	20–40 g/ha	Sunflower
1 mines	CGA–248757	POST	3–15 g/ha	Maize, soybean
Isoxazolidinones	Clomazone	PPI or PRE	0.56–1.7 kg/ha	Soybean
Nitriles	Bromoxynil	POST	0.20–0.56 kg/ha	Wheat, barley, oats, maize, sorghum
	Dichlobenil	POST	4.0–8.0 kg/ha	Forest nurseries
	Ioxynil	POST	0.5–1.0 kg/ha	Wheat, barley
Oxadiazoles	Oxadiazon	PRE	0.5–0.4 kg/ha	Rice, soybean, groundnut, cotton, sugarcane
Oxadiazolidines	Methazole	PPI, PRE or POST	0.5–1.5 kg/ha	Sorghum, maize, wheat, soybean, cotton, sugarcane.
Phenols	Dinoseb	POST	0.75–1.2 kg/ha	Cereal crops, potato

1	2	3	4	5
Phenoxyalkanoic acids				
Phenoxyacetics	2, 4–D	POST	0.26–2.24 kg/ha	Cereal crops, sugarcane, non-crop areas
	MCPA	POST	0.25–1.5 kg/ha	Cereal crops, legumes
Phenoxybuxyrics	2,4–DB	POST	0.20–0.45 kg/ha	Forage legume crops
Aryloxyphenoxy propionics	Fluzifop–P	POST	50–200 g/ha	Wheat, soybean, cotton
N–phenylphthalim-ides	Flumiclorac	POST	30–60 g/ha	Maize, soybean
Phenylpyridazines	Sutfentrazone	PPI or PRE	60–90 g/ha	Soybean
Phthalamates	Naptalam	PRE or POST	2.0–4.0 kg/ha	Groundnut, soybean
Pyrazoliums	Difenzoquat	POST	0.7–1.0 kg/ha	Wild oats
Pyridazinones	Picloram	POST	0.15–1.0 kg/ha	Non–crop area
Pyridines	Dithiopyr	PRE	0.25–0.5 kg/ha	Rice
Pyridinones	Fluridone	–	0.06–0.09 mg/litre	Aquatic weeds
Pyrimidinylthio–benzoates	Pyrithiobac	POST	0.07–kg/ha	Cotton
Quinolinecarboxylic acids	Quinclorac	PRE, late–PRE or early – POST	0.25–.05 kg/ha	Rice
Sulfonylureas	Chlorsulfuron	PRE or POST	10–20 g/ha	Wheat

1	*2*	*3*	*4*	*5*
	Sulfometuron	PRE or POST	50–200 g/ha	Non–crop areas
Thiocarbamates	Butylate	PPI	3.0–6.0 kg/ha	Maize
	Diallate	PPI	2.0–5.0 kg/ha	Wheat, maize, soybean, oilseed crops
	EPTC	PPI	2.0–5.0 kg/ha	Maize, sunflower, cotton
	Thiobencarb	PRE	1.0–1.5 kg/ha	Irrigated rice
Triazines	Ametryn	POST	0.5–2.0 kg/ha	Sugarcane
	Atrazin	PRE	1.0–2.0 kg/ha	Maize, sorghum, sugarcane
	Prometryn	PPI or PRE	0.5–3.0 kg/ha	Cotton, soybean, non-crop areas
	Simazine	PRE	1.0–4.0 kg/ha	Maize, sugarcane
Triazinones	Metribuzin	PPI, PRE or POST	0.25–0.80 kg/ha	Maize, soybean, sugarcane
Triazoles	Amitrole	POST	2.0–10.0 kg/ha	Non-crop areas
Trizolopyrimidine				
Sulfonanilides	Flumetsulam	PRE, PPI or PPS	50–75g/ha	Maize, soybean
Uracils	Bromacil	PRE	1.5–5.0 kg/ha	Non-crop areas
Ureas	Diuron	POST	0.2–0.4 kg/ha	Maize, sorghum
	Fluometuron	PPI or PRE	1.0–1.5 kg/ha	Sorghum, cotton, sugarcane
Unclassified	Bensulide	PPI or PRE	5.6–6.7 kg/ha	Vegetable crops
	Glyphosate	POST	0.5–4.0 kg/ha	Perennial weeds

(PPL–preplanting, PPS–Pre-planting surface, E–PPS–Early pre-planting surface, PPI–Pre-planting incorporated PRE–Pre emergence late–PRE–late pre-emergence, POST–Post-emergence, E–POST–Early post-emergence).

Source : Rao, 2000.

The same herbicides can be made in different formulation. As for examples, 2, 4–D when added to sodium, amine and ethyl ester, it forms soluble solid, soluble concentrate and emulsifiable concentrate respectively.

The herbicides have to be mixed with some inert materials such as water, sand clay etc. for uniform distribution. These materials are called carrier of herbicides. The herbicide of different formulations such as SP, WP, WSC, EC, etc. are mixed with water, a common carrier of herbicide and sprayed on the soil or foliage. The liquid formulation of herbicide such as EC, WSC etc. can be applied on soil after mixing with sand.

Table 8.15 : List of Herbicides Recommended for Marketing in India

Name	*Formulation*
1. Atrazine	50 per cent WP
Simezine	60 per cent WP
2. Paraquat	24 per cent WSC (w/w)
3. MSMA	34 per cent in water solution
4. Fluchloralin	45 per cent EC (w/w)
5. Alachlor	50 per cent EC & 100 per cent G
Butachlor	50 per cent EC & 5 per cent G
Triallate	50 per cent EC
Glyphosate	41 per cent WSC
6. MCPA	40 per cent Amme salt
7. Tribunil	70 per cent WP
8. Dosanex	80 per cent WP (w/w)
9. Nitrofen	25 per cent EC & 8 per cent G
10. Propanil	35 per cent EC
11. Dalapon	85 per cent Na salt & per cent Na salt
12. Dichlormate	4 per cent EC & 4 per cent G
13. Divron 2, 4–D	80 per cent WP
(*a*) Ethyl ester	8 per cent & 34 per cent
(*b*) Amine salt	58 per cent *&* *72* per cent
(*c*) Sodium salt	80 per cent Na salt and 85 per cent Na salt
14. Paraquat dimethyl sulphate	24 per cent WSC

Name	*Formulation*
Growth Regulating Substances	
15. Maleic hydroxide (MH)	30 per cent amine salt
16. Ethephon	39.2 per cent
17. G.A.	100 per cent
18. Chloremequat (CCC) (Cycocel)	50 per cent AS
19. Alpha naphthalene acetic acid (A.N.A.)	45 per cent WS (w/w)

Source : Singh S.S. 2001 Principles and Practices of Agronomy, Kalyani Publishers, New Delhi.

8.14 AQUATIC WEEDS

Aquatic weeds are water inhabitant plants. They grow in water reservoirs and causes much damage in pisciculture.

Aquatic weeds are very much harmful in various reasons as follows :

(*i*) The floating and submerged weeds create obstacle in water current.

(*ii*) Drainage is hampered if weeds grow in drainage channel. As a result water gets stagnant in the field and causes damage to the crop. In the same way, irrigation water also suffers very much.

(*iii*) Aquatic weeds are harmful to pisciculture and they cause damage in pisiculture in various ways as follows :

(*a*) Aquatic weeds take share in the food of fish and they do damage in pisciculture. The foods that are applied for fishes are also useful to the aquatic weeds. As a result, the aquatic weeds take major portion of the food and create deficiency of food of the fishes.

(*b*) Aquatic weeds hamper the production of phyto-plankton, a natural food for the fishes.

(*c*) Algae, *Eichhornia crassipes* etc. when grow in the water reservoirs luxuriantly creates deficiency of light and oxygen. As a result, fishes do not get sufficient oxygen for their respiration and in many times, they die.

(*d*) Aquatic weeds decompose on the bottom of the water reservoirs and toxic gas is created. As a result, the general quality of water is lost and the fishes can not live here freely.

(*e*) Aquatic weeds curtail the space for the fishes. As a result, fishes do not get more space for their movement and health hazard occurs.

(*f*) The predator and unwanted fishes and also predator animal such as snake, frog etc. get shelter in the aquatic weeds and causes damage to the fishes.

(*g*) Aquatic weeds create much difficulty in netting of fishes besides this, they also hamper the breeding of the fishes.

(*h*) Aquatic weeds release carbondioxide during their respiration and it get dissolved in water and creates difficulty in the respiration of fishes.

(*i*) Aquatic weeds create much difficulty in the rearing of duck.

(*iv*) Aquatic weed create difficulty in the cultivation of aquatic crop such as wetland paddy, water chestnut, lotus etc.

(*v*) Aquatic weeds damage the aesthetic values of water reservoirs. Aquatic weed create difficulty in catching of fish, boating, water play etc.

Aquatic Weeds and their Control Measures : Aquatic weeds should be controlled in time to get the benefit of water reservoirs. Aquatic weeds may be controlled in various ways as follows :

1. ***Mechanical Methods***

(*i*) *Use of Implement* : The aquatic weed can be controlled by using some implements.

(*ii*) *Drying of Pond* : It is a good method of controlling submerged aquatic weeds. In this method, pond is dried by lifting the water. As a result, weeds die. In the dry pond tillage is also given. As a result, the weeds are uprooted and die.

(*iii*) *Use of Mower :* The mower is used for cutting the top portion of weeds on the bany of the pond. The flowering and fruiting of the weeds are not possible due to removal of their top portion. As a result, dispersal of weed is checked. But, in this method, weeds can not be controlled permanently.

(*iv*) *Hand Weeding :* Hand weeding is the most efficient and easiest method of controlling aquatic weeds. The aquatic weeds are controlled by pulting out of them by hand.

(*v*) *Rope or Chain Dragging :* This method is very effective to control the floating weeds and also floating and submerged aquatic weeds.

(*vi*) *Burning :* This method is generally employed to control weeds on the bank of water reservoirs. Flaming followed by burning after 10–12 days have been found very effective in controlling the weeds. Besides this, weeds are cut by sharp implement and kept under sun for two to four days, then they are destroyed by burning.

(*vii*) *Cutting :*The weeds are cut above the ground surface, with the help of sickles, axes and saws. The bank side weeds such as brush and trees are controlled by this method.

2. *Biological Method :* Fungus, fishes, snails, insects etc., is used to control aquatic weeds as follows :

(*i*) *Fishes :* The herbivorus fishes like grass carp, common carp, Tellapia etc. are generally employed for destruction of aquatic vegetation.

(*ii*) *Snail :* The submerged weeds like contail, algae etc. can be controlled by fresh water snails.

(*iii*) *Fungus :* Different types of fungus (*e.g. Cercospora rodmani, Acromonium jonatus* etc.) attacks water hyacinth (*Eichhornia crassipes*) and destroys them.

(*iv*) *Insect :* A number of insects help in controlling the aquatic weeds. The azolla and water hyacinth can be controlled by grasshopper and *Gasonulla punctifrons* respectively.

3. *Chemical Method :* The aquatic weeds can be controlled by using chemical herbicides. It is a easy, effective and less costly method. The chemical herbicides create sometimes some problem and do harm to the fishes in the pond. This problem can be avoided by selection of good and safety herbicides. When other method of controlling aquatic weed is not feasible, aquatic weed can be controlled by using chemical herbicides.

Controlling of aquatic weed by using the chemical herbicides have some advantages as follows :

(*i*) The chemical method of controlling aquatic weed is very much effective when the aquatic weed is not possible to controlled by mechanical method.

(*ii*) Use of chemical herbicides once or twice per year is found very effective to control aquatic weeds. On the other hand, mechanical methods need to be followed occasionally.

(*iii*) The plants that are killed due to application of herbicides get deposited on the bottom of water reservoirs and act as a source of food for the fishes.

(*iv*) The chemical method is less costly and time saving process. Some herbicides are used to control aquatic weeds as follows :

(*a*) 2, 4–D (*b*) Paraquat.

(*c*) Diquat. (*d*) Amitrole.

(*e*) Ametryn. (*f*) Dalapon.

(*g*) Simazine. (*h*) Monuron.

(*i*) Diuron. (*j*) Endothal etc.

Calculation of Amount of Herbicides

(*a*) ***Marginal and Floating Weeds***

$$\text{Amount of herbicides} = \frac{a \times b \times c}{10,000 \times 1000}$$

where, a = Area of water reservoir (m).

b = Amount of spray mixture per hectare.

c = Formulation (in percentage).

(*b*) ***Submerged Weed***

$$\text{Amount of herbicides} = \frac{a \times b \times c \times 10.11}{10,000}$$

Where, a = Area of water reservoir (m)

b = Depth of water (m).

c = Amount of herbicide (ppm. W).

The effect of weed control by application of herbicide on fish is as follows :

(*i*) The herbicides kills the algae, which is the food of the fish.

(*ii*) The weeds that are killed by application of herbicides are rotten in the water and make the water polluted.

8.15 WEED CONTROL IN FOREST

Measures to be taken to control weeds in forest in some conditions as follows :

(*i*) To make seedling of forest plant in weed free condition.

(*ii*) To control unwanted plants and remove the seedlings.

(*iii*) To plant some plants as intercrop that grow well under the shade of the tree. *e.g.* Ginger, Turmeric etc.

(*iv*) To plant creeper plant (*e.g.* Beetle vine with Arecanut) by the side of big plant.

(*v*) To cut the tail plant by the side of road or to control their growth.

Method of Weed Control

(*a*) *Mechanical Method*

(*i*) *Cutting of Tree :* The dense and unwanted plant should be cut down. Because dense and unwanted plants do harm to principal plants and no good quality plant is produced from the forest.

(*ii*) *Ringing :* Ringing of the plant is *found* effective for killing of unwanted plants. The plants die quickly if herbicide is applied to the ringing plants.

(*iii*) *Burning :* The killing of unwanted plant by burning is a easy and effective method. Pine and other plants of the forest can not endure the damage caused by firing.

(*iv*) *Grazing :* Goat is very useful animal for destroying the small plants in the forest.

(*b*) *Chemical Method :* The chemical method for controlling shrubs, small tree and other unwanted plants is very effective and economical method. Sodium arsenite, Sodium chlorate and Ammonium sulphates are used to control the timber plants.

The herbicides of forest are used in different ways as follows:

(*i*) Pre-emergence and pre-planting application of herbicide is followed in seed bed and newly planted forest respectively.

(*ii*) Application of herbicides by spraying in the growing seedling.

(*iii*) Application of herbicide by injecting in the plants.

Some herbicides are used to control weeds in the forest as follows :

(*a*) 2, 4–D (*b*) 2, 4, 5–T

(*c*) Paraquat. (*d*) Glysophate.

(*e*) Amitrole etc.

8.16 WEED CONTROL IN NON-CROPPED AREAS

Weed creates problem in non-cropped areas like the cropped areas as follows :

(*a*) Road Side Areas : The weeds that grow in the road side areas do some harms as follows :

(*i*) The main source of the seeds of weed in the crop field is the seeds of weed that grown in road side areas.

(*ii*) The insectpests take shelter in the weeds of road side areas and spread to the crop field from there.

(*iii*) Some road side poisonous weed causes harm to walker.

The road side weeds are controlled by adopting some methods as follows :

1. *Mechanical Method*

(*i*) *Cutting :* Cutting the road side weeds by sharp knife and burning them after 4-5 days is the best method of controlling them.

(*ii*) *Burning :* Burning some weed (*e.g. Eragrostis cynosuroides, Saccharum spontaneum* etc.) is found effective in controlling the road side weeds as they caught fire in green stage.

2. *Chemical Method* : The road side weeds can be controlled by spraying herbicides. The herbicides that are used to control roadside weeds are as follows :

(*i*) 2, 4–D.

(*ii*) Paraquat.

(*iii*) Glysophet etc.

(*b*) Railline and Factory Side : The weeds that grow in railline and factory side cause some problem. So, control of weeds in those places is almost important. The methods of controlling the weeds in those places are same to the method of controlling roadside weeds.

(*c*) Open Space in Village, City and Town : Various types of plants grow in open space that remains vacant and create some problem. Now-a-days, *Parthenium hysterophorus* is found growing here and there. It is very poisonous weeds and causes health hazard. Some measures need to be taken for controlling this notorious weeds as follows :

(*i*) This weed should be destroyed before flowering and it help to check the spread of the weed.

(*ii*) The germination of seed is to be checked by application of herbicide (*e.g.* Paraquate, glysophate etc) in the soil.

(*iii*) Planting of *Cassia sp.* is found effective in controlling this weed as the sap secreted from *Casia sp.* prevents the germination of Parthenium seed.

(*iv*) *Parthenium sp.* causes allergy in human beings, so controlling this weed by mechanical method is not safe. The chemical method is more effective in controlling this weed. The herbicides that are used in controlling this weed are as follows :

(*a*) Mixed herbicides

+

Glysophate : 10–15 ml per litre of water.

2, 4–D : 5 gm per litre of water.

(*b*) Single herbicides

(*i*) Glysophate : 10–15 ml per litre of water.

(*ii*) Paraquate : 5–10 ml per litre of water.

Table 8.16 : Some Parasite and Problematic Weeds and Their Control

S.No.	*Name of weeds*	*Control measures*
1.	Nut sedge (*Cyperus rotundus* L)	(*i*) Deep ploughing and exposure of the underground rhizomes to the sun is very much effecting in controlling this weeds. (*ii*) Mulching with straws of Boro paddy is found effective in controlling this weeds as this weed is sensitive to light and suppressed in shade.

S.No.	*Name of weeds*	*Control measures*
		(*iii*) Removal of rhizome by hand weeding with khurpi, Nirani, spade etc. is very effective method of controlling this weed.
		(*iv*) Pig penning is also good method of controlling this weed.
		(*v*) Growing of dhaincha (*Sesbania aculeata*), a green manuring crop continuously for three years under crop sequence *i.e. Dhaincha* potato wheat is a good method for controlling this weed.
		(*vi*) Application of herbicide (*i.e.* glyphosate @ 1kg ai/ha, Alachlor @ o.s. kg ai/ha Eptam @ 1litre /ha Atrazine @ 4kg a.i/ha etc.) is more effective in controlling of this weed.
2.	Bermuda or Dubgrass (*Cynodon dactylon*)	(*i*) Cultural method such as summer ploughing, hand weeding, puddling etc is most effective in keeping the weed infestation under control.
		(*ii*) Application of herbicide (*i.e.* dalapon, TCA etc. @ 2–4 kg/ha, glyphosphate @ 1 ai kg/ha Alachlor @ 0.5 kg.a.i/ha, Eptam @ 1 litre ai./ha etc.) is found effective in controlling this weed.
3.	Witch weed (*Striga lutea*)	(*i*) Hand weeding and summer ploughing is found effective in controlling this weed.
		(*ii*) Growing trap crop like Soyabean, Cotton, Cowpea, groundnut, Bengal gram, Red gram etc. with catch crop like sorghum, maize and millets control this weed.
		(*iii*) Application of herbicide is effective in controlling this weed as follow: *(a)* pre-Planting soil incorporation : Fenaca 2,3,6–TBA etc. *(b)* Pre-emergence application : Atrazine, simazine etc. *(c)* Post emergence application : 2, 4–D, Paraquat etc. Application of Diesel oil is also effective in controlling this weed.

S.No.	Name of weeds	Control measures
4.	Broomrape (*Orobanche cernua*)	A parasitic weed grow in Tobacco field and it depend entirely on host plant (*i.e.* Tobacco) for its growth and development.
		(*i*) Hand weeding is the quick method of controlling this weed.
		(*ii*) Growing trap crops like sunflower, sesame, cotton, soyabean, finger millet etc. in rotation with Tobacco takes care this weed.
		(*iii*) Chemical weed control method is not very effecting in controlling this weed. But post emergence application of herbicides such as 2, 4–D Sodium sulphate, Alleyl alcohol @ 0.1–0.2 per cent has given high percentage of control.
5.	Bird vine (*Loranthus longiflorus*) A stem parasite mainly grow an mango and other trees.	(*i*) Hand picking this parasitic weed in early stage of its growth is the most common method of control of this weed.
		(*ii*) Spraying of 30–40% diesel oil emulsion in soap water has been found effective in the eradication of this parasite.
6.	Wild carrot or Congress grass (*Parthenium hysterophorus)* Annual weeds and grows throughout the year.	(*i*) Cutting and burning subsequently before flowering is effective method of controlling this weed.
		(*ii*) Spraying 20% common salt is found effective in controlling this weed in fallow land and non-cropped area.
		(*iii*) Sowing of marigold and *Cassia* sp. is found effective in suppression of the growth of this weed.
		(*iv*) This weed can be controlled by application of herbicides as follows: *(a)* Pre-emergence application : Alachlor, Atrazine, Butachlor, Diuron, Metribuzin, Nitrofen, Terbutryne etc. *(b)* Post emergence application : 2, 4–D, Glyphosate 2, 4–D+ ghyphosate, MSMA, MSMA + Paraquat etc. Pre-emergence application of herbicide may prevent the emergence of this weed upto five months.

8.17 PARTICULARS' ABOUT APPLICATION OF HERBICIDE

1. Types of Sprayers

(*a*) ***Hand Operated Sprayer :*** Knapsack sprayer, Logarithmic sprayer etc.

(*b*) ***Power Driven Sprayer :*** Usually mounted on tractor. Full cone jet is used. Hallow cone or fan jet is not used.

Knapsack Sprayer : VMD – 310, NMD – 60

$$\frac{\text{VMD}}{\text{NMD}} = 5.16$$

Herby sprayer : VMD - 230, NMD

$$\frac{\text{VMD}}{\text{NMD}} = 1.2$$

2. Spray Droplet Size

VMD : Volume media diameter.

NMD : Number media diameter.

CDA : Controlled droplet application by micron Herby spray.

3. Spray Drift

(*i*) Spraying should be done when no wind is blowing.

(*ii*) Lowering the spray height from the ground.

(*iii*) Using proper nozzle and reducing the delivery pressure.

4. Nozzle Types

Flat jet, Flat pan, Solid cane, Hollow cane, Flood jet deflector etc.

Nozzle size : WFN–0.04

5. Formulation

(*i*) EC (Emulsified concentration).

(*ii*) SC (Soluble concentration).

(*iii*) WP (Wettable power).

(*iv*) G (Granules).

(G–Less, EC–More, WP/SC-Medium)

6. Mode of Action

(*i*) Contact.

(*ii*) Systemic.

(Contact-less, systemic-more).

7. Time of Application

(*i*) P.P. (Pre-planting).

(*ii*) P.E. (Pre-emergence).

(*iii*) POE (Post-emergence).

(PP–Less, PE–more and POE–Medium).

8. Method of Application

(*i*) Soil application *(SA).*

(*ii*) Foliar application (FA).

(SA–More and FA–Medium).

Plastic hand operated granule application is used for application of granule herbicide.

Herbicide Incorporater : Mixing with soil by spring tooth harrow, Bed conditioner, Rotary hoe etc.

9. Application Technique

(*i*) Backward movement instead of forward movement.

(*ii*) Proper mixing of herbicide with water.

(*iii*) Washing/Cleaning of sprayer and nozzle should be done properly.

10. Spray Volume

450 – 600 litres per hectare. The amounts of spray volume depend on soil type, soil moisture and type of herbicide use.

11. Calculation of the Amount of Herbicide

(*a*) $$Q = \frac{R \times A}{C \times 100}$$

where, Q = Quantity of herbicide in kg/ha.

R = Rate in kg/ha.

A = Area to be sprayed in square meter (sq.m)

C = Concentration of herbicide in Percentage (%).

(*b*) Spray volume (litre/ha)

$$= \frac{\text{Water used in litre} \times 10,000}{\text{Distance covered in meter (m)} \times \text{Swath width in meter (m)}}$$

REFERENCES

Aldrich R.J. and Kremer, R.J. Principles of Weed Management (1977), Panima Publishing Corporation, New Delhi.

Gupta O.P. Modern Weed Management Principles and Practices (2004) Agrobios (India), Jodhpur.

Gupta O.P. Scientific Weed Mangement (1984), Today and Tomorrow's Publishers, New Delhi.

Joyakumar R. and Jaganathan R. Weed Science Principles (2003), Kalyani Publishers, Ludhiana.

Mandal R.C., Weeds, Weedicide and Weed Control : Principles and Practices (2002), Agrobios (India), Jodhpur.

Rao. V.S., Principles of Weed Science (2000), Oxford and IBH Publishing Co., New Delhi.

Singh. S.S., Principles and Practices of Agronomy (2001), Kalyani Publishers, Ludhiana.

Thakur. C., Weed Science (1984), Metropolitan Book Co., New Delhi.

Wallia U.S., Weed Management (2003), Kalyani Publishers, Ludhiana.

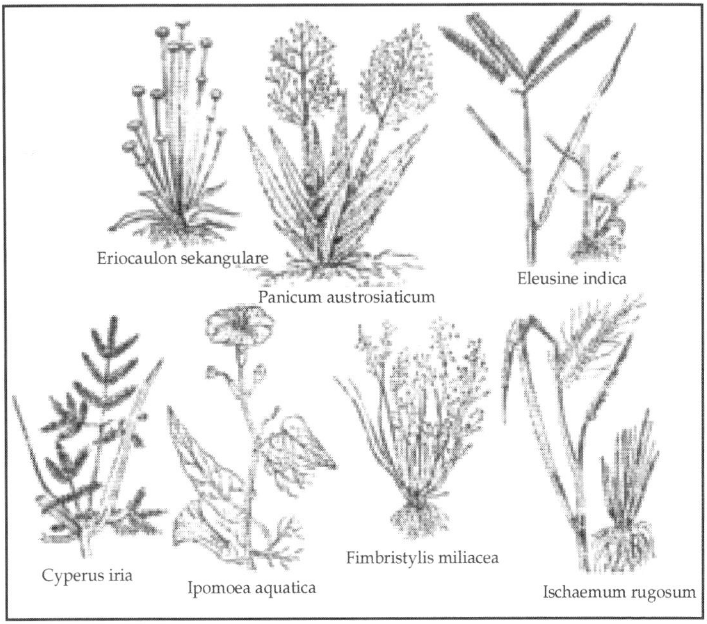

Fig. 8.1 : Some Common Weeds

Echinochloa crusgalli *Echinochloa colonum*

Echinochloa cruspavonis *Setaria barbata*

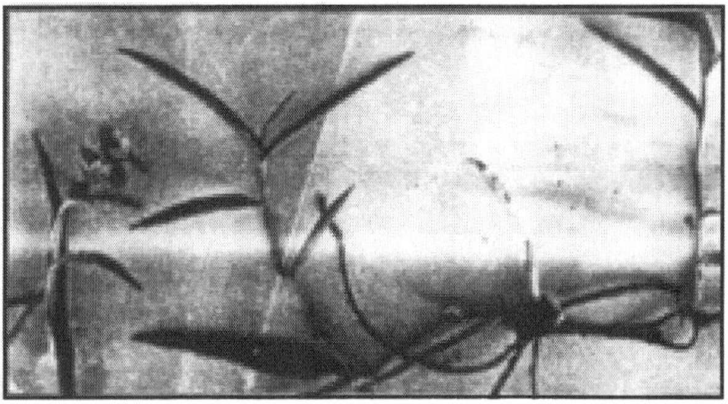

Commelina benghalensis

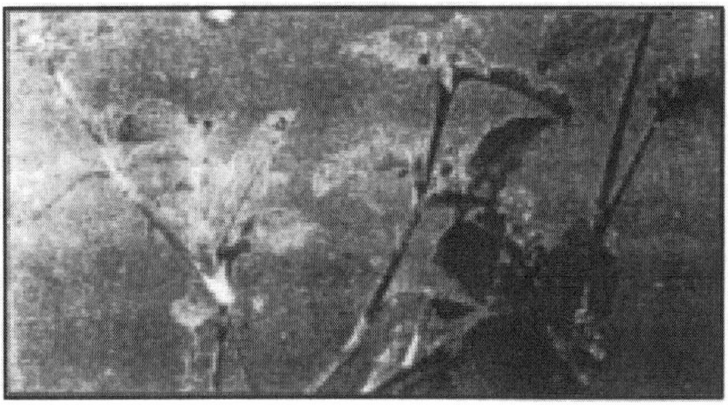

Cyanotis axillaris

Eclipta alba

Ludwigia parviflora

Monochoria vaginalis

Panicum spp.

Dactyloctenium aegyptium

Digitaria sanguinalis

Ischaemum rugosum

9

AGROFORESTRY

9.1 INTRODUCTION

Agroforestry means practice of agriculture and forestry on the same piece of land. Agroforestry is a collective name for land use system and technologies where woody perennials (Shrubs, palms, bamboos, trees etc.) are deliberately used on the same piece of land management units as agricultural crops and/or animals in some form of spartial management or temporal sequences. In Agroforestry system, there are both ecological and economical interactions between different components (Lundergren and Raintree, 1982). In India, the word of Agro-forestry is often used as a synonym for the farm forestry. Farm forestry is defined as the practice of forestry in all its aspect on the farm or village lands generally integrated with other farm operations (NCA, 1976). In India, Agroforestry is sustainable land use system, which integrates growing of agricultural crops and forest trees together in the same pieces of land for maximum production of food, fodder and wood and other products in such a way that the system is economically and ecologicaly, desirable and acceptable to the local population. Vergera (1985) observes that the Agroforestry is a land use system combining Agriculture and tree crops of varying longevity (ranging from annual through biennials and perennial plants) arranged either temporarily (*i.e.* crop rotation) or spartially (*i.e.* inter cropping). At a times, live stocks and fish may be added as components. Under Indian conditions, Agroforestry has two main components, as follows :

(*i*) Growing of agricultural crops including fodder.

(*ii*) Forestry *i.e.* growing of trees and woddy perennials. King (1978) suggests that Agroforestry might be considered to be practiced whenever, trees and agricultural crops are grown in mixture provided that the combined width of the rows of agricultural crops do not exceed height of the forest trees at maturity and provided further that the combined width of rows of forest trees do not exceed the height of tree crops at maturity or at the same selected rotations.

9.2 DEFINITION OF AGROFORESTRY

(*i*) Agroforestry is sustainable land management system which increases the overall yield of land, combines the production of crops (including tree crops) and forest plants and/or animals simultaneously or sequentially on the same unit of land and applied management practices that are compatible with cultural practices of local population.—*King and Chandler* (1978).

(*ii*) Agroforestry as a land use unit of land and applies management practices that are compatible with the cultural practices of local population.—*Nair* (1979).

9.3 COMPONENTS OF AGROFORESTRY

Agroforestry has some components as follows (*Nair*, 1989)

(*i*) It is deliberate growing of woody perennials on the same unit land as agricultural crops and/or animals either in some form of spartial mixture or in sequence.

(*ii*) There must be significant interaction (Positive/negative) between the woody and non-woody components of the system either ecological and/or economical.

(*iii*) This is a production system which tends to harmonise the production of various components and also maximises the production from a given unit of land.

(*iv*) The production and use is sustainable and makes use of modern technologies and traditional local experience and is compatible with the social and cultural life of the local population.

(*v*) It is a long term land management systems and cycles of agroforestry system is more than one year.

(*vi*) Agroforestry is a more complex form of land management both ecologically and economically than other agricultural or forestry systems.

9.4 CLASSIFICATION OF AGROFORESTRY SYSTEMS

Nair (1985) Classified Agroforestry system on the basis of structure, function, socio-economic and ecological status as follows :

1. Structural Classification

(*i*) Agrisilviculture : Agrisilviculture means use of land for concurrent production of agricultural crops and forest crops. Agrisilviculture covers all system in which land is used to produce both forest trees and agricultural crops either simultaneously or alternately (F.A.O. 1978). In India, Agrisilviculture is used for intercropping of forest plantations with agricultural crops in the initial years until the canopy of forest closes. The Agrisilviculture system aims at producing of enough food grains, timber, fodder, fire wood or other products. Important agrisilvicultural may includes the followings :

(*a*) Growing of multipurpose trees on farm land.

(*b*) Homestead plantations and home gardens.

(*c*) Trees with Plantation crops.

(*d*) Multipurpose trees with horticultural crops.

(*e*) Alley cropping or hedgrow cropping.

The Agrisilviculture systems are more productive and sustainable than agriculture. Agrisilviculture system also includes growing tree crops such as Cocoa, Coffee, Oilpalm, Coconut, Citrus, Rubber, Papaya etc. with forest trees. In this system, there is a keen competition between two types of crops for water, nutrients, light and space.

In West Bengal, cultivation of direct seeded paddy is kharif crop and green gram as autumn crop gave better performance. *Acacia tortilis* can be grown in combination with the fodder crop like cowpea, cluster bean or pearl millet with remunerative return.

(*ii*) Agrisilvi Pasture System : Agrisilvi pasture system is a combination of Agrisiliviculture and Silvi pasture system. In this system, the land is managed for the concurrent production of agricultural and forest crops and for grazing by domestic animals. In this system, provision is made for the production of food grains, fodder and wood and for grazing cattle.

The components suitable for mild tropical climate are as follows :

(*a*) ***Field Crops :*** Paddy, (*Oryza sativa*) Maize (*Zea mays* etc.).

(*b*) ***Tree Species :*** *Eucalyptus sp, Erythrina suserosa, Morus leavigata, Putranjiva roxburgi, Robina pseudoacacia* etc.

(*c*) ***Fodder Grass :*** *Cymbopogon citratus, Medicago sativa, Spilanthes, Teosinte, Vetiveria zizanoides* etc.

(*iii*) Silvipastoral System : The silvipastoral system means a land management system in which forests are managed for the production of wood as well as rearing of domesticated animals. In this system, grass or grass-legume mixture is grown along with woody perennials simultaneously on the same unit of land. Soil improvement in terns of pH, EC, organic carbon and total nitrogen was observed under improved silvipasture as compared to pasture alone.

Example : Sesbania grandiflora or *Sesbania sesban* is grown with *Cenchrus ciliaris, Setaria ancepts, Demanthus Sp.* etc.

(*iv*) Multipurpose Forest Tree Plantation System : In this system, forest is managed to yield multiple product (*i.e.*, flower, fruit, leaves, barks, roots, gums, medicine etc.) in addition to wood. This system is practised particularly in areas where tribal peoples live as they derive from forest not only wood but also large number of other product for their livelihood.

2. Functional Classification

(*i*) Productive Agroforestry System : In this system, goods are produced predominantly for meeting the basic needs of the society and measures taken to maximise production of annual and perennial crops. This system may include some system of cultivations as follows :

(*a*) Shifting Cultivation.

(*b*) Intercropping of trees.

(*c*) Home or kitchen gardening.

(*d*) Planting of trees on the boundaries of the field.

(*e*) Cultivation of plantation crop.

(*f*) Planting of plant fuel wood, pulp wood etc.

(*ii*) Protective Agro Forestry System : This system aims at some measures to be taken as follows :

(*a*) Amelioration of climate.

(*b*) Reducing soil erosion by water and wind.

(*c*) Improving the fertility of soil.

(*d*) Providing shelter and other benefits.

(*e*) Providing protection to land.

This system includes some important aspect as follows:

(*a*) Wind break and shelter belt.

(*b*) Some hedgerow intercropping.

(*c*) Plantation for soil conservation.

(*d*) Introduction of cover crop.

(*iii*) Multipurpose Agroforestry System : This system aims at optimising both of the above functions as follows :

(*a*) Product required for meeting the basic needs such as food, fodder, fuel etc.

(*b*) Influences for conservation of environment.

Several other systems may be grouped into this class as follows :

(*a*) Home garden or Kitchen garden.

(*b*) Hedgerow intercropping.

(*c*) Multipurpose tea plantation.

(*d*) Planting trees on agricultural field.

(*e*) Silvipasture and several other system.

3. Socio-economic Classification

(*i*) Subsistence Agroforestry System : This system aims at meeting the basic needs of farming having small holding and managing different activities by the members of the family. This system includes some system of cultivation as follows :

(*a*) Shifting cultivation.

(*b*) Home gardens.

(*c*) Homestead forestry.

(*d*) Planting scattered traces on the field.

(*ii*) Commercial Agroforestry System : In this system, emphasis is given at production of commodities in commercial basis. This system includes plantation crop in combination with agricultural crops.

(*iii*) Intermediate Agroforestry System : This is a system of Agroforestry which is intermediate between subsistence and commercial Agroforestry. This system includes cultivation of fruit trees with agricultural crops.

9.5 ADVANTAGES OF AGROFORESTRY

Agroforestry system have advantages as follows :

1. Agroforestry aids in supplementing food and fodder. Mahua tree (*Madhuca indica*) is most important tree in Agroforestry. This tree yields an edible flower, fruit and seeds. The flower are rich in sugar (65–75%), Calcium (14%) Vitamin B complex and Vitamin–C (Anon 1962). A single tree of Mahua can yield about 50–52 kg of dried mahua flower (Anon, 1962). The tribals and other rural people eat the flower of mahua traditionally. Trees of several species are known to yield palatable green leaf fodder (Sen, *et. al.*, 1978, Gulati *et. al.*, 1982, Singh *et. al.*, 1982) as follows :

(*i*) *Acacia nilotica.*

(*ii*) *Aegle marmelos.*

(*iii*) *Ailanthus excelsa (Gokul).*

(*iv*) *Bauhinia variegata.*

(*v*) *Celtis australis.*

(*vi*) *Dalbergia sissoo.*

(*vii*) *Dendrocalamus strictus.*

(*viii*) *Grewia optiva.*

(*ix*) *Hardwickia bibata.*

(*x*) *Loucama leucocephala.*

(*xi*) *Moringa oleifera.*

(*xii*) *Morus alba.*

(*xiii*) *Zizyphus sp.*

Table 9.1 : Fodder Yields from some Trees (Singh, 1982)

	Names of trees	*Yield of fodder per tree per annum (in kilogram)*
(*i*)	*Acacia nilotica*	50–120
(*ii*)	*Ailanthus excelsa*	100–200
(*iii*)	*Bauhinia variegata*	30–50
(*iv*)	*Grewia optiva*	25–50
(*v*)	*Prosopis cineraria*	25–50

Table 9.2 : Nutritive Value of Some Fodder Trees (Sen *et. al.*, 1978, Gulati *et. al.*, 1982, Singh 1982)

Sl. No.	*Tree Leaves*	*Crude Protein %*	*Digestible Crude Protein %*	*Total Digestible Nutrients %*	*Crude Fibre %*	*Calcium %*	*Phosphorus %*
1	*2*	*3*	*4*	*5*	*6*	*7*	*8*
1.	Khair *Acacia catechu*	12.0–18.7	2.9	46.3	21.9–22.6	1.6–2.7	0.1–0.2
2.	Babul *Acacia nilotica*	7.0–15.0	–	–	20.1–33.3	1.2–2.6	0.1–0.2
3.	Israeli babul *Acacia tortillis*	18.1	–	–	–	3.1	0.2
4.	Haldu *Adina cordifolia*	8.7–15.3	2.7–2.8	50.9	12.1–14.0	1.7–3.2	0.1–0.5
5.	Bel *Aegle marmelos*	15.1–15.3	10.8	56.7	16.5–18.1	4.2–4.8	0.1–0.3
6.	Mahrukh *Ailanthus excelsa*	16.3–19.9	–	–	12.8–21.9	1.5–2.4	0.2–0.3
7.	Ohi *Albizia chinensis*	15.1	4.9	40.2	31.6	1.2	0.1
8.	Siris *Albizia lebbeck*	14.9–29.2	11.6	49.3	25.3–37.5	1.1–2.7	0.1–0.3
9.	Allmania nodiflora	10.3	–	–	25.8	1.0	0.1
10.	Zamin-kand *Amorphophallus campanulatus*	15.3	–	–	14.2	1.5	0.8
11.	Bakli *Anogeissus latifolia*	7.5–11.5	0.6	47.8	16.4–24.2	2.7–4.0	0.1–0.6
12.	Kardahai *Anogeissus pendula*	7.6	–	–	19.0	3.5–3.7	0.1
13.	Kadam *Anthocephalus chinensis*	21.3	–	–	–	2.7	0.3
14.	Kathal *Artocarpus heterophyllus*	11.2–14.2	–	–	18.7–22.8	0.5–2.2	0.1–0.3
15.	Neem *Azadirachta indica*	12.4–18.3	8.4–9.3	42.8–53.3	11.4–23.1	0.9–4.0	0.1–0.3
16.	Bans *Bambusa arundinacea*	18.6	13.5	46.5	24.1	0.6	0.2

1	2	3	4	5	6	7	8
17.	Kachnar *Bauhinia variegata*	10.7–15.9	5.0–9.2	47.9–55.5	20.7–33.0	1.4–4.1	0.2–0.4
18.	Bishop wood *Bischofia javanica*	18.0	–	–	–	2.6	0.2
19.	Buddleia *Buddleia madagascariensis*	13.0	–	–	–	1.1	0.2
20.	Dhak *Butea monosperma*	17.5	–	–	–	3.3	0.2
21.	Kumbhi *Careya arborea*	110.4	0.2	43.1	25.9	1.6	0.3
22.	Amaltas *Cassia fistula*	15.8	–	–	–	1.3	0.2
23.	Kassod *Cassia siamea*	13.5	–	–	–	1.9	0.6
24.	Jangli saru *Casuarina equisetifolia*	9.5	–	–	–	2.1	0.1
25.	Khark *Celus australis*	12.0–15.5	5.1–9.7	41.2–59.4	19.5–22.0	2.8–4.9	0.1–0.2
26.	Lasora *Cordia dichotoma*	12.4–15.1	5.1–5.4	26.9	16.5–26.8	2.4–4.3	0.2–0.3
27.	Shisham *Dalbergia sissoo*	2.7–24.1	3.7–9.1	20.9–52.2	12.5–32.0	2.0–2.3	0.2
28.	Karail Bamboo *Dendrocalanus strictus*	14.2–15.1	9.3	48.9	23.5–15.6	1.1–1.6	0.2–0.3
29.	Vayni *Delonix clota*	25.6–26.5	–	–	7.2	3.5	0.2
30.	Tendu *Diospyros melanoxylon*	7.1	0.0	34.1	25.3	1.8	0.2
31.	Chamror *Ehretia laevis*	13.5	8.5	54.9	11.0	1.5	0.3
32.	Jamun *Syzygium cumini*	8.8–10.2	0.1	43.8	19.8	1.3	0.1–0.2
33.	Bargad *Ficus benghalensis*	7.7–11.5	1.9	44.5	1.3–30.0	1.2–4.1	0.1–0.4

1	2	3	4	5	6	7	8
34.	Rubber *Ficus elastica*	12.0	–	–	–	1.3	0.2
35.	Gular *Ficus glomerata*	11.2–15.2	6.6	53.8	12.3–16.5	1.7–3.0	0.2–0.5
36.	Pakar *Ficus lacor*	7.3–16.0	0.4–6.3	32.2–62.0	12.8–27.7	1.1–3.5	0.1–0.4
37.	Pipal *Ficus religiosa*	9.0–25.6	5.5–7.0	38.2–39.2	14.7–20.0	2.3–6.3	0.2–0.3
38.	Timla *Ficus roxburghii*	12.3–13.4	–	–	7.7–17.8	1.3–2.2	0.2
39.	Fig *Ficus sycomorus*	11.4–15.2	–	–	13.7–27.4	2.6–3.0	0.2–0.3
40.	Bhimal *Grewia optiva*	15.6–20.6	12.4–16.3	57.4–61.7	18.9–22.1	2.7–3.5	0.1–0.3
41.	Dhamni *Grewia tiliifolia*	13.2	–	–	–	1.5	0.1
42.	Anjan *Hardwickia binata*	9.0	–	–	30.4	2.3–3.3	0.1
43.	Marorphali *Helicteres integrifolia*	13.3	9.7	58.4	19.8	2.3	0.3
44.	Kanju *Holoptelea integrifolia*	13.7	5.2	58.1	–	4.3	0.2
45.	Madhavilata *Hiptage benghalensis*	13.2	–	–	–	2.1	0.2
46.	Sausage *Kigelia pinnata*	12.3	–	–	–	2.4	0.2
47.	Pula *Kydia calycina*	11.1–13.6	7.9	45.2	14.5–23.7	3.1–5.2	0.4–0.6
48.	Dhaura *Logersaoernia parviflora*	7.8	0.9	49.5	17.3	2.6	0.3
49.	Jhingan *Lannea coromandelica*	11.4–16.7	4.9	55.2	16.1–19.8	1.8–2.4	0.2–0.3
50.	Subabul *Leucaena leucocephala*	15.2–27.6	12.6–16.4	57.1–70.2	10.2–17.2	2.7–3.1	0.2
51.	Banda *Loranthus falcatus*	14.8	6.2	58.9	17.9	1.8	0.2

1	2	3	4	5	6	7	8
52.	Mahua *Madhuca indica*	9.4–10.0	0.0	37.0	19.5–20.3	1.6	0.1–0.2
53.	Kamila *Mallotus philippinensis*	12.5–23.1	7.9	42.1–16.6	21.6–29.7	1.3–1.6	0.2
54.	Am *Mangifera indica*	9.3	–	–	23.7	1.9–2.2	0.2–0.3
55.	Cassava *Manihot esculenta*	23.0	–	–	24.4	1.6	0.2
56.	Khirni *Manilkara hexandra*	9.4	–	–	23.1	1.5	0.4
57.	Bakain *Melia azedarach*	13.3	–	–	–	2.6	0.2
58.	Gauj *Millettia extensa*	22.7	15.5	44.9	32.5	1.9	0.3
59.	Kaim *Mitragyna parvifolia*	7.7	1.5	49.9	19.6	2.3	0.2
60.	Al *Morinda coreia*	15.1	–	–	22.7	1.7	0.2
61.	Sainjana *Moringa oleifera*	15.3–20.7	11.1	61.5	7.1–17.9	1.7–3.8	0.1–0.5
62.	Mulberry *Morus alba*	15.0–27.6	10.7	59.6	9.1–15.3	2.4–4.7	0.2–1.6
63.	Sandan *Ougeinia oojeinensis*	11.6–15.1	3.7	45.6	21.9–28.5	0.8–3.6	0.2–0.4
64.	Karanja *Pongamia pinnata*	17.6	–	–	–	2.2	0.2
65.	Jand Khejra *Prosopis cineraria*	13.9–15.3	–	–	17.5–22.1	1.9–3.6	0.2–0.5
66.	Tilonj *Quercus dilatata*	9.6	4.2	43.2	29.1	1.6	0.3
67.	Phaliant *Quercus glauca*	9.6	4.6	39.8	29.0	1.9	0.2

1	2	3	4	5	6	7	8
68.	Banj *Quercus leucotrichophora*	8.9–15.9	4.2–5.8	43.2–43.8	13.7–42.1	1.0–1.7	0.1–0.2
69.	Robina *Robinla pseudoacacia*	17.1–25.5	–	–	16.6–23.1	1.1–1.5	0.1–0.8
70.	Gogina Ratendu *Saurauta napaulensis*	12.3	1.2	34.0	18.4	2.7	0.2
71.	Kusum *Schleichera oleosa*	10.4	3.4	47.5	32.3	1.7	0.3
72.	Sal *Shorea robusta*	10.1	0.1	42.7	27.4	0.8	0.1
73.	Padri *Stereospermum personatum*	8.1–11.4	–	–	22.0–28.9	1.1–2.7	0.2–0.6
74.	Imli *Tamarindus indica*	11.2–15.4	–	–	14.2–22.0	1.7–3.2	0.1–0.6
75.	Lauret *Terminalia alata*	8.9–12.7	0.0	34.9	12.4–21.8	2.7–3.2	0.2–0.3
76.	Bahera *Terminalia bellrica*	7.7–17.2	0.9	54.5	7.4–18.6	1.6–3.6	0.2–0.3
77.	Amrita *Tinospora cordifolia*	11.2	–	–	17.5	1.1	0.6
78.	Mored *Ulmus callichama*	18.1	–	–	–	0.7	0.3
79.	Ber *Ziziphus mauritiana*	8.6–16.9	3.1	30.7	13.5–30.1	1.3–3.6	0.2–0.3
80.	Jharber *Ziziphus nummularia*	11.5	5.5	51.1	33.8	1.9	0.3

2. The production is possible to maximise in Agroforestry system. The higher yields of crop have been observed in forest influenced soil than in ordinary soil (Chaturvedi, 1981, Songhad, 1983, Verinumbi, 1987). The International Institute of Tropical Agriculture (IITA) has developed a system of alley cropping :

In this system Leucaena is intercropped with paddy, corn and yam. The tree is cut and prunned in the growing season of the agricultural crop so that they do not shade the nearby crop. The prunned larger branches of the trees are used as fuel and poles and the leaves are used as mulch. In the dry and off season, the trees are allowed to regrow. On infertile sandy soils, alley cropping has given corn yields of more than 3–5 tonnes per hectare in second season (Anon 1984 b).

The reasons for higher production under Agroforestry are as follows :

(*a*) Greater efficiency of perennial crops for photosynthesis, the food manufacturing process of green plants.

(*b*) Tapping required nutrients and water form deeper layer by perennial crops.

(*c*) Creating better environmental conditions for growth of annual crops.

The production can be sustaintially increased by selecting suitable trees and agricultural crops as follows :

(*a*) Fodder cum Fuelwood Species :

- (*i*) Bakphool (*Sesbania grandiflora*).
- (*ii*) Coral tree (*Erythrina indica*).
- (*iii*) Gliricidia (*Gliricidia specium)*
- (*iv*) Shevari (*Sesbania sesban*)

(*b*) Fuel Wood and Timber Species

- (*i*) Akasmoni (*Acacia auriculiformis*).
- (*ii*) Arjun (*Terminalia arjuna*).
- (*iii*) Babul (*Acacia nilotica*).
- (*iv*) Bahera (*Terminalia belerica*).
- (*v*) Bakainneem (*Melia azedarach*).
- (*vi*) Bilati babla (*Prosopis juliflora*).

(*vii*) Capas (*Ceiba pentandra*).

(*viii*) Casurina (*Casuarina equisetifolia*).

(*ix*) Gamar (*Gmelina arborea*).

(*x*) Jarul (*Lagerstroemia flosreginae*).

(*xi*) Kadam (*Anthocephalus cadamba*).

(*xii*) Karanji (*Pongamia pinnata*).

(*xiii*) Krishna chura (*Delonix regia*).

(*xiv*) Lombu (*Dezoxylum binectiforum*).

(*xv*) Minjiri (*Cassia siamea*).

(*xvi*) Neem (*Azadirachta indica*).

(*xvii*) Palas (*Butea monosperma*).

(*xviii*) Piasal (*Pterocarpus marsupium*).

(*xix*) Radhachura (*Peltophorum inerimi*).

(*xx*) Shisum (*Dalbergia sissoo*).

(*xxi*) Siris (*Albizia lebbeck*).

(*xxii*) Sojne (*Moringa oleifera*).

(*xxiii*) White siris (*Albizia procera*).

(*c*) Fuit tree/Vegetable Species

(*i*) Amloki (*Emblica officinalis*).

(*ii*) Arecanut (*Areca catechu*).

(*iii*) Ber (*Ziziphus mauritiana*).

(*iv*) Drum stick (*Moringa oleifera*).

(*v*) Jack fruit (*Artocarpus heterophyllus*).

(*vi*) Jamun (*Syzygium cumini*).

(*d*) Crops to be Grown within the Alleys of Trees :

(*i*) Arhar or pigeon pea (*Cajanus cajan*).

(*ii*) Banana (*Musa sp.*).

(*iii*) Bilter gourd (*Momordica charantia*).

(*iv*) Cowpea (*Vigna sinensis*).

(*v*) Cucumber (*Cucumis sativus*).

(*vi*) Dinanath grass (*Pennisetum pedicellatum*).

(*vii*) Ginger (*Zingiber officinale*).

(*viii*) Gram (*Cicer arietinum*).

(*ix*) Groundnut (*Arachis hypogaea*).

(*x*) Guinea grass (*Panicum maximum*).

(*xi*) Hybrid napier (*Pennisetum purpureum*).

(*xii*) Jowar (*Sorghum vulgare*).

(*xiii*) Kalai (*Vigna mungo*).

(*xiv*) Lady's finger (*Abelmoschus esculentus*).

(*xv*) Maize (*Zea mays*).

(*xvi*) Mustard (*Brassica sp.*).

(*xvii*) Papaya (*Carica papaya*).

(*xviii*) Radish (*Raphanus sativus*).

(*xix*) Sesame (*Sesamum indicum*).

(*xx*) Seed potato (*Solanum tuberosum*).

(*xxi*) Seem (*Dolichos lablab*).

(*xxii*) Snake gourd (*Trichosanthes anguina*).

Leucaena leucocephala, when intercropped with agricultural crops and fodder grasses, the total yield of food grains, fodder and fuel is increased (Tiwari, 1970, Anon 1984 b, Pathak, 1989)

It is indicated in an experiment conducted at IGFRI, Jhansi, that the total yield of fodder is more when fodder grasses are grown with fodder, trees that pure fodder grass cultivation. (DebRoy, 1990).

3. Agroforestry system protects soil from several adverse effects. Most Agroforestry system constitute sustainable land use and help to improves soil in a number of ways as follows (Nair, 1987, Young, 1989) :

(*i*) Reduction of loss of soil as well as nutrients through reduction of run-off. Tree crowns protect the soil from the impact of raindrops. The leaf litter from trees absorbs a large quantity of water which means less quantity of water is available for surface run-off. Soil erosion from vegetated water shed is always less in comparison to cultivated lands.

(*ii*) Addition of carbon and its transformation through leaf, twig, bark fall etc.

(*iii*) Nitrogen enrichment by fixation of Nitrogen fixing trees, shrubs etc.

(*iv*) Improvement of physical condition of soil such as water holding capacity, permeability, drainage etc.

(*v*) Release and recycling of nutrients by affecting bio-chemical nutrient cycling. The leaflitter after decomposition forms humus, releases nutrients and various soil properties. The nutrient content of leaf litter may be per hectare per annum may be as follows:

Nitrogen (N)	150–300 kg.
Phosphorus (P_2O_5)	10–20 kg.
Potassium (K_2O)	75–150 kg.
Calcium (Ca)	100–300 kg.

From the above data, it may mean that nutrient recycling can sustantially reduce the needs of fertilizer.

(*vi*) More microbial associations and addition of more root biomass. Roots in several trees develop a mycorrhizal association which help in nutrition and water absorption.

(*vii*) Moderating effect on extreme condition of soil acidity and soil alkalinity.

(*viii*) Creating more favourable microclimate by wind break and shelter belt effects.

(*ix*) Lowering effect on water table in areas where the water table is high.

Some other beneficial effects for which sufficient research evidence is yet to be confirmed include the followings :

(*i*) Increased precipitation and addition of nutrients.

(*ii*) Uptake of nutrients from deeper layer of soil and deposition on the surface of the soil via leaf litter.

(*iii*) With holding of nutrients which are usually lost by leaching.

(*iv*) Augmenting of soil water availability in arid and semi arid condition by reducing surface evaporation (Dwivedi, 1992).

Most of the beneficial effects of trees on soils are primarily due to their roots, crown, litter fall, the nitrogen fixing ability and the effects depend on the density of trees. Tree crowns protect the soil from impact of rain drop and create specific microclimate (Pradhan, 1973, George, 1978, Ghosh *et. al.*, 1980). The leaf litter from trees absorbs large quantity of water which means less quantity of water is available for surface run-off. Soil erosion from vegetated water shed is always less in comparison to cultivated land (Sinha, 1975, Anon, 1975). The rate of infiltration of soil water is 3 to 5 times more in forested areas in comparison to agricultural land (Ghosh, 1974).

4. Agroforestry have an effect on improved environmental conservation. An Agroforestry system helps to increase the tree cover. It also makes available to the people the required quantity of timber, fuel, fodder etc. for which they traditionally depend on forest. Thus this system help to reduce the pressure of forest. Agroforestry improves the environment by combating the adverse effect of deforestation (such as increase in carbon

dioxide (CO_2), global warming, repeated drought and flood, serious soil losses, serious pollution problem etc.). Trees protect us from different kinds of pollutants. Forest and trees increase precipitation, humidity and decrease temperature (Dabral *et. al.*, 1969, Warren, 1979) and also protect from strong winds and cyclones. Forests are reported to increase 5 to 15 per cent annual precipitation (Martin, 1944, Ranganathan, 1949, Rakhamnov 1966, Nicholson 1960, Warren 1974). It is reported that annual nutrient supply through rainwater is as follows (Ovington, 1965, 1968) :

Nitrogen (N)	0.8–4.9 kg.
Phosphorus (P_2O_5)	0.2–0.6 kg.
Potassium (K_2O)	1.0–10.0 kg.
Calcium (Ca)	3.0–19.0 kg.
Magnesium (Mg)	4.0–11.0 kg.

Agroforestry play a key role in minimizing adverse effect of climatic factor.

5. Agroforestry meets diverse needs (*i.e.* food grains, fibre, sugar, fodder, fuel wood, tea, coffee, spices, rubber, cashewnut etc.) of the people. It aims to maximise production of biomass of trees and agricultural crops. The total production and value of fuel, fodder and small timber in degraded lands are reported to be many times more than the coarse grains usually produced on that land (Chaturvedi, 1981, Gupta and Mohan, 1982, Shyamsunder, 1983). The total production of wood expected from Agroforestry alone is about 350 million m^3 which is enough to the foreseeable demand of wood in near future (Dwivedi and Sharma, 1989).

6. Agroforestry can made successfully in waste land and degraded land, whereas such lands are not being effectively utilised under agricultural land use due to several reasons.

7. Agroforestry system is more appropriate land use system for areas which suffer periodically due to drought and flood.

8. Agroforestry system increases employment opportunities in various ways. The agricultural labour gets alternate opportunities to work in different operations of Agroforestry.

9. Agroforestry gives more income to the farmer per unit of land than pure agriculture or forestry. Silvipastural use consisting of growing trees such as *Acacia sp., Albizia sp., Prosopis sp., Ziziphus sp.* etc.

10. Agroforestry plays an important role in maintaining natural balance.

11. Agroforestry provide some social benefits as follows:

(*i*) Improvement of living standard in rural area from sustained employment and higher income from Agroforestry.

(*ii*) Improvement in nutrition and health due to increased quality and diversity of food outputs of Agroforestry.

(*iii*) Stabilisation and improvement of upland communities through elimination of the need to shift sites of farm activities.

12. In Agroforestry, the production of medicinal plant is increased and it is possible to supply raw materials to medical industry.

9.6 LIMITATION OF AGROFORESTRY

Agroforestry have some negative aspect as follows :

1. Environmental Aspects

(*i*) The trees competes with food crops for space, sunlight, moisture, nutrient etc. As a result, the yield of food crops may reduce.

(*ii*) The trees serve as a potential host to the insectpest that are harmful to food crops.

(*iii*) Food crops get damaged during the harvesting operation of trees.

(*iv*) Rapid regeneration by prolific trees which may displace food crops and take over entire fields.

(*v*) The plant of Eucalyptus when planted in large number creates environmental problem. Because this plant absorbs more water from the soil, decreases the fertility of soil and this plant is not use for fuel and fodder.

2. Socio-economic Aspect

(*i*) Agroforestry needs more labour in different operation which may cause scarcity of labour in other activities of agricultural farm.

(*ii*) In Agroforestry, longer period is required for trees to attain maturity and acquire an economic value.

(*iii*) Farmers fail to displace food crop with trees where land is scarce.

(*iv*) Agroforestry system is more complex and less well understood and difficult to apply compared to single crop farm.

Table 9.3 : Productivity and economic out turn from the Agroforestry System of Cropping (On One Hectare Basis)

Year	*With Casuarina equisetifolia*				*With Leucaena leucocephala*		
	Elephant foot yam a/ha	*Guinea grass q/ha*	*Branches of Casuarina q/ha*	*Wood of Casuarina q/ha*	*Turmeric q/ha*	*Branches of Leucaena q/ha*	*Fuel wood q/ha*
First	**Establishment of tree seedling**						
Second	410		20		86	30	
Third	300		21		99	50	
Fourth		500	25		100	40	
Fifth		700	–	705	106	35	500
Net Income (Rs.)	1,39,434	8,000	50,500		45,161	31,550	
Total Income (Rs.)			1,97,934			76,711	

Source : Maiti, S. (2000) Wasteland development through agroforestry system. IFFCO Agricultural Development Manual.

Table 9.4 : Infiltration, Runoff and Soil Loss from the Water Shed

Type of land use	*Infiltration % of rainfall*	*Runoff % of rainfall*	*Soil loss t/ha*
Forest			
Well managed	25–43	57–75	0.0–0.01
gll managed	10–30	70–90	20–60
Grassland			
Well managed	10–30	70–90	0.0–0.01
gll managed	10–30	70–90	20–60
Agroforestry			
Well managed	20–35	65–80	0.0–2.0
gll managed	10–30	70–90	20–60
Agricultural lands			
Well managed	6–30	70–94	0.5–4.0
gll managed	10–30	70–90	20–60

Source : Hazra. C. R. (1995) Soil and water conservation in relation to land use and biomass production. In Forage production and utilization by R.P. Singh, pp. 286–308.

It is observed from the Table 9.5 that the rates of infiltration was improved and runoff and soil loss were less when land was utilized through agroforestry and pasture production system.

Table 9.5 : Physico-Chemical Properties of Soil After Several Years of Cropping Under *Albizia lebbeck* Based Agroforestry *vis-a-vis* Open Canopy (Without Tree Cropping)

Characters	*Agroforestry*	*Open land*	*Cd%*
Water stable aggregates (0.25%)	29	18	3.11
Bulk density (g.cc)	1.37	1.58	0.12
Infiltration (cm/hour)	5.68	3.12	0.38
Pore space (%)	46.90	38.80	1.55
Field capacity (%)	16.20	14.10	0.90
Wilting point (%)	5.60	5.10	0.22
pH	7.1	7.7	0.12
EC (mmhos/cm)	0.13	0.30	0.04
Organic carbon (%)	0.89	0.46	0.06

Characters		*Agroforestry*	*Open land*	*Cd%*
Available nitrogen (kg/ha)		272.20	189.00	12.30
Available phosphorus (kg/ha)		19.80	10.20	1.51
Cation exchange capacity (me 100/g)		18.70	9.30	1.03
Photosynthetically active radiation ($\mu E/m^2/Sec$)		1168	1682	141
Forage yield (t/ha)				
Maize (*Zea mays*)	Green	32.20	41.70	4.53
	Dry	7.10	9.40	1.02
Oats (*Avena sativa*)	Green	40.60	44.50	NS
	Dry	9.60	10.30	NS

Source : Hazra C.R. (1995) Soil and water conservation in relation land use and Biomass production (in) Forage Production and utilization edited by R.P. Singh, IGFRI, pp. 286–308.

9.7 CHARACTERISTIC FEATURES OF TREES FOR AGROFORESTRY

The success of agroforestry programme depends on the selection of suitable tree spaces. While selecting the tree species for Agroforestry, some criteria should be taken into consideration as follows :

1. Tree species capable of meeting different needs of the people should be selected.
2. Tree should have erect growth habit and factor rate of stem elongation. Tree having fast growing habit should be preferred.
3. Trees having broad leaves should not be selected for Agroforestry.
4. Trees of leguminous family should be preferred as they fix atmospheric nitrogen and enrich the soil. Besides this, leaves of most of leguminous plant are small in size, decompose fact, adding large quality of humus and nutrients to the soil. So the leguminous plant do not compete with agricultural crops for nutrients.

5. Tree should have the ability of tolerance to relatively high incidence of prunning. The prunned branches and leaves are used as fuel wood and fodder respectively.
6. Tree species having copicing ability should be preferred. Because the farmers need not replant after cutting the trees for fuel woods and poles.
7. Trees having the character of erect branches and smaller leaflets should be preferred so that sunlight can penetrate through the canopy.
8. Trees having no allelopathic effect on other crop should be preferred.
9. Tree should produce wood of high caloric value that burns without spays of toxic smoke.
10. Tree should be free from disease pathogen and should not work as alternative host for pests of commercial crops.

9.8 MATERIALS AVAILABLE THROUGH AGRO-FORESTRY

Materials available through Agroforestry are as follows:

(*i*) *Fuel* : Twig of plant, wood etc.

(*ii*) *Wood* : Furniture, agricultural implement, Railway sleeper, Boat, Toys etc.

(*iii*) *Fodder* : Leaves of plant, silkworm, Apiculture etc.

(*iv*) *Wood for Industry* : Plywood, Match box, Paper pulp, Rayon, Lac etc.

(*v*) *Medicine* : Rajan, Tanin, Insecticide, Spices, Gum etc.

(*vi*) *Food* : Fruit, Seed, Oil, Gur, Vegetable etc.

9.9 SUITABLE PLANTS FOR AGROFORESTRY

The plants for Agroforestry are selected on the basis of the position of the land as follows :

1. Fallow and Waste Land : Neem (*Azadirachta indica*), Kadam (*Anthocephalus chinensis*), Sal (*Shorea robusta*), Shishum (*Dalbergia sissoo*), Silver oak (*Grevillea robusta*), Mahogany (*Swietenia macrophylla*) etc.

2. Boarder of the Cropped Land : *Eucalyptus sp.*, Akasmoni (*Acacia auriculiformis*), Silver oak (*Grevillea robusta*), Toddy palm (*Borassus flabellifer*), Arecanut (*Areca catechu*) etc.

3. Bank of Pond and River : Mango (*Mangifera indica*), Neem (*Azadirachta indica*), Coconut (*Cocos nucifera*), Banyan (*Ficus benghalensis*), Jacksfruit (*Artocarpus heterophyllus*), Toddy palm (*Borassus flabellifer*), Arecanut (*Areca catechu*), Karambel (*Dillenia indica*), Phalsa (*Grewia asiatica*), Guava (*Psidium guajava*) etc.

4. Open Space of Cropped Land : Dhaincha (*Sesbania aculeata*), Jute (*Chorchorus* sp.), Arhar (*Cajanus cajan*) Lentil (*Lens esculentus*), groundnut (*Arachis hypogaea*) etc.

5. Plantation Crop of Tea, Coffee, Cocco etc. : Banana (*Musa sp.*) Maize (*Zea mays*), Siris (*Albizia lebbeck*) etc.

6. Homestead Open Space : Mango (*Mangifera indica*), Jamun (*Syzygium cumini*), coconut (*Cocos nucifera*), phalsa (*Grewia asiatica*), guava (*Psidium guajava*), Banana (*Musa sp.*), Neem (*Azadirachta indica*), Mahogany (*Swietenia macrophylla*), Tamarind (*Tamarindus indica*), shishum (*Dalbergia sissoo*), Akasmoni (*Acacia auriculiformis*), karambel (*Dillenia indica*), Jamun (*Syzygium cumini*), Star apple (*Syzygium javanica*), Arecanut (*Areca catechu*), toddy palm (*Borassus flabellifer*) etc.

REFERENCES

Balasubramaniyam P. and S.P. Palaniappan, Principles and Practices of Agronomy (2004), Agrobios (India), Jodhpur.

Dwivedi A.P. Agroforestry, Principles and Practices (1992), Oxford and IBH Publishing Co. Pvt. Ltd. New Delhi.

Jha. L.K. Advances in Agroforestry (1995), A.P.H. Publishing Company, New Delhi.

Nair P.K.R. Classification of Agroforestry System (1985), Agroforestry System 3. 97–128.

Singh S.P. Hand Book of Agroforestry (1994), Agrotech Publishing Academy, Udaipur.

GLOSSARY

Absolute Humidity : Absolute or actual quantity of water vapour present in a given volume of air.

Acaricide : Chemicals used for controlling mites.

Acid Soil : The Hydrogen (H^+) ion is more than Hydroxyl (OH^-) ion in soil solution.

Activated Compost Process : Process of compost making introduced by Fowler and Redge in 1922 at Indian institute of Science, Bangalore.

Adco Process : Process of compost making introduced by Hutchinson and Richard in 1921.

After Tillage The tillage operation that are done in the standing crop after sowing or planting and prior to the harvesting of crop plants.

Agrisilviculture : Land use system for concurrent production of Agricultural crops and forest crops.

Agrisilvipasture System : Combination of Agrisilviculture and silvipasture system.

Agroforestry : Practice of Agriculture and Forestry on the same piece of land.

Agronomy : Important discipline of Agriculture that deals with principles and practices of field management for production of field crops.

Anemometer : An instrument used to indicate the direction of wind.

Annual Weed : The weeds that complete their life cycle within a year or season.

Aquatic Weed : Water inhabitant plant.

Arid Climate : The annual precipitation not exceeding 50 cm.

Atmosphere : Mixtures of gases.

Available Water : The quantity of water present in the soil from the field capacity to permanent wilting point.

Bangalore Process : Process of compost making devised by C.N. Acharya at the Institute of Science, Bangalore.

Bare Fallowing : Keeping the land completely fallow *z.e.* without cultivating any crop and repeated tillage is given with a view to control weeds.

Barometer : The instrument used to record the pressure of the atmosphere.

Bastard Fallowing : Taking one short duration crop, the land is kept fallow and repeated ploughing is given in the rest period with a view to control weeds.

Biennial Weed : The weeds that complete their life cycle in two years or two seasons.

Breeder Seed or Nucleus Seed : The initial amount of pure seed of an improved variety available with the breeder.

Broadcasting : Random scattering of seed by hand on the surface of seed bed.

Bulk Density : The mass (or weight) of a unit volume of dry soil in its natural structure.

Capillary Water : The water that is held by soil due to capillary forces, it is available water to plants.

Carbonation : The chemical combination of carbondioxide (CO_2) with minerals which is present in atmosphere or released by decomposition of organic substances.

Catch Crop : Cultivation of short growing crop in between two main crop in the field.

Certified Seed : The seeds that are actually the progeny of registered seed.

Chaff Cutter : The machine used for chopping the straw of fodder.

Chlorophyll : The green pigment of green plants.

Climate : Average condition of weather in a particular region during comparatively long period (25–30 years or more).

Climatology : The study of climate and weather.

Cloud : Aggregate of minute drops of water suspended in the air at higher altitude.

Compost : Organic manures artificially prepared from vegetable and animal waste products.

Compost Tea : It is prepared by decomposing green leaves or leaves with pungent odour mixed with cowdung or droppings of goat, sheep or poultry.

Compound Fertilizer : Sources of more than one nutrient *e.g.* Gromor 28:28:0.

Consumptive Use of Water : The quantity of water in hectare centimetre utilised by the crop in crop season together with evaporation from the field where the crop is grown, transpiration and metabolic activities of plant.

Cover Crop : The crop that cover the land completely within a short period of time *e.g.* cowpea, sweet potato etc.

Crop Science : The science that deals with mainly the production of field crop.

Crop Rotation : Planned sequence of cropping.

Cultivator : The implements used for tilling the soil between rows of standing crop.

Dapog Method of Nursery : Growing seedlings on concrete floor or raised seedbed covered with polythene sheet or banana leaves.

Density of Soil : The mass (or weight) of a unit volume of soil.

Desuckering : Removal of suckers.

Development : The change over from vegetative stage to reproductive stage leading to differentiation of floral parts and its subsequent growth.

Dibbling : The process of placing the seed in the holes made by convenient implement such as spade, khurpi, nirani etc.

Dormance : The state of rest period of seed in which it does not germinate even they are placed in a condition suitable for germination.

Drainage : Removal of excess water known as free water or gravitational water from the field.

Drilling : The practice of dropping seeds in furrow lines in continuous flow by seed drill and covering them with soil.

Dry Tillage : The tillage operation that is generally practised for the crop to be sown or planted in dry land condition having sufficient moisture for germination of seed.

Drought : Adverse weather condition due to delay in onset of monsoon, prolonged dry spells during crop period and early cessation of monsoon.

Duster : The equipment used for dispersal of pesticides in dust form.

Ecology of Weed : The interrelationship of weeds with climate, soil, crops and other biotic factors.

Environmental Science : It deals with the applied aspect of environment.

Evaporation : Loss of water in the form of vapour.

Evapotranspiration : Evaporation and transpiration.

Fertilizer : The inorganic materials which can supply plant nutrient in available form having the high analytical value and having a definite composition and mostly are industrial product.

Fertilization : The union of male gamets with eggs or female gamets.

Field Capacity : The maximum amount of capillary water remaining in the soil after removing of gravitational water.

Flower : A modified shoot and reproductive organ of plant.

Foundation Seed : Seed produced on government multiplication farm and it is second grade seed in order of its genetic purity.

Fungicide : Chemicals used for controlling fungi.

Fumigant : Chemicals used for fumigation.

F.Y.M. : Organic manures produced from liquid and liquid excreta of cattle mixed with litter used for breeding purposes of cattle.

Glacier : The drainage system of the region of perpetual snow.

Gravitational Water : A free water in excess of capillary water.

Green Manure : The manures obtained by the process of green manuring.

Green Manuring : The practice of turning into the soil undecomposed green plant tissue (Bruckman and Brady, 1980).

Growth of Crop : Permanent and irreversible change in volume or size of an organism accompanied by an increase of dry weight.

Guano : The materials obtained from excreta and dead bodies of sea birds.

Harrow : The implement used for secondary tillage.

Harvesting : Removal of entire plants or economic parts after maturity of the crop from the field.

Hekistotherm : Plants that are restricted only to artic and to alpine regions above 4800 metres in tropic and about 3600 metres in temperate zone of the world.

Hot Water Treatment : Soaking of seed for 15–20 minutes at hot water of 50°C.

Humidity : Water vapour content in the atmosphere.

Hydration : The chemical combination of with other chemical substances.

Hydrolysis : An exchange of constituent part between water and mineral.

Hygrometer : The instrument used for measuring the humidity of surrounding air.

Hygroscopic Co-efficient : The percentage of moisture in the soil when soil is kept in an atmosphere having 100 per cent relative humidity.

Hygroscopic Water : The percentage of water occurs as a thin film (4–5 μ thick, 1 μ or micron = 0.0001 millimeter) surrounding the particles of soil or organic matter.

Indore Process : Process of compost making devised by Howard and Ward at the Institute of Plant Industry, Indore.

Infiltration : Percolation of water.

Inflorescence : The branch system bearing a collection of flower or cluster of flower.

Ionosphere : The lower layer of thermosphere.

Insecticides : Chemicals used for controlling insectpest.

Intercropping : The process of growing a short duration companion crop in between the spaces of widely spaced principal or main crop.

Intercultural Operation : Different agricultural operations adopted in between sowing or transplanting and prior to harvesting.

Intermediate Agroforestry : The system intermediate between subsistence and commercial Agroforestry.

Irrigation : Artificial application of water for the purpose of supplying moisture essential to plant growth.

Irrigation Requirement : The quantity of irrigation to be applied to the field for raising a successful crops.

Jute Decorticator : A machine designed in such a way that it can extract and scrape out barky material from the stem of any bast fibre crop.

Kharif Crop : The crops that grow in kharif season.

Kharif Season : The time between the middle of June to middle of October.

Local or Residual or Indigenous Soil : The soil that is found to exist on the parent rock from which soil is originated.

Macro or Major Nutrient : The nutrients which are used by plants in relatively large quantity (*i.e.,* concentration exceeding 1 ppm)

Manures : The substances which are organic in nature, capable of supplying plant nutrient in available form, bulky in nature having the low analytical value and having no definite composition and most of them are obtained from animal and plant waste product.

Maximum Retentive Capacity or Maximum Water Holding Capacity : The percentage of moisture in the soil when all the pore space are completely filled with water.

Maximum and Minimum Thermometer : The instrument used for recording the maximum and minimum temperature of a day in a particular place.

Megatherm : Plants that require more or less constant high temperature throughout the year for their optimal growth and development.

Mesosphere : Layer of atmosphere which lies above stratosphere.

Mesotherm : Plants that are capable of enduring considerable lower temperature during some period of the year.

Meteorology : The study of physical phenomena which takes place in the atmosphere.

Micro or Minor Nutrient : The nutrients that are used by plant in very small quantity (*i.e.* less than 1 ppm).

Microtherm : Plants that demand much lower temperature for their growth and development.

Mimicry : One plant resembles with other plant. *e.g. Echinochloa colonum* resembles Paddy (*Oryza sativa*) plant.

Minimum Tillage : The reducing tillage to the minimum necessary for ensuring a good seed bed, rapid germination, a satisfactory stand and a favourable growing condition.

Mower : The implements mainly used for harvesting of grasses and forage crop.

Mulching : The process of conserving soil moisture by the use of mulch *i.e.* Straw, dry grass, paper, polythene etc.

Negative Photoblastic : Seeds that do not germinate when exposed to light.

Nematocide : Chemicals used for controlling nematodes.

Nitrogenous Fertilizer : The fertilizers that supply only Nitrogen.

Nipping : Removal of shoot tips to encourage branching and flower formation.

Nonphotoblastic : Seeds that germinate either in light or in dark.

Non Selective Herbicides : The herbicides that kill all vegetation in which they come in contact with irrespective of weeds and crop plants.

North-East Monsoon : Winds that start blowing during winter and cold wind blows from north east direction.

Particle Density : The weight per unit volume of solid portion of soil.

Percolation : Downward movement of water.

Periodic Wind : With that blow during a particular period in definite direction.

Perennial Weed : The weeds that live for more than two years.

Phosphatic Fertilizer : The fertilizer that supplies only phosphorus.

Phosphorus Fixation : The conversion of phosphorus to iron and aluminium phosphate.

PF : The logarithmic value of soil moisture tension expressed in centimetre height of water column.

Photosynthesis : Food in manufacturing process of green plants.

Ploughing : The opening of compact soil with the help of different ploughs.

Pollination : The transfer of pollen grain from anther to stigma.

Pore Space : The portion of soil occupied by air and water.

Positive Photoblastic : Seeds that require light for inducing germination.

Postemergence Herbicide : The herbicides that are applied after emergence of crop and weed on soil surface.

Precipitation : The falling of any type of condensed moisture to the ground surface.

Potassic Fertilizer : The fertilizer that supply only potassium.

Pre-emergence Herbicide : The herbicides that are applied after sowing of the crop but before a crop or weed has emerged or germinated.

Pre-Planting Herbicide : The herbicides that are applied before planting of the crop in the field.

Preparatory Tillage : Tillage operations that are carried out to prepare the land for raising crops from the harvest of a crop to the sowing of next crops.

Primary Dormancy : The seeds that are incapable of germination after ripening if they are provided suitable condition of germination.

Primary Tillage : The tillage operations that are done after the harvest of the crop to bring the land under cultivation.

Productive Agroforestry System : In this system goods are produced predominantly for meeting the basic needs of the society and measures taken to maximise production of annual and perennial crops.

Rabi Crop : The crops that grow in rabi season.

Rabi Season : The time between the middle of October to middle of February.

Rainfall : Precipitation in the form of liquid drops longer than 0.5 mm diameter falling an earth.

Raingauge : The instrument used for measuring the amount of rainfall.

Reaper : An improved implement used for cutting and threashing of the crop in the field.

Registered Seed : The seeds multiplied either the breeder seed or foundation seed.

Relative Humidity (R.H) : The ratio between the amount of water vapour present in the air and the amount of water required for saturation at a particular temperature and pressure.

$$\textbf{R.H.} = \frac{\text{Water vapour present in the air}}{\text{Water vapour required for saturation}} \times 100$$

Rodenticide : Chemical used for controlling rodents.

Saline Soil : The soil that contains high percentage of soluble salt to interfere the growth of most plant.

Saline-Alkali Soil : The soils which contain appreciable quantities of neutral soluble salt and enough absorbed sodium ions to effect the plant growth.

Scarification : The process of rubbing the seed coat.

Secondary or Induced Dormancy : The seeds that are capable of germinating under suitable condition just after ripening become incapable of germination when they are stored under unfavourable condition even for a few days.

Secondary Tillage : The tillage operations that are carried out on the soil after primary tillage to bring the good soil tilth.

Seed : Mature, ovule and it is the protectors and propagator of its kind.

Seed Drill : The machine that is used for sowing seeds in rows at uniform rate and controlled depth.

Seed Coat : Outer covering of seed which shields the embryo and endosperm and protect the seed.

Seepage : Lateral subsurface movement of water.

Seed Treatment Chemicals : Chemicals used for treating the seed before sowing.

Selective Herbicide : The herbicides that are more toxic to one plant than another plant.

Silvipastoral System : A land management system in which forests managed for production of wood as well as rearing of domestic animals.

Semi Arid Climate : The annual rain fall belowing 125cm.

Soil : The upper weathering layer of earth crust.

Soil Consistence : The manifestation of physical forces of cohesion and adhesion acting within the various level of moisture content.

Soil Fertility : The inherent capacity of the soil to supply plant nutrient in adequate amount and suitable proportion and free from toxic substances.

Soil Moisture Tension : The total force with which water is held in the soil.

Soil Productivity : The ability of soil to produce crop per unit area.

Soil Science : It deals with soils and their origin and impact of crop production etc.

Soil Sterilant : The chemicals that prevent the growth of living organism when present in the soil.

Soil Structure : The arrangement of soil particles in the soil mass.

Soil Texture : The relative size of soil particles.

South West Monsoons : Winds that start blowing in India from the month of June and lasts upto the first week of October.

Specific Humidity : The ratio of mass of water vapour actually in the air to the unit mass of air including its moisture.

Sprayer : The equipment used for dispersal of pesticides in liquid form.

Stratospause : The stratosphere and mesosphere are separated by narrow traditional layer.

Stratosphere : Layer of atmosphere which lies above the tropopause.

Subsistence Agroforestry : This system aims at meeting the basic need of the family having small holding and managing different activities by the members of the family.

Sugarcane Crusher : The machine used for extraction of juice from sugarcane.

Superfluous Water : The quantity of water in between maximum relative capacity and field capacity.

Stubble Mulch Tillage : Keeping the soil protected at all times by leaving crop residues or stubbles on the surface of soil during fallow period.

Temperature : Measurement of intensity of heat.

Terrestrial Weed : The weeds that grow in crop field.

Threhing : The process of separating the seeds from the ears of cereal crops.

Thermosphere : Outer most layer extending from the mesosphere where atmospheric densities are low.

Tillage : The physical manipulation of soil with tools and implement to result in good tilth for better germination of seed and subsequent growth of crop (Reddy *et. al.,* 1993).

Tilth : The physical condition of soil in its relation to plant growth.

Topping : Removal of terminal bud with or without top leaves before and after emergence of flower.

Transpiration : Process of release of water vapour from all aerial parts of plants especially from leaves and it is the biological process.

Transpiration Ratio : The ratio of quantity of water used to the quantity of dry matter produced.

Transplanting : The process of planting the seedlings in the main field for further growth, development and production.

Transported Soil : The soil that is transported from original place by air or by other means.

Transplanter : The implement used for transplanting of paddy seedlings.

Trashing : Removal of loosely adhering dried and drying leaves from sugarcane.

Troposphere : Lower layer of atmosphere extending upto height of 8–18 kilometers from the earth.

Water Application Efficiency : The ratio between the quantity of water stored in the root zone and water delivered to the field.

Water Convenience Efficiency : The losses of water that occurs during convenience from the source of supply to field.

Water Distribution Efficiency : The measurement of water distribution within the field.

Water Requirement of Crop : The ratio of water transpired during the growth of the crop to weight of dry matter produced.

Water Storage Efficiency : It is the ratio between water stored in the root zone and water needed in the root zone.

Water Use Efficiency : It is the ratio between crop yield and water used in evaporation.

Weather : The condition of the atmosphere in a place at a certain time or over a certain period.

Weedicide/Herbicide : Chemical used for controlling weed.

Wet or Puddling Tillage : The tillage operation that is done in a land with standing water.

Wind : Air in motion in a stream along the surface of the earth.

(*a*) ***Permanent Wind or Prevailing Wind :*** Wind that blows in the same direction all the year round.

(*b*) ***Periodic Wind :*** Wind that blows during a particular period in definite direction.

Wind Van : The instrument used to measure the velocity of wind.

Zaid Crop : The crops that grow in zaid season.

Zaid Season : The time between the middle of February to the middle of June.

APPENDIX

1. Land and Its Measurement

1 Bigha	= 20 katha.
1 Bigha	= 33 cent (40 cent in some areas).
1 Acre	= 3 Bigha or 2.5 Bigha.
1 Acre	= 100 cent or 43560 sq. ft or 0.40 hectare.
1 Hectare	= 2.471 Acre or 10,000 sq. metre.
1 Katha	= 726 sq. ft.

2. Weight : Its Unit and Conversion

1 Ounce (O_2)	= 28.3495 grams (gm).
1 Pound (lb)	= 0.454 kilogram (kg).
1 Tola	= 11.6638 grams (gm).
1 Chhatak	= 58.319 grams (gm).
1 Maund (40 Secrs)	= 33.3242 kilogram (kg) or 0.37 quintal.
1 Metric ton	= 1000 kilograms (kg).
1 Quintal	= 100 kilogram or 2.679 maunds.

3. Length : Its Unit and Conversion

1 Foot	= 12 inches.
1 Yard	= 3 feet or 0.9 metre.
1 Inch	= 2.54 centimetre (cm.).
1 Foot	= 30.48 centimetre (cm.).
1 Metre	= 39.3701 inches.

1 Kilometre	=	1093.61 yard or 0.62137 mile.
1 sq. cm	=	0.155 sp. inch.
1 sq. ft	=	92.903 sq.cm.
1 sq. yard	=	0.48 sq. metre.

4. Liquid : Its Unit and Conversion

1 gallon	=	4.593 litre.
1 litre	=	0.22 gallon.
1 kerosin tin	=	18 litres.

5. Weight Per Acre into Kilogram Per Hectare

1 Maund per acre	=	0.9223 kg per hectare.
1 Pound per acre	=	1.1209 kg per hectare.
1 Ounce per acre	=	70.05 gm per hectare.
1 Cwt per acre	=	125.6 kg per hectare.

6. Fertilizers and Its Relation with Plant Nutrient

1 kg. Nitrogen (N)	=	5 kg Ammonium Sulphate (20% N).
		or 2.2 kg. Urea (45% N).
		or 4 kg. Calcium Ammonium Nitrate (CAN, 25% N).
		or 4 kg. Ammonium chloride (25% N).
		or 3.84 kg. Ammonium Sulphate Nitrate (26% N).
1 kg. Phosphorus (P_2O_5)	=	6.250 kg Super Phosphate, (16% P_2O_5)
		or 2.220 kg Super Phosphate, (45% P_2O_5).
1 kg Potassium K_2O).	=	1.660 kg. Muriate of Potash (60%

(K_2O)		or 2.0 kg Muriate of Potash (50% K_2O).
1 kg. Nitrogen +		5 kg Suphala 20:20:0
1 kg. Phosphorus	=	or 3.570 kg gromor 28:28:0. or Paras 28:28:0.

1 kg. Nitrogen + 1 kg. Phosphorus + 1kg. Potassium = 6.660 kg Suphala 15:15:15.

7. Area Coverage By Some Agricultural Implements

(1) A country plough covers 33–50 cent of land in a working day of 8 hours.

(2) A mould board plough cover 50–66 cent is a working of 8 hours.

(3) A hand seed drill covers 1 acre of land for Aus paddy and Jute or 0.66 acre of land for wheat in a working day of 8 hours.

(4) A wheel hoe covers 0.33–0.50 acre in a working day of 8 hours.

(5) A knap sack sprayer covers 1 acre of land in a working day of 8 hours.

(6) A micronet sprayer covers 12–14 acres of land in a working day of 8 hours.

(7) A hand sprayer covers 1 acre of land in working day of 8 hours.

(8) A hand duster covers 3 acres of land in a working day of 8 hours.

(9) A power sprayer covers 10–12 acres of land in a working day of 8 hours

(10) A farmer and bullocks walks 8.25 mile to plough an acre of land.

8. Soil pH and Its Range for Some Crops

Names of Crop	*Suitable pH Range*
Paddy	5.0–6.5

Wheat, Maize, Oat, Barley etc.	5.5–7.5
Names of Crop	*Suitable pH Range*
Pulse crops (Gram, Pea, Kalai etc.)	6.0–7.5
Mustard	6.0–7.5
Jute	6.0–7.5
Potato	5.0–6.5
Sugarcane	6.0–8.0
Vegetable crops	5.5–7.5
Onion	6.0–8.0
Groundnut	5.5–6.5
Tomato	6.0–7.0
Tobacco	5.5–7.5
Tea	4.0–5.5
Banana	6.0–7.5
Pine apple	5.0–6.5
Coconut	6.0–8.0

9. Crops and Its Unusable Product (Ratio) :

Straw : Paddy = 1 : 1 (High yielding variety).

= 1.5–1.7 : 1 (Local Aman Paddy)

Paddy : Rice = 3 : 2

Sugarcane : Jaggery (gur) = 8 : 1

Jaggery : Sugar = 10 : 1

Jute : Fibre = 100 : 5–6

Groundnut (Whole) : Seed = 3 : 2

10. Different Oil Seed Crops and Their Oil Content (Percentage)

Mustard : 30 – 48

Sesamum : 38 – 54

Linseed : 35 – 45

Groundnut : 40 – 50

Sunflower : 35 – 45

Safflower : 25 – 35

Castor : 40 – 55

Coconut : 50 – 75

Soyabean : 18 – 25

11. Land Use in India

Uses of Land	*Amount of Land (Crore Hectare)*	*Percent Amount*
Geographical area	32.87	—
(*i*) Cultivated land	14.24	43.32
(*ii*) Forest area	6.83	20.77
(*iii*) Non-agricultural land	3.94	12.96
(*iv*) Fallow land	2.20	6.69
(*v*) Other non cultivable land	3.25	9.88
Total Usuable land	30.47	92.70

12. List of ICAR Research Institute Dealing with Crops and Related Activities

1. Central Arid Zone Research Institute, Jodhpur, Rajasthan.
2. Central Institute for Agricultural Engineering, Bhopal, Madhya Pradesh.
3. Central Institute for Cotton Research, Nagpur, Maharashtra.
4. Central Research Institute for Dryland Agriculture, Saidabad P.O. Hyderabad, Andhra Pradesh.
5. Central Institute for Research on Cotton Technology, Mumbai, Maharashtra.
6. Central Institute for Sub-Tropical Horticulture, Lucknow, Uttar Pradesh.
7. Central Plantation Crop Research Institute, Kasargod, Kerala.
8. Central Potato Research Institute, Shimla, Himachal Pradesh.
9. Central Research Institute for Jute and Allied Fibre, Barrackpur, West Bengal.

10. Central Rice Research Institute, Cuttack, Orissa.
11. Central Soil Salinity Research Institute, Karnal, Haryana.
12. Central Soil and Water Conservation Research and Training Institute, Dehradun, Uttar Pradesh.
13. Central Tobacco Research Institute, Rajahmundry, Andhra Pradesh.
14. Central Tuber Crop Research Institute, Trivandrum, Kerala.
15. Directorate of Biological Pest Control, Bangalore, Karnataka.
16. Directorate of Cropping System Research, Modipuram, Meerut, Uttar Pradesh.
17. Directorate of Oil Seed Research, Rajendranagar, Hyderabad, Andhra Pradesh.
18. Directorate of Poultry Improvement, Andhra Pradesh Agricultural University, Rajendranagar, Hyderabad, Andhra Pradesh.
19. Directorate of Rice Research, Hyderabad, Andhra Pradesh.
20. Directorate of Water Management Research, Rahuri, Maharashtra.
21. Directorate of Wheat Research, Karnal, Haryana.
22. ICAR Research Complex for NEH Region, Meghalaya.
23. ICAR Research Complex in Goa.
24. Indian Agricultural Research Institute (Deemed University), Pusa, New Delhi.
25. India Agricultural Statistical Institute, New Delhi.
26. Indian Grassland and Fodder Research Institute, Jhansi, Uttar Pradesh.

27. Indian Institute of Horticultural Research, Bangalore, Karnataka.
28. Indian Institute of Pulse Research, Kanpur, Uttar Pradesh.
29. Indian Institute of Soil Science, Bhopal, Madhya Pradesh.
30. Indian Institute of Sugarcane Research, Lucknow, Uttar Pradesh.
31. Indian Lac Research Institute, Ranchi, Bihar.
32. Jute Technological Research Laboratory, Kolkata, West Bengal.
33. National Biotechnological Centre for Crop Science, Pusa, New Delhi.
34. National Bureau of Plant Genetics Resources, New Delhi.
35. National Bureau of Soil Survey and Land use Planning Nagpur, Maharashtra.
36. National Centre for Agricultural Economics and Policy Research, Library Avenue, New Delhi.
37. National Centre for Integral Pest Management, Faridabad, Haryana.
38. National Centre for Mushroom Research and Training, Solan, Himachal Pradesh.
39. National Research Centre for Agroforestry, Jhansi, Uttar Pradesh.
40. National Research Centre for Arid Horticulture, Bikaner, Rajasthan.
41. National Research Centre for Cashew, Tuttur, Karnataka.
42. National Research Centre on Citrus, Nagpur, Maharashtra.

43. National Research Centre for Groundnut, Junagadh, Gujarat.
44. National Research Centre for Oil Palm, Pedavegi, West–Godavari, Andhra Pradesh.
45. National Research Centre for Onion and Garlic, Pune.
46. National Research Centre on Rapeseed and Mustard, Bharatpur, Rajasthan.
47. National Research Centre on Soyabean, Indore, Madhya Pradesh.
48. National Research Centre for Sorghum, Hyderabad, Andhra Pradesh.
49. National Research Centre for Spices, Calicut, Kerala.
50. National Centre for Weed Science, Jabbalpur, Madhya Pradesh.
51. Nuclear Research Laboratory, Pusa, New-Delhi.
52. Plant Quarantine Regional Station, National Bureau of Plant Genetics Resources, Hyderabad, Andhra Pradesh.
53. Project Directorate for Cropping System Research, Meerut, Uttar Pradesh.
54. Sugarcane Breeding Institute, Coimbatore, Tamil Nadu.
55. Vivekananda Parvatia Krishi Anusandhana Sala, Almora, Uttar Pradesh.
56. Water Technology Centre for Eastern Region, Bhubaneswar, Orissa.

13. List of Agricultural Universities in India

1. Andhra Pradesh Agricultural University, Rajendra-Nagar, Hyderabad, Pin. 500030.
2. Assam Agricultural University, Jorhat, Assam, Pin. 785013.

3. Bidhan Chandra Krishi Vishwavidyalaya, Mohanpur, Nadia.
4. Birsa Agricultural University, Kanke, Ranchi, Pin. 834006.
5. Central Agricultural University, Iroisemba, Imphal, Manipur, Pin. 795001.
6. Chandrasekhar Azad University of Agriculture and Technology, Kanpur, Uttar Pradesh, Pin. 208002.
7. Chaudhury Charan Singh Haryana Agricultural University, Hissar, Haryana, Pin. 125004.
8. Dr. Yaswant Rao Parmar University of Horticulture and Forestry, Solan, Himachal Pradesh, Pin. 173230.
9. Govind Ballabh Pant University of Agriculture and Technology, Pantnagar, Uttarakhand, Pin. 263145.
10. Gujarat Agricultural University, Sardar Krishi Nagar, Banaskantha, Gujarat, Pin. 385506.
11. Himachal Pradesh Krishi Vishwavidyalaya, Palampur, Himachal Pradesh, Pin. 176062.
12. Indira Gandhi Krishi Viswavidyalaya, Krishak Nagar, Raipur, Chhatisgarh, Pin. 492006.
13. Jawaharlal Nehru Krishi Viswavidyalaya, Jabalpur, Madhya Pradesh, Pin. 482004.
14. Kerala Agricultural University, Vellanikara, Thrissur, Kerala, Pin. 680654.
15. Konkan Krishi Vidhapeeth, Dapoli, Maharashtra, Pin. 431712.
16. Mahatma Phule Agricultural University, Rahuri, Ahmednagar, Maharashtra, Pin. 413722.
17. Marathwada Agricultural University, Parbani, Maharashtra, Pin. 431402.
18. Narendra Deva University of Agriculture and Technology, Faizabad, Uttar Pradesh, Pin. 224001.

19. North Eastern Hill University, School of Agricultural Science and Rural Development, Medziphema Nagaland.
20. Orissa University of Agriculture and Technology, Bhubneswar, Orissa, Pin. 751003.
21. Punjab Agricultural University, Ludhiana, Punjab, Pin. 141994.
22. Punjabrao Krishi Vidyapeeth, Krishinagar, Akola, Maharashtra, Pin. 444104.
23. Rajasthan Agricultural University, Bikaner, Rajasthan, Pin. 334002.
24. Rajendra Agricultural University, Pusa, Samastipur, Bihar, Pin. 848125.
25. Shere-e-Kashmir, University of Agricultural Science and Technology.

(*a*) Dalgate, Srinagar, Pin 191001 (In Summer : May-October)

(*b*) Gandhinagar, Jammu-Tawi, Pin 180009 (In Winter: November-April).

26. Tamilnadu Agricultural University, Coimbatore, Tamil Nadu, Pin. 691003.
27. Anand Agricultural University, Anand, Gujarat, Pin. 388110.
28. University of Agricultural Science, Bangalore, Karnataka, Pin. 560065.
29. University of Agricultural Science, Dharwad, Karnataka, Pin. 580005.
30. Uttar Banga Krishi Viswavidyalaya, Pundibari Coochbehar, West Bangal, Pin. 736165.
31. Allahabad Agricultural Institute. Allahabad, Uttar Pradesh, Pin. 211007.

32. Junagadh Agricultural University, Junagadh, Gujarat, Pin. 362001.
33. Maharana Pratap University Agriculture and Technology, Udapur, Rajasthan Pin. 313001.
34. Navsari Agricultural University, Navsari, Gujarat, Pin. 396450.
35. Rajmata V. Simdia Krishi viswavidyalaya, Gwalior, M.P., Pin. 474002.

12. List of Publications by Author (Dr. A.K. Ghosh)

1. Agronomy and Crop Science 162, 342–346 (1989). 1989 Paul Parey Scientific Publishers Berlin and Hamburg ISSN 0931–2250. Monocarpic Senescence in Relation to Yield of *Sesamum indicum* During Source–Sink Alternation, A.K. Bisws and A.K. Ghosh.
2. *Indian J. Plant Physiol.*, Vol. XXXII, No. 2, pp. 172–174 (June, 1989).

   Monocarpic Senescence of *Arachis hypogaea* : Nutrient Withdrawal Vs. Migration of Senescence Signal, A.K. Ghosh and A.K. Biswas.
3. Indian Journal of Experimental Biology. Vol. 28 May 1990, pp. 492–493.

   Whole plant senescence in *Ophyioglossum vulgatum:* Influence of sporangiferous spike, kinetin and abscisic acid. Arun Kumar Biswas,* Ashis Kumar Ghosh, Swapan Kumar Mandal and Udayan Sarkar.
4. *Indian J. Plant Physiol.,* vol. XXXIV No. 1, pp. 25–29 (March, 1991).

   Source : Sink Relationship During Ageing and Senescence of *Solanum tuberosum* L. Ashis Kumar Ghosh and Arun Kumar Biswas.
5. Agri. Biol. Res. 7 (2) 132–138 (1991).

Source : Sink Relationship During Monocarpic Senescence of *Raphanus sativus*. A.K. Ghosh and A. Biswas.

6. An article the Botanica, Vol 42, pp. 20–23 (1992) Magazine of Delhi Univ. Botanical Society. Inhibitory Effect of Fishes Exposed to some Cultivated, Wild plant from the Bangiya West Rardh (Bankura district). A.K. Ghosh.

7. Indian Journal of Forestry, Vol. 16(3) 201–203. 1993. Intact leaf senescence in some polycarpic forest trees in relation to their reproductive development. Ashis Kumar Ghosh and Arun Kumar Biswas.

8. *Geobios New Reports* 12 : 94–96, 1993. Herbal folk medicines of Bankura district, West Bengal. D. Namhata and A. Ghosh.

9. *Geobios New Reports* 12 : 96–99, 1993. Effect of polluted air of Durgapur on the vegetation. Ashis Kumar Ghosh.

10. *Revista Biol (Lisboa)* 15 : 63–68, 1994. Monocarpic senescence in *Amaranthus tricolor L. : Dual action of kinetin.* A.K. Ghosh and A.K. Biswas.

11. Indian Journal of Experimental Biology. Vol. 32. Novemeber 1994, pp. 807–811. Correlative senescence in *Tagetes patula and Chysanthemum coronarium* during reproductive development. Ashis Kumar Ghosh and Arun Kumar Biswas.

12. J. Agronomy and Crop Science 175–202 (1995). (c) 1995 Blakwell Wissenshafts Verlag, Berlin ISSN 0931–2250. Regulation of Correlative Senescence in *Arachis hypogaea* L. by Source–Sink. Alteration through Physical and Hormonal Means. A.K. Ghosh and A.K. Biswas.

13. J. Econ Taxon Bot additional Series, 12. Scientific Publishers, Jodhpur (India), 1996, pp. 318–320. Ethnomedicine in Bankura district, West Bengal. Ashis Ghosh, Sarathi Maity and Malaty Maity.

14. J. Econ. Taxon. Bot. Vol. 23 No. 2 (1999), pp. 557–560. Herbal Veterinary medicine from the Tribal Area of Bankura District, West Bengal, A.K. Ghosh.

15. Pak J. Sci. Ind Res 2002 45 (3) 212. Mechanism of Monocarpic Senescence of *Trichosanthes dioica* Roxb (Cucurbitaceae). Ashis Ghosh.

16. Ghosh Ashis Kumar (2002). "Herbal Veterinary Medicine from the Tribal Areas of Bankura and Midnapur District." In : Series Recent Progress in Medicinal Plants. Vol. 1, Ethnomedicine and Pharmacognosy (Eds : V.K. Singh, J.N. Govil and Gurdip Singh). Sci Tech Pub., USA 233–237 pp.

17. Indian Journal of Traditional Knowledge, Vol. I (i) October, 2002, pp. 93–95. Ethnoveterinary medicines from the tribal areas of Bankura and Medinipur districts, West Bengal, Ashis Ghosh.

18. J. Econ. Taxon. Bot. Vol. 27 No. 3 (2003). Scientific Publishers (India), pp. 573–575. Herbal veterinary medicine from the tribal areas of Midnapur and Bankura district, West Bengal, Ashis Kumar Ghosh.

19. J. Econ. Taxon. Bot. Vol. 27 No. 4 (2003). Scientific Publishers (India), pp. 825–826. Traditional vegetable dyes from central West Bengal, Ashis Ghosh.

20. Indian Journal of Traditional knowledge, Vol. 2 (4) October, 2003, pp. 393–396. Herbal folk remedies of Bankura and Medinipur districts, West Bengal, Ashis Ghosh.

21. Natural Product Radiance Vol 3 (2), p. 91 March-April 2004. Plant and Clay dyes used by Weavers and Potters in West Bengal, Ashis Ghosh.

22. Natural Product Radiance Vol. 3(3), p. 170 May-June 2004. Apiphilic plants in agro-forestry. Ashis Ghosh.

23. Natural Product Radiance Vol. 3 (3), p. 173 May-June 2004. Natural biocides and biofertilizers. Ashis Ghosh.

24. Natural Product Radiance Vol. 3 (6), p. 426 Nov.-Dec. 2004 Home made baby food. Ashis Ghosh and Jhuma Karan.

25. Pak J. Sci Ind Res. 2005 48(1) 55–56. Mechanism of *Monocarpic Senescence* of *Momordica dioica*. Source : Sink Regulation by Reproductive Organs. Ashis Ghosh.

26. Non-Wood News, No. 12, p. 30 March 2005, Rome, Italy, Plant and Clay Dyes. Ashis Ghosh.

27. Ageing and Society; The Indian Journal of Gerontology, Vol. 15 (3&4), pp. 75–81, 2005. How one can survive for a long time in the earth. Ashis Ghosh.

28. Ghosh Ashis (2006). Identification of veterinary Medicinal plants through participatory approach. In : Advances in Medicinal Plants (Eds. N.D. Prajapati, T. Prajapati and S. Jajpura) Vol. 2 pp. 97–105. Asian Medicinal Plants and Health Care Trust, Jodhpur, Rajasthan.

29. Natural Product Radiance Vol. 5 (4) p. 260. July-Aug. 2006. Healthy Pan-drink. Ashis Ghosh.

30. J. Econ. Taxon. Bot. Vol. 30 (Suppl.), pp. 233–238 (2006). Medicinal plants used for treatment of diabetes by the tribals of Bankura, Purulia and Medinipur of W. Bengal. Ashis Ghosh.

31. Natural Product Radiance Vol. 7(5), 761–765, Sept.-Oct. 2008. Ethnobotanical Survey in West Rarrh for natural health care and green belt movement. Ashis Ghosh.

32. Geobios 35 : 286–288, 2008. Source : Sink relationship during monocarpic senescence of *Pachyrhizus angulatus* (Papilionaceae) and *Papaver somniferum* (Papaveraceae). Ashis Ghosh.

33. J. Phytol. Res. 21(1) : 153, 2008. Tailoring of haulms serve as a natural remedy against Late Blight of Potato and natural resource for biofertilizer–An Observation. Ashis Ghosh.

34. J. Phytol. Res. 21(1) : 155, 2008. Ancient indigenous abolishing paddy varieties of Bengal. An observation. Ashis Ghosh.

35. Natural Product Radiance Vol. 8(5), p. 477 Sept.-Oct 2009. How to get honey in off season from detached plants? Ashis Kumar Ghosh.

36. Natural Product Radiance Vol. 8(5), p. 477 Sept.-Oct. 2009. Allelopathic susceptibility of some climbers. Ashis Kumar Ghosh.

37. J. Econ. Taxon. Bot. Vol. 33(2), pp. 321/323 (2009). Comperative study of monocarpic senescence of a monocot (*Commelina benghalensis* L.) and a dicot (*Oxalis corniculata* L.) : Migration of senescence signal Vs. nutrient drainage. Ashis Kumar Ghosh.

38. J. Econ. Taxon. Bot. Vol. 33(2), pp. 333-334 (2009). A note on quick regeneration of stem cuttings of bamboos as influenced by planting posture and IAA. Ashis Kumar Ghosh.

Books

1. Regulation of Senescence in Various Plant by A.K. Biswas and A.K. Ghosh, Emkay Publication, Delhi. Page 329, Table 87, Figure 86, Price Rs. 400.00 Size. Demy octavo.

2. Paribesh O Udvid, Rs. 50.00 – Ashis Ghosh.

3. Prani Rahasya, Rs. 50.00 – Ashis Ghosh.

4. Ethnobiology : Therapeutics and Natural Resources. Ashis Ghosh, Daya Publishing House, New Delhi, Rs. 595.00.

5. Ethnomedicine for Human and Veterinary Development. Ashis Ghosh, Daya Publishing House, New Delhi, Rs. 295.00.

6. Agronomy : Facts and Approahes. Dr. A.K. Ghosh and Dr. P. Das, Daya Publishing House, New Delhi.

INDEX

SUBJECT INDEX

PLANT INDEX

C

H

I

L

M

N

O

P

ANIMAL INDEX